MW00762947

LITHIUM AND THE CELL

Lithium and the Cell:

Pharmacology and Biochemistry

Edited by

Nicholas J. Birch

*Biomedical Research Laboratory, School of
Health Sciences, Wolverhampton Polytechnic, UK*

ACADEMIC PRESS
Harcourt Brace Jovanovich, Publishers
London   Boston   San Diego   New York
Sydney   Tokyo

ACADEMIC PRESS LIMITED
24/28 Oval Road,
London NW1

United States Edition published by
ACADEMIC PRESS INC.
San Diego, CA 92101

This book is printed on acid-free paper

British Library Cataloguing in Publication Data
Lithium and the cell.
1. Man. Metabolism. Effects of lithium compounds
I. Birch, Nicholas J.
615.739

ISBN 0-12-099300-7

Typeset by P & R Typesetters Ltd, Salisbury, Wilts, UK
and printed in Great Britain by St Edmundsbury Press Ltd, Bury St Edmunds, Suffolk.

Contributors

Joseph Attias Echanges Cellulaires, Faculté des Sciences et Techniques de l'Université de Rouen-Haute Normandie, BP 118, F 76134 Mont-Saint-Aignan Cedex, France

Sophia Avissar Beer Sheva Mental Health Centre, Ida and Solomon Stern Psychiatry Research Unit, Ben Gurion University, P.O. Box 4600, Beer Sheva, Israel

Robert H. Belmaker Beer Sheva Mental Health Centre, Ida and Solomon Stern Psychiatry Research Unit, Ben Gurion University, P.O. Box 4600, Beer Sheva, Israel

Nicholas J. Birch Biomedical Research Laboratory, School of Health Sciences, Wolverhampton Polytechnic, Lichfield St, Wolverhampton WV1 1DJ, UK

Alexander Buchan Department of Medical Microbiology, The Medical School, University of Birmingham, Birmingham B15 2TJ, UK

De-Maw Chuang Biological Psychiatry Branch NIMH, Bldg 10, Room 3N 212, 9000 Rockville Pike, Bethesda, MD 20892, USA

David R. Crow School of Applied Sciences, Wolverhampton Polytechnic, Wolverhampton WV1 1LY, UK

Robert J. Davie Biomedical Research Laboratory, School of Health Sciences, Wolverhampton Polytechnic, Lichfield St, Wolverhampton WV1 4DJ, UK

Vincent S. Gallicchio Lucille P. Markey Cancer Center, University of Kentucky Medical Center, Lexington, KY 40536-0084, USA

David A. Hart Department of Microbiology and Infectious Diseases, University of Calgary HSC, 3330 Hospital Drive NW, Calgary, Alberta, Canada T2N 4N1

Christopher E. Hartley   Department of Medical Microbiology, The Medical School, University of Birmingham, Birmingham B15 2TJ, UK

David F. Horrobin   Efamol Research Institute PO Box 818, Kentville, Nova Scotia, Canada B4N 4H8

Mark S. Hughes   Biomedical Research Laboratory, School of Health Sciences, Wolverhampton Polytechnic, Lichfield St, Wolverhampton WV1 1DJ, UK

Roy P. Hullin   Regional Metabolic Research Unit, High Royds Hospital, Leeds LS29 6AQ, UK

Sarah Lancaster   Department of Medical Microbiology, The Medical School, University of Birmingham, Birmingham B15 2TJ, UK

Michael J. Levell   Department of Chemical Pathology, University of Leeds, Leeds LS2 9JT, UK

Jonathan D. Phillips   Biomedical Research Laboratory, School of Health Sciences, Wolverhampton Polytechnic, Lichfield St, Wolverhampton WV1 1DJ, UK

Robert M. Post   Biological Psychiatry Branch NIMH, Building 10, Room 3N 212, 9000 Rockville Pike, Bethesda, MD 20892, USA

Sharon Randall   Department of Medical Microbiology, The Medical School, University of Birmingham, Birmingham B15 2TJ, UK

Frank G. Riddell   Department of Chemistry, The University, St Andrews, Fife KY16 9ST, UK

Mogens Schou   The Psychiatric Hospital, 2 Skovagervej, DK-8240 Risskov, Denmark

Gabi Schreiber   Beer Sheva Mental Health Centre, Ida and Solomon Stern Psychiatry Research Unit, Ben Gurion University, P.O. Box 4600, Beer Sheva, Israel

Michael Shalmi   Department of Pharmacology, University of Copenhagen, Juliana Maries Vej 20, 2100 Copenhagen Ø, Denmark

William R. Sherman   Washington University School of Medicine, Department of Psychiatry, St Louis, MO 63110, USA

Gordon R.B. Skinner   Department of Medical Microbiology, The Medical School, University of Birmingham, Birmingham B15 2TJ, UK

Michel Thellier   Echanges Cellulaires, Faculté des Sciences et Techniques de l'Université de Rouen-Haute Normandie, BP 118, F 76134 Mont-Saint-Aignan Cedex, France

Klaus Thomsen   Psychopharmacology Research Unit, University of Aarhus, Denmark

Jean-Claude Wissocq   Biologie Animale et Cellulaire, Faculté des Sciences de l'Université de Picardie, F 80039 Amiens Cedex, France

Preface

This book was conceived when it became apparent that there was a plethora of recent lithium work, particularly in the pharmacological and biochemical areas, which had not yet found application in clinical medicine. It was clear also that a number of burgeoning areas were represented and that a wider audience might be interested in lithium, not only for its most common use in psychiatry, but because it has effects on a variety of systems such as the phosphoinositide signalling systems, viral replication, immunology, dermatology and leukocyte metabolism. This new audience might not be aware of the vast range of knowledge already existing in unrelated areas. Indeed it is common, for instance, to read totally distorted accounts of the psychopharmacology of lithium in chemical manuscripts to justify an elegant synthesis of a novel lithium ionophore which has no real relevance to the psychiatric use of lithium. In contrast, biological psychiatrists need to be aware of the variety of lithium action in order to pick up further clues to the biochemical basis of the diseases with which they are concerned. This, then, is an attempt to provide a broad-based, but not necessarily comprehensive, view of current lithium knowledge.

The book opens with a review of the present psychiatric status of lithium by Professor Mogens Schou, who is, above all others, the person who made lithium the useful psychotropic drug that it has become over the last 20 years. Without Mogens' persistence lithium, as a drug, would have failed: there are, after all, no large pharmaceutical companies which would have funded the research necessary to prove the safety of the drug since a simple inorganic compound is not a major potential revenue source. The subsequent chapter by Dr Wissocq describes the breadth of effects of lithium in biological systems.

The book then progresses from the molecular level to the whole body and reverses the more usual 'microscopic' hierarchy in which the broad view leads to ever more magnified, but limited, views. This approach has been deliberate to try to emphasize the general applicability of the molecular approach to subsequent biochemical chapters and these, in turn, to the physiology of the tissues and to the whole body. I am much indebted to my friend and colleague David Crow who, in discussion of this chapter order, pithily extended a quotation from Professor J.C. Polkinghorne, the eminent mathematician, by saying that 'Since chemistry is clearly physics writ large, biology might be considered to be physics writ larger still!'

It is my belief that we must recognize the value of a multidisciplinary approach to problems in medicine and it is important that we should strive to become 'multilingual' in the various biomedical sciences. Only in this way will we be able to recognize the breadth of information which is available to use and use it for the benefit of patients.

Lithium is currently in use for the prophylaxis of recurrent affective disorders by some 40 000 patients in the United Kingdom. One can perhaps extrapolate to provide an estimate of 200 000 patients in Europe as a whole and 300 000 world-wide. The potential usage, simply for this disease, must be five times that number. In the United Kingdom alone the savings in hospital cost represented by those 40 000 patients is £23 million per year. New uses for lithium, for instance in the treatment of aggression, herpes simplex (including genital herpes), leukocytopenias of various origins, inflammatory diseases and dermatoses appear frequently in the literature. The broad distribution of loci of lithium actions suggests that more uses yet will be found. In all these applications an appreciation of the basic pharmacology and biochemistry of the ion, the lightest solid element, is essential for the effective design of its mode of administration and optimization of dosage. The fascination of lithium is that this 'simple' ion has such varied and widespread effects: whatever its locus of action, it must be at a fundamental level of organization.

This book is based in part on the widespread clinical use of lithium in psychiatry. It is interesting to note that Belmaker *et al.,* in an early chapter, comment that: 'Theories of lithium action are more varied, more contradictory and more short-lived than in almost any other area of psychopharmacology'. In the closing chapter, Levell and Hullin remark that transient hypomania is a common experience for all humans and that this may last for '... only seconds, minutes or for the duration of the party.' I hope that the ideas expressed in this volume may prove durable and survive intellectually at least '... for the duration of the party.' It is in the nature of science, however, that they will be replaced ultimately and the value of this book will be judged by the quality of ideas which derive from it.

The publishers, Academic Press, have been most cooperative and helpful during all stages of the preparation of this volume. Particularly I thank Dr Carey Chapman, who has guided me around the pitfalls of the editing task. I wish to thank my friends and associates in the Biomedical Research Laboratory for their efforts and for their patience with me regarding other unfinished tasks, shelved in the obsession to complete this book, and my colleagues in the School of Health Sciences for their tolerance. Especial thanks are due to Mrs Sue Sydenham, who has struggled with tape and manuscript for my text and has kept control of the communications with my co-authors. Finally I should like to thank my wife, Vance, for all her support and encouragement despite the household repairs not completed, lost manuscripts and other crises.

Wolverhampton   N.J. Birch

Contents

Contents

1 Clinical Aspects of Lithium in Psychiatry

MOGENS SCHOU

The Psychiatric Hospital, 2 Skovagervej,
DK-8240 Risskov, Denmark

Lithium was introduced into modern pharmacotherapy in 1949. It has been estimated that at the present time about one person out of every thousand in the population is undergoing lithium treatment in most industrialized countries. Although employed against various somatic diseases (neutropenia, cluster headache, the on–off phenomenon in Parkinsonism), lithium is used primarily in psychiatry for treatment and relapse-prevention in manic-depressive illness. In order to understand its special therapeutic and prophylactic properties, one must know something about this widespread and devastating mental disease. Detailed information in non-technical language is provided in a small book I have written for patients and relatives (Schou, 1989).

Manic-depressive Illness

We usually distinguish between two major psychoses, schizophrenia and manic-depressive illness. Schizophrenia typically takes an irregularly progressive course, and the disease affects the patient's contact with reality and with other people. Sufferers develop delusions and hallucinations and become autistic. Manic-depressive illness typically shows an episodic course with symptom-free intervals between the manic and depressive episodes. Manias are characterized by elation and increased mental speed, anger is easily aroused, and sometimes the sufferer becomes violent. Manic patients are excessively self-confident and lack self-criticism. They commit ill-considered actions, may destroy their marriages, spoil their reputations, and ruin themselves financially. Depressions, on the other hand, are characterized by sadness, a lack of self-confidence, and lowered mental speed. The patients are inhibited and lack initiative. Weighed down by feelings of hopelessness and guilt they may consider suicide; about one-fifth actually kill themselves. Some manic-depressive patients may suffer a single or a few episodes during a lifetime,

LITHIUM AND THE CELL
ISBN 0-12-099300-7

but much more often episode follows episode in regular or irregular sequence. One distinguishes between a bipolar disease course with both manias and depressions and a unipolar course with depressions only.

Manic-depressive episodes have been designated 'endogenous', i.e. originating without (discernible) cause, but in fact they may, especially in the early phase of the illness, be precipitated by various kinds of stress. It is, however, characteristic that the further course of the episode and the illness are largely independent of external events.

Manic-depressive illness occurs with increased frequency in some families, and hereditary factors presumably play a role. What is inherited is not the disease as such, but rather an increased vulnerability to, as yet unknown, somatic or psychological factors which may perturb those regulatory mechanisms which normally maintain stability of mood, activity, and mental speed.

Lithium Treatment of Manic-depressive Illness

Lithium exerts therapeutic action against mania; during treatment the patients lose their talkativeness, overactivity, and inflated self-confidence.

Lithium also exerts therapeutic action against depressive phases of manic-depressive illness, but its effect is not great, and usually treatment with antidepressant drugs is preferred. It has been suggested that addition of lithium to antidepressants in patients who do not respond to the latter drugs alone may lead to rapid recovery, but final proof of this has still to be provided.

The main use of lithium is in relapse prevention or (secondary) prophylaxis. When given as a maintenance treatment lithium significantly ameliorates or prevents further episodes, and in patients who previously suffered from frequent and severe attacks, lithium treatment can produce an almost miraculous change of life course. Lithium prophylaxis is equally good in women and men and in young and old. Lithium prevents manias and depressions with the same efficacy, and the prophylactic effect is as good in unipolar patients, who never experience a mania, as in bipolar patients who suffer both manias and depressions.

Lithium treatment is not completely effective. It has been estimated that out of 10 treated patients one or two may not respond at all, and of the remainder one half may respond partially and one half completely. These figures vary with patient selection, treatment management, definition of response, etc.

The therapeutic and prophylactic effects of lithium are less pronounced when atypical clinical features of schizophrenic or neurotic character are present. Nevertheless, lithium has occasionally been used with good effect for the treatment of schizoaffective illness, i.e. the simultaneous occurrence of schizophrenic and manic-depressive features. Lithium treatment has been tried in various other psychiatric conditions, but the only solidly documented effect is a stabilizing and antiaggressive action in patients who are subject to periodic emotional instability or bouts of explosive behaviour.

Effects on Normal Mental Functions

In view of its marked effect on episodes of abnormal mood and activity, lithium exerts remarkably little action on normal mental functions. Most patients in lithium treatment feel normal and function normally. Some patients may complain of impaired memory and lowered reaction speed, and some feel that life under lithium is less colourful and interesting than before. It is difficult to decide to what extent these patients miss the emotional lift of previous mild manias and to what extent the phenomena are caused by a direct action of lithium on mental functions.

Side Effects and Intoxication

Among fairly common side effects of lithium treatment may be mentioned hand tremor (which may yield to treatment with beta-receptor blocking agents), weight gain, and increased urine production and thirst. Side effects are strongly dependent on treatment intensity and can often be alleviated through a moderate lowering of lithium doses and serum levels.

Lithium intoxication may be caused by overdosage, for example as a result of self-administration with suicidal intent, or by lowered elimination rate. The latter can be secondary to kidney disease, or it may be induced by a negative sodium or water balance. Lithium intoxications have developed in patients who went on salt-free diets or crash slimming cures, or who used diuretics concurrently with lithium. During physical illness with fever lithium treatment should be discontinued.

A fully developed lithium intoxication resembles cerebral haemorrhage; severe poisoning may lead to death or permanent cerebellar damage. Lithium intoxication is treated with correction of salt and water balance and in more pronounced cases with haemodialysis in an artificial kidney.

Administration and Elimination

Lithium cannot be given by injection or be administered against the patient's own wish; the drug is consumed in the form of tablets or capsules. In principle,

M. Schou

any soluble lithium salt can be used, provided the preparation has constant release properties and all lithium is released during passage through the gastrointestinal tract.

Lithium is eliminated almost exclusively through the kidneys. After filtration through the glomerular membrane lithium is reabsorbed, together with and to the same extent as sodium and water, in the proximal kidney tubules; little or no lithium is reabsorbed in the distal parts of the nephron. In healthy persons the renal lithium clearance is about 25% of the glomerular filtration rate, i.e. in the order of 20 ml min^{-1}. The elimination half-time of lithium is about 20–30 h, and with constant intake steady-state conditions are reached in approximately a week.

Serum and Tissue Concentrations

Table 1 shows steady-state serum and tissue lithium concentrations. During prophylactic treatment the lithium concentration in blood serum is maintained at levels of the order of 0.3–1.0 mmol l^{-1}. This is in blood samples drawn 12 h after the last intake of lithium and corresponds roughly to the mean concentration over the 24 h of the day. The serum concentration fluctuates in relation to the intake, being 2–3 times higher than average at absorption peaks and about half average at absorption troughs. In saliva the concentration is about double and in cerebrospinal fluid about half that in serum.

Lithium is distributed unevenly in the tissues. At steady-state the concentration is lower than that in serum in such tissues as liver and erythrocytes and higher in, for example, kidneys, thyroid and bone. The concentration in whole brain is approximately the same as in serum; there are regional differences, but no brain area accumulates lithium to a high concentration.

The concentrations in Table 1 may be worth keeping in mind when experimental work is planned. Studies in which all or a large part of the medium's sodium content is replaced by lithium are hardly of clinical relevance.

Oral administration of lithium to animals with food gives less fluctuation of the serum concentration than does administration by injection and is more related to the clinical situation. Administration with drinking water has the

TABLE 1   Serum and tissue lithium concentrations.

	Treatment	Intoxication
Serum (mmol l^{-1})	0.3–1	2–10
Tissues (mmol kg^{-1} wet wt)	0.2–5	1–15

disadvantage that intake increases if the animals develop lithium-induced polyuria/polydipsia.

Time Factors

The full antimanic action of lithium is reached in about a week. This is slower than with neuroleptics but less unpleasant for the patients. Full antidepressant effect may be achieved in 2–3 weeks, i.e. about the same time as with antidepressants and electroconvulsive therapy. When lithium is used prophylactically, some patients may require treatment for weeks or months to achieve full protection against relapses.

From a clinical point of view one is less interested in acute experiments with administration of a single or a few doses of lithium than in studies where lithium is given for days or weeks or months. Acute and chronic administration of lithium may produce entirely different results.

Experimental Work: A Research Strategy Based on Clinical Considerations

Experimental research concerning the mode of action of psychiatric treatments is notoriously difficult and fraught with frustration. The main obstacle is the difficulty of knowing whether the changes we produce in our experimental systems, be they healthy rats or brain cortex slices or isolated enzymes, are in any way connected with what happens when the same drugs or procedures are used to heal the sick human mind. The distance between the former systems and the latter is so large that attempts at extrapolation or deduction seem far-fetched and even ridiculous. I believe, however, that scientists who are interested in treatments that have an effect on manic-depressive illness could gainfully adopt a strategy based on clinical experience.

There is admittedly much in psychiatry that is less than solid, but few would dispute that one can group treatment modalities in categories which are reasonably distinct as regards their effect on manic-depressive illness. In the first category one might place electroconvulsive therapy (ECT) and lithium, both treatments of well-documented efficacy, with a good deal of diagnostic specificity, and acting on both the manic and the depressive manifestations of the disease. Carbamazepine, which also has shown prophylactic action against both manic and depressive relapses, should possibly be placed in the same group. The antidepressant drugs would constitute a second category. They share with the treatments in the first group efficiacy and specificity, but they act on depressions only, not on manias. A third category might then contain the neuroleptic drugs and drugs with hypnotic-sedative-anxiolytic action. These agents can alleviate some of the symptoms of mania and depression (violence, restlessness, agitation, anxiety), but they are not specific for manic-depressive illness.

Now if we expose our experimental systems to appropriate representatives of all the groups or categories, we might find that none of them has any effect, and then there would be no evidence of a connection with affective disease. Or we might find that the experimental system is affected equally or similarly by all the groups, and in that case evidence of diagnostic specificity would be lacking. But we might also be lucky enough to find that the experimental system is affected by lithium and ECT (and perhaps carbamazepine), that it is or is not affected by the antidepressant drugs, and that the system is unaffected by neuroleptics and barbiturates and benzodiazepines. In that case it would not seem entirely absurd to assume that the phenomenon under investigation might in some way, however indirectly, be related to mood normalization in manic-depressive illness. Different drugs may clearly exert similar actions through different mechanisms, but somewhere in the chain of reactions there is likely to be a common pathway, the discovery of which would be of great value.

My suggestion is therefore that rather than exposing our experimental systems to ten different antidepressants, whether of the same or different generations, we should investigate the effects of different categories of psychotropic agents and procedures and then compare clinical similarities and differences with experimental similarities and differences. This might give us information about the clinical relevance of experimental data and clues to further progress.

Reference

Schou, M. (1989). 'Lithium Treatment of Manic-Depressive Illness: A Practical Guide', 4th edn. Karger, Basel, London and New York.

2 Exotic Effects of Lithium

JEAN-CLAUDE WISSOCQ[1], JOSEPH ATTIAS[2] and
MICHEL THELLIER[2]

[1] *Biologie Animale et Cellulaire, Faculté des Sciences de l'Université de Picardie, F 80039 Amiens Cedex, France*
[2] *Echanges Cellulaires, Faculté des Sciences et Techniques de l'Université de Rouen-Haute Normandie, BP 118, F 76134 Mont-Saint-Aignan Cedex, France*

Introduction

Following the work of Cade (1949) and Schou (1957), the efficacy of lithium salts for the treatment of manic depressive psychosis in man is now generally accepted. One important characteristic of this therapeutic effect of lithium lies in the low values of the active concentrations which range from 0.5 to 1 mM in blood serum; this is about one tenth of potassium ion concentration and less than one hundredth that of sodium ions. However, its clinical action is far from being the only biological effect of lithium. In the present contribution, we describe lithium effects interfering with a variety of physio-logical processes and occurring in essentially all classes of organisms from higher plants to viruses. These effects are quite different from a general toxic effect since (1) most often they are observed at fairly low lithium concentrations, and (2) they do not usually decrease the overall growth rate of a given biological system but will rather affect qualitatively its physiological behaviour.

Meiotic Maturation of Oocytes

Invertebrates

Picard and Dorée (1983) reported that LiCl microinjections into starfish oocytes inhibited reversibly the hormone-induced meiotic maturation. Meiosis arrested at the prophase stage can be reinitiated by the hormone 2-methyl adenine (Kanatani *et al.*, 1969). This hormone acts on oocyte membrane receptors by producing a cytoplasmic maturation promoting factor (or M-phase promoting factor, MPF) which stimulates meiotic maturation. Lithium prevents meiosis reinitiation, even when it is microinjected at the end of the hormone-dependent period when the presence of the hormone is necessary for meiosis to proceed. The transfer of MPF-containing cytoplasm into oocytes did not trigger the reinitiation of meiosis if the recipient oocytes were previously injected with Li^+. These results suggested that lithium interfered both with the activity and the amplification of MPF at the level of the nuclear envelope. The possibility that Li^+ could inhibit MPF activity by an early increase of free cytoplasmic Ca^{2+} as well as the hypothesis of an inhibition of the Na^+/K^+ pump was ruled out. Indeed, ouabain, a specific inhibitor of the Na^+/K^+ pump, had no effect on the prevention of meiosis reinitiation by Li^+. Besides, the simultaneous microinjection of K^+, Na^+ or EGTA, a specific chelator of calcium ions, did not prevent the Li^+-dependent inhibition of oocyte meiosis (Picard and Dorée, 1983).

Vertebrates

Mammalian oocytes are cells arrested at the first meiotic prophase. Meiosis is normally resumed after a stimulus by gonadotropins and arrests at metaphase II. *In vitro*, the spontaneous resumption of meiosis of uncoated oocytes is prevented by forskolin, an activator of adenylate cyclase. Gavin and Shorderet-Slatkine (1988) reported that Li^+ reversed the inhibitory effect of forskolin in uncoated mouse oocytes in a concentration-dependent relationship. Li^+ acted on adenylate cyclase only when the enzyme had been previously activated by forskolin, thus inducing the reinitiation of meiosis. However, Li^+ ions did not seem to modify the phosphoinositide cell-signalling pathway by limiting the production of *myo*-inositol because the addition of exogenous *myo*-inositol, simultaneously with Li^+, did not reverse the effects of Li^+. A possible action of Li^+ on G proteins has also been suggested.

Embryogenesis

The interference of lithium with embryogenesis was reported almost a century ago by Herbst (1893) in his studies on sea-urchin eggs. The sea-urchin embryo was later adopted as a model for the study of lithium effects in invertebrates, while the amphibian egg was used in studies on vertebrates. Lithium has turned out to be a powerful tool for investigating the mechanisms of induction or of cell differentiation during embryogenesis.

Sea-urchin

Herbst (1893) was the first to report that the addition of iso-osmotic concentrations of lithium to seawater induced the formation of anomalous embryos with a preferential formation of the endoderm (primitive gut) at the expenses of the ectoderm (external epithelium). In some cases the endoderm did not invaginate normally, but formed an external vesicle (exogastrula). A few decades later, Runnström (1928) and Hörstadius (1936) introduced the concept of two distinct gradients occurring during sea-urchin differentiation termed 'vegetalizing' and 'animalizing' gradients. These gradients orient the formation of distinct anatomic parts of the embryo. Several reports (Von Ubisch, 1925; Hörstadius, 1936; Lindahl, 1936) demonstrated that lithium exerted primarily a vegetalizing action on sea-urchin embryos by favouring the formation of the endoderm and of the mesoderm (larval skeleton). The degree of vegetalization, as observed by the presence of endodermal and mesodermal formations, increased with lithium concentration and with the length of treatment for doses compatible with the embryo's survival.

The vegetalizing effects of lithium were antagonized by K^+ and Rb^+. The absence of Ca^{2+} or the lowering of Na^+ concentration enhanced the vegetalizing effects of lithium, whereas uramildiacetic acid, a Li^+-chelator, suppressed these effects (Lallier, 1979). Cell adhesion of vegetalized embryo was increased by lithium, but this effect was in part suppressed by Zn^{2+} (Fujisawa and Amemiya, 1982).

Invertebrates other than echinoderms

Reliable data are still few. Lithium appears to be active mainly at the beginning of development, during the cleavage stage, as observed with the annelid *Eisenia foetida* (Devries, 1976) and with the gastropod mollusc *Lymnaea* (Raven, 1952). Again, in *Lymnaea*, Li^+ was shown to decrease the rate of cell division (Verdonk, 1968; Van den Biggelaar, 1971). Lithium-induced endodermization and exogastrulation have been frequently observed, particularly with the polychaete worm *Orphryotrocha* (Emanuelsson, 1971) with *Lymnaea* (Raven, 1952) and with *Ascidia* (Nieuwkoop, 1953). The latter effect may correspond to lithium disturbing induction messengers among cells, as a result of the decrease of cleavage rate. The presence of lithium was also shown to cause cephalic abnormalities of the nervous tissues and of the sensory organs in *Lymnaea* (Raven, 1976; Verdonk, 1968). As with sea-urchin, the lithium-induced decrease of cleavage rate could be due to the disturbance of the cellular exchanges of Ca^{2+}.

Amphibians

Early reports (Lehmann, 1937; Pasteels, 1945; Backström, 1954; Lallier, 1955) showed that lithium disturbed the development of amphibian embryos at the cleavage and the gastrulation stages. Lithium prevented the installation of the chordamesoderm, the precursor of the future vertebral axis (Lehmann, 1937). When ectodermal cells were grown *in vitro* in the presence of lithium, mesodermal structures were formed instead of ectodermal ones (Masui, 1961). Further evidence for this mesodermizing effect of lithium was obtained from graft experiments and from the use of heterogeneous inductors like guinea-pig liver extracts (Engländer and Johnen, 1967). According to Nieuwkoop (1970), lithium could increase indirectly the inductive effects of the endoderm on the formation of the mesoderm.

More recently, Kao *et al.* (1986) have shown that early *Xenopus* embryos of 32 cells treated with lithium displayed malformations at the dorsal-anterior level. UV-irradiated, fertilized *Xenopus* eggs produced 'ventralized' embryos deprived of dorsal structures. However, if these embryos were exposed to lithium in the early stages, they recovered their dorsal structures. When administered during cleavage, lithium increased the volume of the embryos' head structures and suppressed the formation of the trunk (Cooke and Smith, 1988; Kao and Elinson, 1989; Regen and Steinhardt, 1988). During the formation of head structures (cephalization) in the presence of lithium, the number of neurones was reported to be increased by two-fold (Breckenridge *et al.*, 1987). More recent observations (Klein and Moody, 1989) indicate that lithium has an indirect effect by modifying the spatial interactions between the ectoderm and the chordamesoderm during gastrulation, thus orienting the determination of head structures towards the posterior parts of the embryo. Lithium was also reported to interfere with a number of inducing substances such as the fibroblast growth factor (Slack *et al.*, 1988) and the mesoderm inducing factor XTC-MIF (Cooke *et al.*, 1989) in the determination of mesodermal structures.

As regards the later events of embryogenesis, lithium was shown to induce mainly malformations of the brain and sensory systems of *Xenopus* and *Pleurodeles* (Pasteels, 1945; Backström, 1954; Signoret, 1960).

Vertebrates other than amphibians

Studies of vertebrates other than amphibians have been very few and have remained rather descriptive. To our knowledge, the effect of lithium on reptile embryogenesis has not been investigated at all.

In fishes, lithium is responsible for a significant decrease in size of the front brain and for severe malformations of the optic vesicles causing synophthalmia, anophthalmia or cyclopia (Stockard, 1906; Vahs and Zenner, 1964). With *Salmo irideus*, Vahs and Zenner (1964) demonstrated the existence of two distinct periods of sensitivity to lithium. The first period takes place at the beginning of the cleavage stage: lithium causes the endodermization of the embryo, with a significant reduction of the notochord and the evolution of muscular areas into connective tissues. During the second period, taking place at the gastrulation stage, lithium affects more specifically the front brain areas, causing proenkephalon degeneration and cyclopia. The other alkaline cations Na^+ and K^+ are much less efficient than Li^+ in disturbing the development of *Salmo irideus*; they do not antagonize the effects of lithium—on the contrary, Ca^{2+} is antagonistic to Li^+ (Vahs and Zenner, 1965).

With birds, lithium-induced cyclopia has also been observed (Rogers, 1963). Using *in vitro* cultures of lithium-treated chicken embryo explants, Nicolet (1965) and Duvauchelle (1965) have described the development of brain malformations and alterations of the sensory vesicles.

A few experiments have been performed on the chicken to study the effect of lithium at the cellular level. Lithium was shown to arrest the onset of anaphase in cultured chick fibroblasts and myoblasts (Chèvremont-Comhaire, 1953). According to Rogers (1964), protein and RNA biosyntheses were reduced, mainly at the level of the lateral parts of the optic vesicles, in both the dienkephalon and the telenkephalon. De Bernardi *et al.* (1969) reported that the Li^+ doses which caused cyclopia also inhibited the binding of ^{14}C-labelled amino acids to the ribosomes. With chick embryonal lymphoid cells, it was observed that the bursal lymphocytes were much more sensitive to lithium than the granulocytes and the erythrocytes, and thymus, spleen, liver and bone marrow were only slightly affected (Jankovic and Jankovic, 1981).

Schou and Amdisen (1970, 1971) reported that the number of birth malformations in humans was not significantly different between children born from lithium-treated or non-treated mothers. It is still recommended, however, to exercise caution when administering lithium to pregnant women, taking into account that (1) the placental barrier is permeable to lithium (Johansen and Ulrich, 1969) and (2) there may be a risk of teratogenic effects of lithium on the cardiovascular system (Weinstein and Goldfield, 1975).

No lithium-induced malformations were observed in the rat by Johansen and Ulrich (1969), Rider and Hsu (1976) or in the mouse by Tuchmann-Duplessis and Mercier-Parrot (1973). In contrast, significant abnormalities affecting mainly the palate, eyes, bones, enkephalon and the spinal cord were reported by Szabo (1970), Wright *et al.* (1971), Smithberg and Dixit (1982), Marathe and Thomas (1986) and Jurand (1988). Fetus resorption was also reported to occur (Wright *et al.*, 1971). These discrepancies between published results may originate from differences in the estimated lithium concentrations, possibly due to differences in the doses and in the mode of administration of lithium—*per os* or by injection.

The imaging of lithium has been performed using neutron-capture radiography in mouse embryos from mothers with 0.2 mM lithium concentrations in plasma (Wissocq *et al.*, 1985). In these studies lithium regionalization was very clear-cut, the ectodermal tissues being significantly more loaded with lithium than the endodermal ones; lithium concentration (mmol kg^{-1}) was 0.9 at the retina, 0.6–0.7 in the brain, 0.5 in the liver and 0.4 in the heart.

Molecular effects

Lithium was reported to diminish protein synthesis in sea-urchin embryos (Spiegel *et al.*, 1970; Wolcott, 1981). In amphibian embryos previously treated with lithium, protein synthesis is decreased (Leonardi Cigada *et al.*, 1973). Concerning *Xenopus* embryos, lithium reduced the rate of synthesis of specific proteins produced during the cleavage stage (Klein and King, 1988). Lithium was also reported to modify the activity of mitochondrial enzymes of the sea-urchin embryo (Gustafson and Hasselberg, 1951; De Vicentiis and Runnström, 1967). Ranzi (1962) showed that, *in vitro*, lithium exerted a protective effect on fibrillar proteins against proteolytic enzymes like pronase or trypsin, and suggested that a similar effect could occur *in vivo* during embryogenesis against endogenous cellular proteases (Ranzi, 1959; Lallier, 1969).

Concerning the effects on nucleic acid synthesis, in sea-urchin eggs lithium was suggested to interfere with transcription during embryogenesis (Markman and Runnström, 1970; Runnström and Immers, 1970). However, another report (Lallier, 1963) concluded that it rather acted on translation. A decrease of RNA synthesis was reported (Flickinger *et al.*, 1967) when fragments of frog embryos at early stages of development were treated with lithium and [^{14}C]uridine. Moreover, lithium decreased the rate of cell division and increased the length of the interphase, allowing an increased rate of synthesis of RNA compared with DNA (Flickinger *et al.*, 1970; Osborn *et al.*, 1979). In *Xenopus* embryos the rate of cell division and the incorporation of [^{3}H]thymidine were reduced in the presence of lithium (Smith *et al.*, 1976).

Hagström and Lönning (1967) reported that lithium slackened the chromosomes' migration in the mitotic spindles during the cleavage stage of the sea-urchin egg. Forer and Siller (1987), studying the effects of lithium during the first mitosis in fertilized sea-urchin eggs, concluded that the alteration of anaphase could result from an effect on the phosphoinositide pathway. These effects are reversed by *myo*-inositol but not by its stereoisomer *scyllo*-inositol or other cyclitols. Lithium could also interfere at the level of the release of intracellular calcium. Busa and Gimlich (1989) reported that the effect of lithium on frog eggs embryogenesis was reversed by injections of *myo*-inositol and concluded that lithium could interfere with the transduction of the inositol phosphate/protein kinase C signal in some cellular groups during cleavage.

Therefore, the most consistent scheme of the effect of lithium on embryogenesis would be: (1) lithium disturbs the phosphoinositide pathway, thus decreasing the release of Ca^{2+}; (2) as a consequence the rate of mitosis is reduced, which disturbs cell arrangement and movements; (3) this changes the spatial relationships between the different interacting groups of cells which (4) finally induce vegetalization in sea-urchin and increase and alter mesodermization and neurodermization in amphibian.

Regeneration

Ciliates (Protozoans)

The main studies concern the effects of lithium on stomatogenesis (i.e. mouth formation) and regeneration in a few ciliate genera and species.

Fauré-Frémiet (1949) reported that in *Leucophris piriformis* lithium caused the appearance of microstomal forms (i.e. reduced mouth structures), which usually appeared only upon starvation, and stomatogenesis was reversibly repressed.

Lithium diminished the motility of *Stentor coeruleus* and its ability to catch food (König, 1967). Cell division was reduced or arrested, depending on lithium concentration. The membranelles, cilia normally present at the peristome (mouth), may not form and modifications of body morphology were observed (Tartar, 1957). Membranelle morphogenesis was retarded and stomatogenesis was modified with the formation of anomalous peristomal forms.

Schweickhardt (1966) reported that lithium could interfere with a system of morphogen gradients similar to those existing in embryogenesis. Two gradients seem to be involved in the course of cell division or regeneration, affecting the differentiation pattern of the oral region formed by the peristome and the membranelles. One of these gradients is directed longitudinally with a maximal intensity at the basal level. The second gradient is circular and controls the width of a series of longitudinal bands located on both sides of a 'contrast line' of the body of *S. coeruleus*. These two gradients are localized close to the cytoplasmic side of the plasma membrane (cortical cytoplasm) but their chemical nature is still unknown. By limiting the regeneration of membranelles, lithium would strengthen the basal tendency of the longitudinal gradient (Schweickhardt, 1966; König, 1967). Besides, by favouring the appearance of a ventral line it could also act on the circular gradient. Sodium thiocyanate, an antagonist of lithium, tends to decrease the basal tendency of the gradient in favour of its apical tendency (apicalization) (Schweickhardt, 1966; König, 1967).

Similar observations have been reported for the ciliate *Spirostomum* (Schwartz, 1967) and seem to confirm the hypothesis of the strengthening of the basal pole by lithium.

In the ciliate *Tetrahymena pyriformis*, lithium was reported to inhibit both DNA and RNA synthesis, whereas their synthesis was activated by sodium thiocyanate. Protein synthesis does not seem to be affected by lithium (Volm *et al.*, 1970). In *Spirostomum*, however, synthesis of DNA and RNA seems to be modified differently in the different parts of the cell (Struttmann, 1973). In *Paramecium bursaria*, a strong decrease of macronucleus DNA by lithium was observed during the interphases (Struttmann, 1973).

Hydra

Lithium disturbs the regeneration of the distal part (head) of the *Hydra*. After amputation of this region and treatment with lithium, the extent of regeneration is somewhat inhibited and an increased number of regenerated tentacles is observed (Ham and Eakin, 1958). Chemicals like chloretone and xylocain, known as nervous system depressors, have an effect similar to that of lithium in increasing the number of regenerated tentacles. On the basis of the antagonistic effect exerted by Na^+ ions towards lithium, the authors consider that lithium could act at the level of the nerve cells. Indeed, NaCl prevents the decrease of distal regeneration and the increase of regenerated tentacles (Ham and Eakin, 1958). K^+, Mg^{2+} and Ca^{2+} have no antagonistic effects towards lithium (Yasugi, 1974).

Body fragments of *Hydra* obtained by a double amputation at mid-body and near the mouth normally regenerate a head and a foot. However, in the presence of lithium two heads were regenerated (Yasugi 1974). Lithium seems to interfere with a morphogen gradient and presumably with that of a cephalic activator protein isolated by Schaller and Bodenmüller (1981).

Interesting results have been reported on *Hydra attenuata* by Hassel and Berking (1988), although they do not concern directly regeneration. LiCl at a concentration of 20 mM inhibited the determination of stem cells into nerve cells, whereas at 1 mM, it inhibited their commitment into nematocytes, a kind of urticating tentacle cell. In *Hydra* exposed for 10 days to LiCl, the density of stem cells was reduced by at least 90% of its initial value, and nematocytes were almost completely absent. However, if the animals were fed prior to lithium treatment, the determination into nerve cells was strongly stimulated within the first three hours following a pulse treatment with LiCl.

Prolonged treatment with 1 mM LiCl blocked the self-renewal of stem cells as well as their determination into nematocytes. However, lithium had no effect on the differentiation of cells which had already started to differentiate into nematocytes. Thus, LiCl could interfere with both the control mechanisms of differentiation and of pattern formation in *Hydra*.

Planarians

Although published results concern various species and genera, some common features can be outlined. After decapitation, planarians regenerate a new head quite rapidly. However, if lithium is present, head regeneration is generally inhibited (Rulon, 1948; Teshirogi, 1955) and supernumerary eyes are often observed (Bronsted, 1942; Kanatani, 1958). With mid-body fragments, the presence of lithium results in the regeneration of duplicated heads and tails (Teshirogi, 1955). More recent results on *Dugesia* sp. (Bustuoabad *et al.*, 1980) show that in the presence of lithium, tail fragments regenerate anomalous heads displaying microcephaly, duplicated heads, cyclopia or supernumerary eyes. As in *Hydra*, lithium also seems to interfere with morphogen gradients.

Annelids

A few results on the effects of lithium have been reported for *Nereis diversicolor* and *Eisenia foetida*. Essentially, after the removal of caudal segments, an incomplete regeneration of these segments is observed (Massaro and Schrank, 1959; Stephan-Dubois and Morniroli, 1967). Lithium seems to cause mainly an inhibition of caudal regeneration in these organisms.

Differentiation of Slime Moulds

Slime moulds are relatively simple organisms and are well known at the level of the signals regulating their differentiation and the expression of some particular genes. During their normal development *Dictyostelium discoideum* cells aggregate into multicellular slugs displaying an antero-posterior (longitudinal) pattern of prestalk and prespore cells which further differentiate into stalk cells and spores. Early studies showed that *D. discoideum* cells growing in the presence of LiCl did not sporulate and differentiated into stalk-like cells after 48 h incubation (Maeda, 1970; Sakai, 1973). The extracellular signal for prestalk and prespore differentiation was shown to be cyclic AMP, which acts as a morphogenetic signal. Van Lookeren-Campagne *et al.* (1988) established that

Li^+ gave rise to a rupture of the prestalk/prespore pattern in the slugs by inducing an almost complete redifferentiation of prespore cells into prestalk cells. LiCl inhibited the induction of expression of the prespore-specific gene by cyclic AMP, whereas in the absence of a cyclic AMP stimulus, it promoted the induction of prestalk-associated gene expression. LiCl had no effect on the binding of cyclic AMP to cell-membrane receptors, but interfered with the transduction of the morphogenetic signals involved in the regulation of gene expression.

Peters *et al.* (1989) reported elsewhere that Li^+ interfered with the inositol phosphate metabolism. The stimulatory effect of LiCl on the expression of the CP2 gene, which encodes for the prestalk-associated cysteine-proteinase 2 (CP2), was not due to an increased production of inositol trisphosphate (InsP3). In contrast, an inhibition of prespore gene expression could result from the inhibitory effect of LiCl on the cyclic AMP-induced production of InsP3. The regulation of the CP2 gene was also suggested to be under the control of Ca^{2+} ions released from intracellular stores.

Viruses

In vitro and *in vivo* studies have shown that lithium has an inhibitory effect on the replication of *Herpes simplex* virus types 1 and 2 (HSV_1 and HSV_2) (Skinner *et al.*, 1980; Trousdale *et al.*, 1984). DNA synthesis is inhibited in the presence of LiCl although the presence of several viral proteins can still be detected. Lithium was also reported to inhibit the replication of other DNA viruses such as poxvirus and adenovirus, but not that of RNA viruses like influenza virus or encephalomyocarditis virus (Buchan *et al.*, 1988). In *in vitro* cultures of endothelial cells, LiCl reduces the synthesis of viral polypeptides; however the synthesis of host-cell proteins is maintained to a high level (Ziaie and Kefalides, 1989). Earlier results from Bach and Specter (1988) suggested that the antiviral effect of Li^+ on HSV could be due to competition between Li^+ and Mg^{2+} considered as a cofactor of many enzymes involved in the biosynthesis of viral polypeptides and nucleic acids. The simultaneous addition of Li^+ and Mg^{2+} showed a marked decrease of the inhibitory effect of Li^+, and conversely, this effect of Li^+ was enhanced in the presence of the divalent-cation chelator EDTA. Indeed, the antiviral action of LiCl was strengthened by using adjuvant chelating agents. More recently, a possible antiviral effect of lithium carbonate on HSV was reported (Amsterdam *et al.*, 1990). Another possibility which has been raised is that the *in vivo* inhibition of HSV infectivity by LiCl could be due to a stimulatory effect of Li^+ on the host immune system.

Bacteria

Escherichia coli

The first reported effects of lithium on bacteria concern mainly effects on morphology. In the presence of LiCl, the normal rod-like *Escherichia coli* cells are converted into spheroplasts which are protoplast-like forms of bacterial cells (Gamaleia, 1900). In addition to their distinct morphology, the spheroplasts show a slow rate of multiplication and a lower sensitivity to streptomycin (Pitzurra and Szybalski, 1959). However the mechanisms of these lithium-induced changes are not still clear.

In contrast with animal cells, in which the movements of K^+ and Na^+ are mediated by ATP-driven pumps (Na^+/K^+ ATPases), ion transport in bacteria is mediated by specific transport systems. In *E. coli*, transport of K^+ was shown to be driven by two ATP-requiring systems, Kdp and TrkA (Sorensen and Rosen, 1980). These authors, using Tl^+ and Li^+ ions, concluded that the transport of K^+ mediated by the Kdp and the TrkA systems involved the participation of Na^+.

In animal cells, transport of carbohydrates and amino acids is coupled with the transport of monovalent cations and mediated by specific membrane-carriers. The energy required for transport is stored as transmembrane electrochemical potentials of the ions which provide the driving force. Most cells use Na^+ gradients for cotransport of substrates across cell membranes, whereas bacteria generally use proton gradients. Some microorganisms, however, are known to use Na^+ for the cotransport of carbohydrates and amino acids.

Li^+ added to the growth medium of *E. coli* cells was shown to stimulate the transport of proline, while Na^+ and K^+ have a slight effect, and Rb^+, Cs^+ and NH_4^+ have no effect at all (Kawasaki and Kayama, 1973). Proline transport was synergistically stimulated by Li^+ in the presence of glucose and succinate as carbon sources. Lithium seemed to have no effect on the transport rate of other amino acids (Kayama and Kawasaki, 1976). Further experiments (Kayama-Gonda and Kawasaki, 1979) showed the presence of an H^+/Li^+ antiport, with Li^+ replacing Na^+. Proline would enter into the cell by a Li^+/Pro symport and induce a decrease of the membrane potential by an efflux of protons. A stimulation of proline transport by lithium was also reported on reconstituted membrane vesicles of *Mycobacterium phlei*; however, Li^+ was less effective than Na^+ in stimulating transport (Hirata *et al.*, 1974).

Glutamic acid and melibiose are cotransported with Na^+ in *E. coli* cells. Using a non-metabolizable analogue of melibiose, thiomethyl β-galactoside (TMG), Lopilato *et al.* (1978) and Tsuchiya *et al.* (1978) showed that Na^+ and Li^+ stimulated the melibiose transport system of *E. coli*. However, Li^+

inhibited cell growth when melibiose was used as sole carbon source. No
inhibition of cell growth by Li^+ occurred when cells were grown on glucose,
galactose, lactose or glycerol. Li^+ blocked the transport of melibiose but had
no effect on the induction of the melibiose operon or on the activity of
α-galactosidase, which cleaves melibiose to glucose and galactose. Tsuchiya
and Wilson (1978) reported that the inward cotransport of Li^+ and TMG
induced an efflux of protons, suggesting an electrogenic $Li^+/$melibiose
cotransport. More recently, the demonstration of a $Li^+/$TMG cotransport
was achieved with the use of Li^+-selective microelectrodes which allowed the
direct measurement of intracellular concentrations of lithium (Tsuchiya *et al.*,
1983). Mutant *E. coli* strains which were able to grow on melibiose in the
presence of Li^+ showed a high increase in Li^+ uptake. Because of their altered
cation-coupling system, these mutants were no longer dependent on Na^+ or
protons but only on Li^+ for the cotransport of melibiose (Niyia *et al.*, 1982;
Shiota *et al.*, 1985).

Other bacteria

In *Salmonella typhimurium*, as in *E. coli*, Na^+ and Li^+ stimulate the uptake of
TMG mediated by the melibiose carrier (Stock and Roseman, 1971).
Membrane vesicles of *S. typhimurium* showed a TMG-dependent uptake of
Na^+ when an outward diffusion potential of K^+ was induced across the
vesicles with valinomycin. Wild-type strains of *S. typhimurium* can grow on
media containing proline as sole carbon source and supplemented with NaCl,
but grow poorly when NaCl is replaced by LiCl. The main pathway for
proline uptake in *Salmonella* occurs via the proline permease encoded by the
putP gene (Ratzkin and Roth, 1978). Proline transport is coupled with Na^+
or Li^+ cotransport. Mutant strains which were able to grow on proline and
Li^+ showed low rates of $Li^+/$Pro and $Na^+/$Pro cotransports compared with
wild-type strains (Myers and Maloy, 1988). The location of each mutation
was determined by deletion maps of the putP gene.

Marine bacteria like *Vibrio alginolyticus* require both Na^+ and K^+ for the
active transport of amino acids. Lithium has been used to investigate the
transport systems of Na^+ and K^+ in relation to the active transport of amino
acids. When iso-osmotic concentrations of LiCl are used, the efflux of K^+
and the concomitant influx of Li^+ are enhanced by an increase of the
extracellular pH (Nakamura *et al.*, 1982). It has been demonstrated that K^+
is necessary for both the establishment of a pH gradient and of an electrochemical gradient of Na^+.

Vibrio parahemolyticus, the agent of gastroenteritis, grown on synthetic media
in the presence of different salts, shows optimal growth with Li^+ compared
with K^+, NH_4^+ and Rb^+ (Morishita and Takada, 1976). Other investigations

were carried on *Actinomyces viscosus* and *Streptococcus mutans*, oral microorganisms related to the cariogenicity of dental plaque in humans. Fluoride was found to increase the production of extracellular polysaccharides, whereas lithium tended to reverse this effect (Treasure, 1981). In *S. mutans*, lithium reduced bacterial growth and enhanced the properties of adhesion of the bacterial cells.

Plants: General Toxic Effects and Specific Effects on Signal Transduction

Fungi and algae

The effect of lithium on yeasts has been studied by Asensio *et al.* (1976), whose work includes references to previous publications. Under normal conditions of nutrition, no lithium was detected in yeast cells, and there is no indication that even low doses of Li^+ may be requisite or favourable to yeast growth. When lithium was added to the medium, a wide range of tolerance was found among 12 different yeast strains. At pH 6.5, the growth of some strains was inhibited by 4 to 60 mM Li^+, whereas in others, growth was still observed at 1.3 M Li^+. Increasing K^+ concentration from 0.5 to 30 mM enhanced lithium tolerance of the susceptible strains but not that of the tolerant ones. Lithium sensitivity proved to depend on the pH of the nutrient medium. However, the interpretation of such effects was not straightforward since some strains were less sensitive to lithium at low pH than at neutral pH, whereas others were more sensitive. Apart from the inhibitory effect on growth, lithium interfered with the sporulation process. When strain 361 of *Saccharomyces cerevisiae* was grown in the presence of K^+ at pH 6.5, under which conditions growth was inhibited by Li^+ concentrations above 1 M, at 10 mM Li^+ sporulation was delayed but the number of asci was not affected. Both a delay and a decrease in the number of asci occurred at 50 mM Li^+, and sporulation was seriously impaired at 100 mM Li^+. Q_{CO_2} and Q_{O_2} were not affected during the first two hours following Li^+ addition under conditions which inhibited growth, but incorporation of $[^3H]$leucine and of $[^{14}C]$adenine was stopped in less than 50 min after the addition of 100 mM Li^+ at pH 6.5 in the sensitive strains. It is therefore possible that inhibition of RNA synthesis is among the primary effects of Li^+. In these experiments, the external concentration of Li^+ had no effect on the cellular concentrations of K^+ and Na^+. Under the same external conditions, the cellular concentration of Li^+ was lower in the tolerant strains than in the sensitive ones, ranging from approximately 5 to 100 mM, against 100 mM for the external concentration. In the case of strain 767, which is sensitive to Li^+ at pH 6.5 but not at pH 3.5, the cellular Li^+ concentration was similar to the external one at pH 6.5 but was much lower at pH 3.5. These results suggest that tolerance to Li^+ depends on the ability of yeast cells to resist to Li^+ invasion under existing experimental conditions.

When mature cells of *Chara ceratophylla* were placed in solutions containing LiCl in concentrations ranging from 0.05 to 100 mM, the amount of Li^+ taken up by the cells increased only about three-fold when the external concentration was increased 10-fold (Collander, 1939). From the more diluted solutions, Li^+ was absorbed so abundantly, even in the dark, that the final concentration in the sap surpassed several times that in the outer medium. The absorption of Li^+ was stimulated by light, but the pH of the external solution, ranging from 5.0 to 8.4, had no great influence. When cells enriched with Li^+ over a period of about two days were placed in lithium-free solutions for two days, no significant exosmosis of Li^+ could be detected. The author has given no indication concerning a possible effect of the absorbed lithium on growth or on any other physiological process in *Chara* cells. In *Chara fragilis* (Strauss, 1968), the natural concentration of Li^+ recalculated by reference to the plant water content was of the order of 40 μM, whereas that in the external medium was less than 10 μM. When lithium was added to the external medium up to 2 mM, cellular lithium concentration referred to plant water was increased to 1.5 mM in the older parts and almost to 6 mM in the younger parts. External Li^+ concentrations above 0.7 mM inhibited plant growth. The addition of Na^+ antagonized the inhibitory effect of Li^+. In contrast, the addition of K^+ increased the toxicity of Li^+, although K^+ stimulated *Chara* growth in the absence of Li^+.

The unicellular marine diatom *Skeletonema costatum* shows a circadian rhythm of several parameters including biomass growth-rate. The growth of the algae was inhibited by lithium doses higher than 2 mM, but 0.5 mM Li^+ was enough to lengthen the circadian period (Ostgaard *et al.*, 1982).

Higher plants

The effect of lithium in higher plants has already been discussed by Desbiez and Thellier (1987). Except for a few desert plants which can accumulate lithium up to 5 mmol per kg dry weight or even more (Romney *et al.*, 1977), under usual natural conditions, higher plants contain practically no lithium, but they can absorb Li^+ ions added to their growth medium. To make comparisons easier between data of different authors, we have recalculated all the values of lithium concentrations in mmol per litre of tissue water. Kent (1941), adding LiCl to the soil of growing wheat plantlets, observed Li^+ concentrations in the first leaf close to 20, 40, 80 and 130 mM after 3, 10, 27 and 35 days of treatment, respectively. Lithium stored in the oldest leaves was apparently immobile, but when the lithiated plants were repotted in untreated soil, lithium was rapidly lost from the roots to the soil. For bean plants grown in a solution containing 0.5 mM Li^+ (Wallace *et al.*, 1977), the estimated Li^+ concentrations were close to 0.2 mM in the leaves, 0.8 mM in the shoots and 0.4 mM in the roots. Also for *Phaseolus* (Millet *et al.*, 1978), after 8 h of contact with a solution containing 8 mM Li^+, the mean concentration of lithium in the plants was close to 3.5 mM. For *Lemna gibba* (G3) grown in the presence of 0.05, 0.5 and 2 mM $LiNO_3$, lithium concentrations in the tissue water after 144 h were close to 1.6, 5.3 and 14.4 mM, respectively, whereas they were approximately half these values in the presence of 10 to 30 μM Na^+ in the external medium (Kondo, 1984).

Bielenski *et al.* (1984) have used the stable isotopes ^6Li and ^7Li to perform the compartment analysis of lithium in *Lemna gibba* (G1) grown in the presence of 1 mM Li^+. Lithium concentrations were found to be about 6 mM in the cell wall, 1.4 mM in the cytoplasm and 2.4 mM in the vacuole. The flux densities, entering and leaving, expressed in pmol $m^{-2} s^{-1}$, were 760 and 750 between outside and wall, 23 and 17 at the plasmalemma and 9.9 and 4.3 at the tonoplast. The results were consistent with the existence of a passive Li^+ leak with a permeability coefficient of the order of 10^{-13}–10^{-12} m s^{-1} both at the tonoplast and at the plasmalemma, and with an active pumping of Li^+ out of the cytoplasm both towards the wall and the vacuoles. Increasing the concentration of Ca^{2+} in the medium inhibits almost completely the cellular absorption of lithium by barley roots (Epstein, 1960; Jacobson *et al.*, 1960; Waisel, 1962). The presence of Li^+ represses the cellular absorption of K^+, but the addition of Ca^{2+} antagonizes the effect of Li^+ (Jacobson *et al.*, 1960).

Although Li^+ was shown to interfere with the cellular transport of glucose in *Lemna gibba* (G1), no direct interaction of Li^+ with the H^+-pump or with the glucose/H^+ symport could be detected (Hartmann *et al.*, 1983). Given the immobility of lithium in the old leaves (Kent, 1941), and noting the more recent studies by Hinz and Fisher (1976) with *Saxifraga*, it was concluded that Li^+ is practically phloemimmobile and that the long-distance transport of lithium in a plant occurs only via the xylem.

The presence of lithium in the medium can have a direct toxic effect on seed germination and plant growth. According to Gupta (1974), the percentage of germination was not affected up to 15 mM of external Li^+ in wheat and barley, whereas it was affected at 1.5 mM in rice; moreover, the growth of roots and shoots of the seedlings tended to be affected above 1 mM. However, there is a much more striking effect of lithium: at low external concentrations, Li^+ can affect the transduction of a variety of signals in plants, with practically no general effects of toxicity.

Lithium can disturb plant motility. The rates of opening and closure of *Pelargonium* stomata, following changes of light and of the CO_2 concentration of the atmosphere, were affected only by relatively high doses of Li^+ (Louguet and Thellier, 1976), but the spontaneous rhythm of stomata opening and closure was fairly lithium-sensitive (Brogärdh and Johnsson, 1974). Some plants exhibit nyctinastic movements, i.e. open their leaves during the day and close them at night. With *Cassia fasciculata* (Gaillochet, 1981), 0.3 mM Li^+ inhibited specifically the closure of folioles and 3 mM enhanced their opening, these Li^+ effects being reversed by Na^+ and K^+. With *Oxalis regnelli* (Johnsson *et al.*, 1981), Li^+ disturbed the nyctinastic rhythm both in period and phase, and the concentration in the pulvini (i.e. the motor organs of the leaves) was then much above that in the leaf limb. Lithium slackens the circadian rhythm of the petal movements of *Kalanchoe* flowers (Engelmann, 1972).

Although lithium has not been reported to exert a permanent effect on geotropism, application of Li^+ ions to root tips of *Zea mays* shifted the direction of initial orthogeotropic curvature from downward to upward (Millet and Pickard, 1988). The circumnutation of plants such as beans or *Ipomoea* (i.e. plants with twining shoots) is quite sensitive to lithium. With the bean *Phaseolus vulgaris*, Millet *et al.* (1978) observed that 5 mM Li^+ caused the plant to tend to grow straight, instead of circumnutating, whereas K^+ antagonized this effect. When growing shoots of *Ipomoea purpurea* are prevented from twining by insertion in a narrow glass tube, they usually exhibit a rapid and transient increase of their content in gibberellic acid, a plant 'hormone', at the site where twining would have occurred with non-immobilized plants. Li^+ ions, at concentrations ranging from 0.01 to 1 mM, inhibit specifically gibberellic response without affecting the rate of enlargement of the shoots (Claire, 1982).

The induction of plants to flower frequently depends on the relative lengths of light and darkness per day: long-day plants will flower when left under long-day/short-night conditions, and the reverse is true for short-day plants. In the two species of *Lemna*, *L. gibba* and *L. perpusilla*, the former is a long-day plant whereas the latter is a short-day one. When the plants were given 1 mM LiCl under long-day conditions, flower production was inhibited in *L. gibba* whereas it was stimulated in *L. perpusilla*, i.e. both plants tended to behave as if they were kept under short-day conditions. In this case Li^+ was assumed to act as an antagonist to K^+ (Kandeler, 1970). The induction to flower has been correlated with changes in peroxidase activities. When basal leaves of spinach plants were given flashes of red or far-red light, changes in the activity of basic peroxidases were observed a few minutes later in all the plant leaves, including those left in the dark (Penel, 1982). Lithium did not affect the peroxidase changes in the leaves which had been illuminated, but it inhibited those in the non-illuminated leaves (Karege *et al.*, 1982).

When a plant is wounded, local responses occur rapidly. For instance, a transient release of ethylene can be detected after 30 min, whereas changes of various protein activities (enzymes, transmembrane carriers) are observed some hours later. The effects of lithium on these reactions have been investigated mainly by sampling discs of mature leaves or fragments of storage organs, and by studying their behaviour within the hours following excision ('ageing' of the specimen under consideration). According to Laties (1959), Li^+ inhibits the stimulation of respiration which is normally observed during ageing of fragments of *Cichorium* roots and Li^+ in the range 1–10 mM inhibits the increase of the rate of cellular absorption of methylglucose which is normally observed during ageing of leaf discs of *Pelargonium*; at similar concentrations Na^+ has no effect. The inhibitory effect of Li^+ can be reversed by K^+ but not by Na^+. The Li^+ doses which inhibit completely the increase of the absorption rate reduce the incorporation of labelled amino acids in leaf proteins by not more than 10–15% (Carlier and Thellier, 1979). Results comparable to those achieved with the ageing processes were also obtained with 'filtration stress' (or 'gas shock') of cultured sycamore cell suspensions. Following the harvesting of the cells by filtration under vacuum, the rate of absorption for various substrates such as sugars, amino acids, phosphate or sulphate falls almost to zero and is then progressively restored after 10–20 h; moreover there are significant differences between the peroxidases released during normal cell growth and those released during the recovery of cells from filtration stress. The addition of 1 mM LiCl completely prevented restoration, although it reduced by only 30% the incorporation of tritiated leucine in the total proteins. NaCl had no effect, but K^+ antagonized the effect of Li^+ (Thoiron et al., 1980; Thellier et al., 1980; Li et al., 1990).

Small mechanical or traumatic signals may significantly affect plant morphogenesis. Rubbing one internode of young *Bryonia dioica* plants causes this internode to stop elongating without altering the other internodes ('thigmomorphogenetic response'). Lithium treatment (1 mM LiCl) prevents the inhibition of elongation due to rubbing; Li^+ treatment also suppresses the release of ethylene and of a specific cathodic isoperoxidase characteristic of rubbed plants (Boyer et al., 1979, 1983). Puncturing one or two cotyledons of six-day-old plantlets of *Bidens pilosa* caused a significant reduction of hypocotyl growth; the effect was particularly large in plants previously transferred into deionized water or 20 μM NaCl, but not with those transferred into normal nutrient medium or 20 μM LiCl. Ethylene metabolism and peroxidase activities were significantly increased in the hypocotyl within one to a few hours after puncturing, whereas Li^+, but not Na^+, inhibited these biochemical responses (Desbiez et al., 1987). With slightly older plants of *Bidens pilosa*, puncturing only one cotyledon of each plant causes the bud opposite the damaged cotyledon to start to grow faster than the other, after decapitation (i.e. ablation of the apex) of the plantlet. Since the normal plantlet is bilaterally symmetrical, this is termed 'breaking the symmetry of the plant' (Desbiez et al., 1984). When the punctured plantlets were not decapitated, nothing happened, both buds at the axil of the cotyledons appearing to remain quiescent. However, if the plants were later decapitated, up to a fortnight after the puncturing, the bud opposite the pricked cotyledon was again the first to start to grow. This means that the symmetry-breaking signal may remain latent, i.e. 'stored' in the plant, for at least two weeks. Even with decapitated plants, the expression of the asymmetry is not compulsory; it depends on the reception by the plant of other signals, e.g. mechanical or thermic, which enable this plant to 'retrieve' the stored symmetry-breaking message and express it in its morphogenesis. The whole process, including storage and retrieval of the symmetry-breaking information, depends very much on the ionic balance of the plant (Thellier and Desbiez, 1989). Lithium, even at micromolar concentrations, could inhibit the symmetry-breaking effect of the asymmetrical pricking treatments, whereas Na^+, at similar or higher concentrations, had no such effects (Desbiez and Thellier, 1975).

Tentative Interpretation of Lithium Effects in Plants

At least in part, the effect of lithium is via alteration of cell division (cytokinesis). The mitotic events in stamen hair-cells of *Tradescantia* are altered dramatically by treatment with 50 μM to 1 mM LiCl; depending on the moment of lithium treatment, the cells may be arrested in prophase, metaphase or anaphase. If cells are treated during late prophase, more than 80% fail to enter metaphase, while in cells treated in mid-metaphase, sister chromatids split at the normal time but further chromosome cleavage is arrested (Wolniak, 1987). When 10 mM LiCl is added to the same cells prior to early anaphase, cell-plate formation occurs by coalescence of Golgi vesicles, but the structure subsequently disperses and the resultant cells are binucleate (Chen and Wolniack, 1987). With meristemal onion root cells, 10–30 mM LiCl also induced binucleate cells by inhibiting cell-plate formation, whereas Na^+, K^+, Ca^{2+} and Mg^{2+} antagonized the lithium-induced inhibition of cytokinesis (Becerra and Eucina, 1987).

Several hypotheses have been made concerning the primary site of Li^+ action in plants. Since lithium and caffeine have additive effects on the induction of binucleate cells, Becerra and Eucina (1987) have suggested that Li^+ may compete with Mg^{2+} for the binding sites of ATPases. It has already been shown that Li^+ affects several activities, mainly in respiration, cellular transport, enzymes controlling ethylene metabolism and peroxidases. At least in part, this is via the biosynthesis of the corresponding polypeptide chains (Espejo *et al.*, 1982), and not only by modifying the activity of pre-existing proteins. Moreover, it is a specific effect of Li^+ on a restricted number of proteins, not a general effect of Li^+ on protein biosynthesis, since it has been emphasized that the incorporation of labelled amino acids in the bulk of the plant proteins was not altered very much by lithium treatment. These effects have appeared to be sensitive to the presence of other ions, mainly K^+. They are also quite sensitive to the tissue water potential (Carlier and Thellier, 1979). Kent (1941) had already noticed that in wheat seedlings lithium treatment depressed the net gain in water, dry weight and ash by the leaf during its growth, whereas depression was highest for water and lowest for ash. This interference of water and ionic characteristics with the lithium-sensitive processes may be quite important, although it is still not clearly understood. Inositol-phosphate phosphatase activity from *Lilium longiflorum* pollen is sensitive to high Li^+ concentrations (Loewus and Loewus, 1982). Callus formation from leaf discs of *Brassica oleracea* was blocked by 80% with 40 mM LiCl, whereas this inhibition was partially reversed by the addition of *myo*-inositol in the medium (Bagga *et al.*, 1987). In the cases where lithium altered cytokinesis, as described above, the addition of 10 μM *myo*-inositol or of 100 μM $CaCl_2$ could reverse Li^+ effects (Chen and Wolniak, 1987; Wolniak, 1987). In carrot cells inositol-bisphosphate phosphatase was partially inhibited by 10–50 mM Li^+; even 50 mM Li^+ had no effect on inositol-trisphosphate metabolism in either the microsomal or the soluble fractions, but Li^+ inhibited inositol-bisphosphate dephosphorylation in the soluble fraction by approximately 25% (Memon *et al.*, 1989).

The hypothesis according to which Li^+ may affect primarily the phospho-inositide metabolism in plants would give a unifying interpretation of lithium effects in animals and plants. There is, however, the difficulty that the effects of lithium on the phosphoinositide pathway in plants have been observed at relatively high lithium concentrations, in the range 10–50 mM in most cases. These, we have seen, are concentrations where lithium is indeed toxic to plant growth, whereas the specific effects of lithium on signal transduction (twining plants, induction to flower, thigmomorphogenesis, breaking of symmetry of bud growth, etc.) generally occur at much lower concentrations. Therefore it cannot be excluded that the effect of lithium on the phosphoinositide pathway only reflects the general toxicity of lithium at high concentrations, and not the specific effects observed at low concentrations. Moreover, Joseph *et al.* (1989) have denied that inositol-phosphate phosphatase of cultured tobacco cells was lithium-sensitive. We think that a clear answer to the question of whether Li^+ effects in plants are via inhibition of inositol phosphatases can be given only by quantitatively mapping lithium in the samples at the subcellular level, and by correlating the locally observed effects to the corresponding local concentrations of lithium. The extension to biological material of the physical methods of imaging such as neutron-capture radiography, nuclear probe or secondary-ion emission mass spectrometry, might be very helpful for this purpose (Thellier, 1984).

References

Amsterdam, J.D., Maislin, G. and Rybakowski, J. (1990). A possible antiviral action of lithium carbonate in herpes simplex virus infections. *Biol. Psych.* **27**, 447–453.

Asensio, J., Ruiz-Argüeso, T. and Rodriguez-Navarro, A. (1976). Sensitivity of yeasts to lithium. *Antonie van Leeuwenhoek* **42**, 1–8.

Bach, R.O. and Specter, S. (1988). Antiviral activity of the lithium ion with adjuvant agents. *In* 'Lithium Inorganic Pharmacology and Psychiatric Use' (ed. N.J. Birch), pp. 91–92. IRL Press, Oxford.

Backström, S. (1954). Morphogenetic effects of lithium on the embryonic development of *Xenopus*. *Ark. Zool.* **6**, 527–536.

Bagga, S., Das, R. and Sopory, S.K. (1987). Inhibition of cell proliferation and glyoxalase-1 activity by calmodulin inhibitors and lithium in *Brassica oleracea*. *J. Plant Physiol.* **129**, 149–154.

Becerra, J. and Eucina, C.L. (1987). Biological effects of lithium. Experimental analysis in plant cytokinesis. *Experientia* **43**, 1025–1026.

Bielenski, U., Ripoll, C., Demarty, M., Lüttge, U. and Thellier, M. (1984). Estimation of cellular parameters in the compartmental analysis of Li^+ transport in *Lemna gibba* using the stable isotopes ^6Li and ^7Li as tracers. *Physiol. Plant* **62**, 32–38.

Boyer, N., Chapelle, B. and Gaspar, T. (1979). Lithium inhibition of thigmomorphogenetic response in *Bryonia dioïca*. *Plant Physiol.* **63**, 1215–1216.

Boyer, N., Desbiez, M.O., Hofinger, M. and Gaspar, T. (1983). Effect of lithium on thigmomorphogenesis in *Bryonia dioïca*, ethylene production and sensitivity. *Plant Physiol.* **72**, 522–529.

Breckenridge, L.J., Warren, R.L. and Warner, A.E. (1987). Lithium inhibits morphogenesis of the nervous system but not neuronal differentiation in *Xenopus laevia*. *Development* **99**, 353–370.

Brogärdh, T. and Johnsson, A. (1974). Effects of lithium on stomatal regulation. *Z. Naturforschung* **29c**, 298–300.

Bronsted, H.V. (1942). Experiments with LiCl on the regeneration of Planarians. *Ark. Zool.* **34b**, 3, 1–8.

Buchan, A., Randall, S., Hartley, C.E., Skinner, G.R.B. and Fuller, A. (1988). Effect of lithium salts on the replication of viruses and non-viral microorganisms. *In* 'Lithium Inorganic Pharmacology and Psychiatric Use' (ed. N.J. Birch), pp. 83–90 IRL Press, Oxford.

Busa, W.B. and Gimlich, R.L. (1989). Li^+ induced teratogenesis in frog embryos prevented by a polyphosphoinositide or diacylglycerol analogue. *Devl. Biol.* **132**, 315–324.

Bustuoabad, O.D., Fiocchi, M.G. and Matteucci, I.M. (1980). Alteracion en la regeneracion de estructuras neurales por chloruro de litio. *Medicina (Buenos Aires)* **40**, 547–552.

Cade, J.F. (1949). Lithium salts in the treatment of psychotic excitement. *Med. J. Aust.* **36**, 349–352.

Carlier, G. and Thellier, M. (1979). Lithium perturbation of the induction of a methyl-glucose transport during aging of foliar disks of *Pelargonium zonale* L. Aiton. *Physiol. Veg.* **17**, 13–26.

Cernescu, C., Popescu, L., Constantinescu, S. and Cernescu, S. (1988). Antiviral effect of lithium chloride. *Rev. Roum. Med. Virol.* **39**, 93–101.

Chen, T.L. and Wolniak, S.M. (1987). Lithium induces cell plate dispersion during cytokinesis in *Tradescancia. Protoplasma* **141**, 56–63.

Chèvremont-Comhaire, S. (1953). Action du lithium sur la croissance et la mitose dans les cultures de fibroblastes et myoblastes. *Arch. Biol.* **65**, 295–299.

Claire, A. (1982). Augmentation de l'activité gibbérellique chez les tiges volubiles d'*Ipomoea purpurea* au cours du mouvement révolutif. *Physiol. Vég.* **20**, 11–22.

Collander, R. (1939). Permeabilitätsstudien an Characeen. III Die Aufnahme und Abgabe von Kationen. *Protoplasma* **33**, 215–257.

Cooke, J. and Smith, E.J. (1988). The restrictive effect of early exposure to lithium upon body pattern in *Xenopus* development studied by quantitative anatomy and immunofluorescence. *Development* **102**, 85–99.

Cooke, J., Symes, K. and Smith, E.J. (1989). Potentiation by the lithium ion of morphogenetic responses to a *Xenopus* inducing factor. *Development* **105**, 549–558.

De Bernardi, F., Cigada, M., Maci, R. and Ranzi, S. (1969). On protein synthesis during the development of lithium-treated embryos. *Experientia* **25**, 211–213.

Desbiez, M.O. and Thellier, M. (1975). Lithium inhibition of the mechanically induced precedence between cotyledonary buds. *Plant Sci. Lett.* **4**, 315–321.

Desbiez, M.O. and Thellier, M. (1987). Rôle des ions minéraux dans la morphogenèse végétale. in 'Le développement des végétaux: aspects théoriques et synthétiques' (ed. H. Le Guyader) pp. 173–197.

Desbiez, M.O., Kergosien, Y., Champagnat, P. and Thellier, M. (1984). Memorization and delayed expression of regulatory messages in plants. *Planta* **160**, 392–399.

Desbiez, M.O., Gaspar, T., Crouzillat, D., Frachisse, J.M. and Thellier, M. (1987). Effect of cotyledonary prickings on growth, ethylene metabolism and peroxidase activity in *Bidens pilosus. Plant Physiol. Biochem.* **25**, 137–143.

De Vicentiis, M. and Runnström, J. (1967). Studies on controlled and released respiration in animalized and vegetalized embryos of the sea urchin *Paracentrotus lividus. Exp. Cell Res.* **45**, 681–689.

Devries, J. (1976). Action du lithium sur l'embryon d'*Eisenia foetida* (Lombricien). *Arch. Biol.* **87**, 225–243.

Duvauchelle, R. (1965). Rol del ion litio sobre el desarrolo de blastodiscos de embriones de pollo cultivados 'in vitro'. *Biologica (Santiago)* **39**, 56–67.

Emanuelsson, H. (1971). Effects of lithium chloride on embryos of *Ophryotrocha labronica. W. Roux's Arch. Entw. Mech. Org.* **168**, 10–19.

Engelmann, W. (1972). Lithium slow down the *Kalanchoë* clock. *Z. Naturforschung* **27**, 477.

Engländer, H. and Johnen, A.G. (1967). Die morphogenetische Wirkung von Li-ionen auf gastrula-ektoderm von *Ambystoma* und *Triturus. W. Roux's Arch. Entw. Mech. Org.* **159**, 346–356.

Epstein, E. (1960). Calcium–lithium competition in absorption by plant roots. *Nature* **185**, 704–706.

Espejo, J., Thoiron, B., Thoiron, A. and Thellier, M. (1982). Effect of Li^+ on the incorporation of labelled leucine in membrane fractions of *Acer pseudoplatanus* cells. *In* 'Plasmalemma and Tonoplast: Their Function in the Plant Cell' (ed. D. Marmé *et al.*), pp. 155–162. Elsevier Biomedical Press, Amsterdam.

Fauré-Frémiet, E. (1949). Action du lithium sur la stomatogenèse chez les Ciliés. *Belg. Nederl. Cyto. Embryol. Dag. Gent,* 100–102.

Flickinger, R.A., Miyagi, M., Moser, C.R. and Rollins, E. (1967). The relation of DNA synthesis to RNA synthesis in developing frog embryos. *Devl. Biol.* **15**, 414–431.

Flickinger, R.A., Lauth, M.R. and Stambrook, P.J. (1970). An inverse relation between the rate of cell division and RNA synthesis per cell in developing frog embryos. *J. Embryol. Exp. Morph.* **23**, 571–582.

Forer, A. and Siller, P.J. (1987). The role of the phosphatidylinositol cycle in mitosis in sea urchin zygotes. *Exp. Cell Res.* **170**, 42–55.

Fujisawa, H. and Amemiya, S. (1982). Effects of zinc and lithium ions on the strengthening cell adhesion in sea urchin blastulae. *Experientia* **38**, 852–853.

Gaillochet, J. (1981). Effects of the lithium chloride on the leaf movements of *Cassia fasciculata. Planta* **151**, 544–548.

Gamaleia, N. (1900). Heteromorphismus der Bakterien unter dem Einfluss von Lithiumsalzen. *In* 'Elemente der allgemeinem Bakteriologie', pp. 204–215. Verlag von A. Hirschwald, Berlin.

Gavin, A.C. and Schorderet-Slatkine, S. (1988). The interaction of lithium with forskolin-inhibited meiotic maturation of denuded mouse oocytes. *Exp. Cell Res.* **179**, 298–302.

Gupta, I.C. (1974). Lithium tolerance wheat, barley, rice and gram at germination and seedling stage. *Indian J. Agric. Res.* **8**, 103–107.

Gustafson, T. and Hasselberg, I. (1951). Studies on enzymes in the developing sea urchin egg. *Exp. Cell Res.* **2**, 642–672.

Hagström, B.E. and Lönning, S. (1967). Cytological and morphological studies of the action of lithium on the development of the sea urchin embryo. *W. Roux's Arch. Entw. Mech. Org.* **158**, 9–11.

Ham, R.G. and Eakin, R.E. (1958). Time sequence of certain physiological events during regeneration in hydra. *J. Exp. Zool,* **139**, 33–53.

Hartmann, A., Bielenski, U., Lüttge, U., Garrec, J.P., Thoiron, A. and Thellier, M. (1983). Measurements of unidirectional fluxes of lithium: application to the study of Li^+/H^+ interactions with the transmembrane exchanges of *Lemna gibba* G1. *In* 'Physical Chemistry of Transmembrane Ion Motions' (ed. G. Spach), pp. 591–597. Elsevier, Amsterdam.

Hassel, M. and Berking, S. (1988). Nerve cell and nematocyte production in *Hydra* is deregulated by lithium ions. *W. Roux's Arch. Dev. Biol.* **197**, 471–475.

Herbst, C. (1893). Experimentelle Untersuchungen über den Einfluss der veränderten chemischen Zusammensetzung des umgebenden Mediums auf die Entwicklung der Tiere. II. Teil. *Mitt. Zool. Stn. Neapel* **11**, 136–220.

Hinz, U. and Fischer, H. (1976). Transport von Lithium und Caesium durch die Stolonen von *Saxifraga sarmentosa* L. *Z. Pflanzenphysiol.* **78**, 283–292.

Hirata, H., Kosmakos, F.C. and Brodie, A.F. (1974). Active transport of proline in membrane preparations from *Mycobacterium phlei. J. Biol. Chem.* **249**, 6965–6970.

Hörstadius, S. (1936). Weitere Studien über die Determination im Verlaufe der Eiachse bei Seeigeln. *W. Roux's Arch. Entw. Mech. Org.* **135**, 40–68.

Jacobson, L., Moore, D.P. and Hannapel, R.J. (1960). Role of calcium in absorption of monovalent cations. *Plant Physiol.* **35**, 352–358.

Jankovic, B.D., and Jankovic, D.L. (1981). The effect of lithium cation on the developing chick embryo. *Iugoslav. Physiol. Pharmacol. Acta* **17**, 239–242.

Johansen, K.T. and Ulrich, K. (1969). Preliminary studies of the possible teratogenic effect of lithium. *Acta Psych. Scand.* (Suppl.) **207**, 91–95.

Johnsson, A., Johnsen, P.I., Rinnan, T., and Skrove, D. (1981). Basic properties of the circadian leaf movements of *Oxalis regnellii* and period change due to lithium ions. *Physiol. Plant* **53**, 361–367.

Joseph, S.K., Esch, T. and Bonner, W.D., Jr (1989). Hydrolysis of inositol phosphates by plant cell extracts. *Biochem. J.* **264**, 851–856.

Jurand, A. (1988). Teratogenic activity of lithium carbonate: an experimental update. *Teratology* **38**, 101–111.

Kanatani, H. (1958). Effect of environment on the occurrence of supplementary eyes induced by lithium in the Planarian. *J. Fac. Sci. Tokyo IV* **8**, 245–251.

Kanatani, H., Shirai, H., Nakanishi, K. and Kurokawa, T. (1969). Isolation and identification of meiosis inducing substance in starfish *Asterias amurensis*. *Nature* **221**, 273–274.

Kandeler, R. (1970). Die Wirkung von Lithium und ADF auf die Phytochromsteverung der Blütenbildung. *Planta* **90**, 203–207.

Kao, K.R. and Elinson, R.P. (1989). Dorsalization of mesoderm by lithium. *Devl. Biol.* **132**, 81–90.

Kao, K.R., Masui, Y. and Elinson, R.P. (1986). Lithium-induced respecification of pattern in *Xenopus laevis* embryos. *Nature* **322**, 371–373.

Karege, F., Penel, C. and Greppin, H. (1982). Rapid correlation between the leaves of spinach and photocontrol of a peroxidase activity. *Plant Physiol.* **69**, 437–441.

Kawasaki, T. and Kayama, Y. (1973). Effect of lithium on proline transport by whole cells of *Escherichia coli*. *Biochem. Biophys. Res. Commun.* **55**, 52–59.

Kayama, Y. and Kawasaki, T. (1976). Stimulatory effect of lithium ion on proline transport by whole cells of *Escherichia coli*. *J. Bacteriol.* **128**, 157–164.

Kayama-Gonda, Y. and Kawasaki, T. (1979). Role of lithium ions in proline transport in *Escherichia coli*. *J. Bacteriol.* **139**, 560–564.

Kent, N.L. (1941). Absorption, translocation and ultimate fate of lithium in the wheat plant. *New. Phytol.* **40**, 291–298.

Klein, S.L. and King, M.L. (1988). Correlations between cell fate and the distribution of proteins that are synthesized before the midblastula transition in *Xenopus*. *W. Roux's Arch. Devl. Biol.* **197**, 275–281.

Klein, S.L. and Moody, S.A. (1989). Lithium changes the ectodermal fate of individual frog blastomeres because it causes ectopic neural plate formation. *Development* **106**, 599–610.

Kondo, T. (1984). Removal by a trace of sodium of the period lengthening of the potassium uptake rhythm due to lithium in *Lemma gibba* G3. *Plant Physiol.* **75**, 1071–1074.

König, K. (1967). Wirkung von Lithium- und Rhodanid-Ionen auf die polare Differenzierung und die Morphogenese von *Stentor coeruleus* E. *Arch. Protistenk.* **110**, 179–230.

Lallier, R. (1955). Recherches sur le problème de la détermination embryonnaire chez les Amphibiens et chez les Echinodermes. *Arch. Biol.* **66**, 223–402.

Lallier, R. (1963). Effets de l'actinomycine D sur le développement de l'oeuf de l'oursin *Paracentrotus lividus*. *C. R. Soc. Biol.* **257**, 2159–2162.

Lallier, R. (1969). Recherches sur les modifications expérimentales de la différenciation de la larve d'oursin par les enzymes protéolytiques. *C. R. Soc. Biol.* **163**, 2028–2032.

Lallier, R. (1979). Relations ioniques au cours du développement de l'oeuf de l'oursin *Paracentrotus lividus* et implications sur l'activité morphogène du lithium. *Arch. Biol.* **90**, 43–58.

Laties, G. (1959). The development and control of coexisting respiratory systems in slices of chicory roots. *Arch. Biochem. Biophys.* **79**, 378–391.

Lehmann, F.E. (1937). Mesodermisierung des präsumptiven Chordamaterials durch

Einwirkung von Lithium chlorid auf die Gastrula von *Triturus alpestris*. *W. Roux's Arch. Entw. Mech. Org.* **136**, 112–146.

Leonardi Cigada, M., De Bernardi, F.L. and Scarpetti Bolzern, A.M. (1973). Sintesi di proteine in embryo di *Xenopus laevis* trattati con LiCl. *Rend. Acad. Naz. Lincei Sci. Fiz.* **8**, 93–96.

Li, Z.S., Attias, J. and Thellier, M. (1990). Filtration stress-induced variations of peroxidase activity in cell suspension cultures of sycamore (*Acer pseudoplatanus*) cells. *Physiol. Plant* **78**, 22–28.

Lindahl, P.E. (1936). Zur Kenntnis der physiologischen Grundlagen der Determination im Seeigelkeim. *Acta Zool.* **17**, 179–365.

Loewus, M.W. and Loewus, F.A. (1982). *Myo*-inositol-1-phosphatase from the pollen of *Lilium longiflorum*. Thunb. *Plant Physiol.* **70**, 765–770.

Lopilato, J., Tsuchiya, T. and Wilson, T.M. (1978). Role of Na^+ and Li^+ in thiomethylgalactoside transport by the melibiose transport system of *Escherichia coli. J. Bacteriol.* **134**, 147–156.

Louguet, P. and Thellier, M. (1976). Influence du lithium sur le degré d'ouverture et les vitesses d'ouverture et de fermeture des stomates chez le *Pelargonium hortorum. C. R. Acad. Sci. Paris Série D* **282**, 2171–2174.

Maeda, Y. (1970). Influence of ionic conditions on cell differentiation and morphogenesis of the cellular slime molds. *Dev. Growth Differ.* **12**, 217–226.

Marathe, M.R. and Thomas, G.P. (1986). Embryotoxicity and teratogenicity of lithium carbonate in Wistar rat. *Toxicol. Lett.* **34**, 115–120.

Markitziu, A., Friedman, S., Steinberg, D. and Sela, M.N. (1988). The *in vitro* effect of lithium on growth and adherence of *Streptococcus mutans* 6715. *J. Trace Elem. Electrolytes Health Dis.* **2**, 199–203.

Markman, B. and Runnström, J. (1970). The removal by actinomycin D of the effect of endogenous or exogenous animalizing agents in sea urchin development. *W. Roux's Arch. Entw. Mech. Org.* **165**, 1–7.

Massaro, E. and Schrank, A.R. (1959). Chemical inhibition of segment regeneration in *Eisenia foetida. Physiol. Zoöl.* **32**, 185–196.

Masui, Y. (1961). Mesodermal and endodermal differentiation of the presumptive ectoderm of *Triturus* gastrula through influence of lithium ion. *Experientia* **17**, 458–459.

Masui, Y. (1966). pH-dependence of the inducing activity of lithium ion. *J. Embryol. Exp. Morph.* **15**, 371–386.

Memon, A.R., Rincon, M. and Boss, W.F. (1989). Inositol triphosphate metabolism in carrot (*Daucus carota* L.) cells. *Plant Physiol.* **91**, 477–480.

Millet, B. and Pickard, B.G. (1988). Early wrong-way response occurs in orthogravitropism of maize roots treated with lithium. *Physiol. Plant* **72**, 555–559.

Millet, B., Melin, D., Bonnet, B., Tavant, H. and Mercier, J. (1978). Contrôle expérimental de la durée de la période du mouvement révolutif du haricot. Rôle des ions Li^+ et K^+ dans le mécanisme de ce phénomène périodique. *Physiol. Vég.* **16**, 805–815.

Morishita, H. and Takada, H. (1976). Sparing effect of lithium ion on the specific requirement for sodium ion for growth of *Vibrio parahaemolyticus. Can. J. Microbiol.* **22**, 1263–1268.

Myers, R.S. and Maloy, S.R. (1988). Mutations of put P that alter the lithium sensitivity of *Salmonella typhimurium. Mol. Microbiol.* **2**, 749–755.

Nakamura, T., Tokuda, H. and Unemoto, T. (1982). Effects of pH and monovalent cations on the potassium ion exit from the marine bacterium, *Vibrio alginolyticus*, and the manipulation of cellular cation contents. *Biochim. Biophys. Acta* **692**, 389–396.

Nicolet, G. (1965). Action du chlorure de lithium sur la morphogenèse du jeune embryon de poule. *Acta Emb. Morph. Exp.* **8**, 32–85.

Nieuwkoop, P.D. (1953). The influence of the Li ion on the development of the egg of *Ascidia malaca* Pub. *Staz. Zool. Napoli* **24**, 101–141.

Nieuwkoop, P.D. (1970). The formation of mesoderm in urodelean amphibians. III—The vegetalizing action of the Li ion. *W. Roux's Arch. Entw. Mech. Org.* **166**, 105–123.

Niiya, S., Yamasaki, K., Wilson, T.M. and Tsuchiya, T. (1982). Altered cation coupling to melibiose transport in mutants of *Escherichia coli*. *J. Biol. Chem.* **257**, 8902–8906.

Osborn, J.C. and Stanisstreet, M. (1977). Comparison of cell division and cell sizes in normal embryos and lithium-induced exogastrulae of *Xenopus laevis*. *Acta Embryol. Exp.* **3**, 283–293.

Osborn, J.C., Wall, R. and Stanisstreet, M. (1979). Quantitative aspects of RNA synthesis in normal and lithium-treated embryos of *Xenopus laevis*. *W. Roux's Arch. Dev. Biol.* **187**, 269–282.

Ostgaard, K., Jensen, A. and Johnsson, A. (1982). Lithium ions lengthen the circadian period of growing cultures of the diatom *Skeletonema costatum*. *Physiol. Plant* **55**, 285–288.

Pasteels, J. (1945). Recherches sur l'action du LiCl sur les oeufs des Amphibiens. *Arch. Biol. Paris* **56**, 105–183.

Penel, C. (1982). Communications rapides chez les plantes et floraison. *In* 'Les mécanismes de l'irritabilité et du fonctionnement des rythmes chez les végétaux' (ed. H. Greppin and E. Wagner), pp. 125–135. Editions du Centre de Botanique, Genève, Suisse.

Peters, D.J.M., Van Lookeren Campagne, M.M., Van Haastert, P.J.M., Spek, W. and Schaap, P. (1989). Lithium ions induce prestalk-associated gene expression and inhibit prespore gene expression in *Dictyostelium discoideum*. *J. Cell Sci.* **93**, 205–210.

Picard, A. and Dorée, M. (1983). Lithium inhibits amplification or action of the maturation-promoting factor (MPF) in meiotic maturation of starfish oocytes. *Exp. Cell Res.* **147**, 41–50.

Pitzurra, M. and Szybalski, W. (1959). Formation and multiplication of spheroplasts of *Escherichia coli* in the presence of lithium chloride. *J. Bacteriol.* **77**, 614–620.

Ranzi, S. (1959). On the stability of the structures of animalized and vegetalized sea urchin embryos. *Biol. Bull.* **117**, 436–437.

Ranzi, S. (1962). The proteins in embryonic and larval development. *Adv. Morphogen.* **2**, 211–257.

Ratzkin, B. and Roth, J. (1978). Cluster of genes controlling proline degradation in *Salmonella typhimurium*. *J. Bacteriol.* **133**, 744–754.

Raven, C.P. (1952). Morphogenesis in *Limnaea stagnalis* and its disturbance by lithium. *J. Exp. Zool.* **121**, 1–77.

Raven, C.P. (1976). Morphogenetic analysis of spiralian development. *Amer. Zool.* **16**, 395–403.

Regen, C.M. and Steinhardt, R.A. (1988). Lithium dorsalises but also mechanically disrupts gastrulation of *Xenopus laevis*. *Development* **102**, 677–686.

Rider, A.A. and Hsu, J.M. (1976). Effect of lithium ingestion and water restriction on rat dam and offspring. *Nutrition Rep. Int.* **13**, 567–577.

Rogers, K.T. (1963). Experimental production of perfect cyclopia in the chick by means of LiCl with a survey of the literature on cyclopia produced experimentally by various means. *Devl. Biol.* **8**, 129–150.

Rogers, K.T. (1964). Radioautographic analysis of the incorporation of protein and nucleic acid precursors into various tissues of early chick embryos cultured in toto on medium containing LiCl. *Devl. Biol.* **9**, 176–196.

Romney, E.M., Wallace, A., Kinnear, J. and Alexander, G.V. (1977). Frequency distribution of lithium in leaves of *Lycium andersonii*. *Commun. Soil. Sci. Plant. Anal.* **8**, 799–802.

Rulon, O. (1948). The control of reconstitutional development in planarians with sodium thiocyanate and lithium chloride. *Physiol. Zoöl.* **21**, 231–237.

Runnström, J. (1928). Zur experimentellen Analyse der Wirkung des Lithiums auf den Seeigelkeim. *Acta Zool.* **9**, 365–424.

Runnström, J. and Immers, J. (1970). Heteromorphic budding in lithium treated sea urchin embryos. A study of gene expression. *Exp. Cell Res.* **62**, 228–238.

Sakai, Y. (1973). Cell type conversion in isolated prestalk and prespore fragments of the cellular slime mold *Dictyostelium discoideum*. *Dev. Growth Differ.* **15**, 11–19.

Schaller, H.C. and Bodenmüller, H. (1981). Isolation and aminoacid sequence of a morphogenetic peptide from *Hydra*. *Proc. Natl. Acad. Sci. USA* **78**, 7000–7004.

Schou, M. (1957). Biology and pharmacology of lithium. *Pharmacol. Rev.* **9**, 17–58.

Schou, M. and Amdisen, A. (1970). Lithium in pregnancy. *Lancet*, 1391.

Schou, M. and Amdisen, A. (1971). Lithium teratogenicity. *Lancet*, 1132.

Schwartz, V. (1967). Reaktionen der polaren Differenzierung von *Spirostomum ambiguum* auf Lithium- and Rhodanidionen. *W. Roux's Arch. Entw. Mech. Org.* **158**, 89–102.

Schweikhardt, F. (1966). Zytochemisch-entwicklungsphysiologische Untersuchungen an *Stentor coeruleus* Ehrbg. *W. Roux's Arch. Entw. Mech. Org.* **157**, 21–74.

Shiota, S., Yamane, Y., Futai, M. and Tsuchiya, T. (1985). *Escherichia coli* mutants possessing an Li^+-resistant melibiose carrier. *J. Bacteriol.* **162**, 106–109.

Signoret, J. (1960). Céphalogenèse chez le triton *Pleurodeles waltlii* Michah après traitement de la gastrula par le chlorure de lithium. *Mem. Soc. Zool. France* **32**, 1–117.

Skinner, G.R.B., Hartley, C., Buchan, A., Harper, L. and Galimore, P. (1980). The effect of lithium chloride on the replication of herpes simplex virus. *Med. Microbiol. Immunol.* **168**, 139–148.

Slack, J.M.W., Isaacs, H.V. and Darlington, B.G. (1988). Inductive effects of fibroblast growth factor and lithium ion on *Xenopus* blastula ectoderm. *Development* **103**, 581–590.

Smith, J.L., Osborn, J.C. and Stanisstreet, M. (1976). Scanning electron microscopy of lithium-induced exogastrulae of *Xenopus laevis*. *J. Embryol. Exp. Morph.* **36**, 513–522.

Smithberg, M. and Dixit, P.K. (1982). Teratogenic effects of lithium in mice. *Teratology* **26**, 239–246.

Sorensen, E.N. and Rosen, B.P. (1980). Effects of sodium and lithium ions on the potassium ion transport systems of *Escherichia coli*. *Biochemistry* **19**, 1458–1462.

Spiegel, M., Spiegel, E.S. and Meltzer, P.S. (1970). Qualitative changes in the basic protein fraction of developing embryos. *Devl. Biol.* **21**, 73–86.

Stephan-Dubois, F. and Morniroli, C. (1967). Effet du chlorure de lithium sur la régénération postérieure de *Nereis diversicolor* (Annélide Polychète). *Bull. Soc. Zool. France* **92**, 335–344.

Stock, J. and Roseman, S. (1971). A sodium-dependent sugar cotransport system in bacteria. *Biochem. Biophys. Res. Commun.* **44**, 132–138.

Stockard, C.R. (1906). The development of *Fundulus heteroclitus* in solutions of lithium chloride. *J. Exp. Zool.* **3**, 99–120.

Strauss, R. (1968). Sur l'absorption de lithium et les interactions entre ions alcalins chez *Chara fragilis* Desvaux. *C. R. Acad. Sci. Paris* **226**, 2249–2252.

Struttmann, C. (1973). Uber Reaktionen des Nukleinsäurehaushaltes und der polaren Differenzierung von Ciliaten auf Lithium- und Rhodanidionen. *Arch. Protistenk.* **115**, 271–323.

Szabo, K.T. (1970). Teratogenic effect of lithium carbonate in the foetal mouse. *Nature* **225**, 73–75.

Tartar, V. (1957). Reactions of *Stentor coeruleus* to certain substances added to the medium. *Exp. Cell Res.* **13**, 317–332.

Teshirogi, W. (1955). The effects of lithium chloride on head frequency in *Dugesia gonocephala. Bull. Mar. Biol. Station Asamushi Tohoku Univ.* **7**, 141–146.

Thellier, M. (1984). From classical to modern methods for element localization in plant systems. *Physiol. Vég.* **22**, 867–886.

Thellier, M. and Desbiez, M.O. (1989). "Stockage" et "rappel" d'information chez un végétal. La vie des Sciences, *C. R. Acad. Sci., Série Générale* **6**, 289–303.

Thellier, M., Thoiron, B., Thoiron, A., Le Guiel, J. and Lüttge, U. (1980). Effects of lithium and potassium on recovery of solute capacity of *Acer pseudoplatanus* cells after gas-shock. *Physiol. Plant.* **49**, 93–99.

Thoiron, B., Espejo, J., Thoiron, A., Le Guiel, J. and Thellier, M. (1980). Effect of alkaline ions on the control of permeation systems of *Acer* cells. *In* 'Plant membrane transport' (ed. R.M. Spanswick *et al.*), pp. 629–630. Elsevier North Holland, Biomed Press, Amsterdam.

Treasure, P. (1981). Effects of fluoride, lithium and strontium on extracellular polysaccharide production by *Streptococcus mutans* and *Actinomyces viscosus. J. Dent. Res.* **60**, 1601–1610.

Trousdale, M.D., Gordon, Y.J., Peters, A.C.B., Gropen, T.I., Nelson, E. and Nesburn, A.B. (1984). Evaluation of lithium as an inhibitory agent of Herpes simplex virus in cell cultures and during reactivation of latent infection in rabbits. *Antimicrob. Agents Chemother.* **25**, 522–523.

Tsuchiya, T. and Wilson, T.H. (1978). Cation-sugar cotransport in the melibiose transport system of *Escherichia coli. Memb. Biochem.* **2**, 63–79.

Tsuchiya, T., Lopilato, J. and Wilson, T.H. (1978). Effect of lithium ion on melibiose transport in *Escherichia coli. J. Membrane Biol.* **42**, 45–59.

Tsuchiya, T., Oho, M. and Shiota-Niiya, S. (1983). Lithium ion-sugar cotransport via the melibiose transport system in *Escherichia coli. J. Biol. Chem.* **258**, 12765–12767.

Tuchmann-Duplessis, H. and Mercier-Parrot, L. (1973). Influence du lithium sur la gestation et le développement prénatal du rat et de la souris. *C. R. Soc. Biol.* **167**, 183–186.

Vahs, W. and Zenner, H. (1964). Der Einfluss von Lithium ionen auf die Embryonalentwicklung meroblastisher Wirbeltiere (*Salmo irideus*). *W. Roux's Arch. Entw. Mech. Org.* **155**, 632–636.

Vahs, W. and Zenner, H. (1965). Der Einfluss von Alkali und Erdalkisalzen auf die Embryonalentwicklung der Regenbogenforelle. *W. Roux's Arch. Entw. Mech. Org.* **156**, 96–100.

Van den Biggelaar, J.A.M. (1971). Development of division asynchrony and bilateral symmetry in the first quartet of micromeres in eggs of *Limnea. J. Embryol. Exp. Morph.* **26**, 393–399.

Van Lookeren-Campagne, M.M., Wang, M., Spek, W., Peters, D. and Schaap, P. (1988). Lithium respecifies cyclic AMP-induced cell-type specific gene expression in *Dictyostelium. Dev. Gen.* **9**, 589–596.

Verdonk, N.H. (1968). The determination of bilateral symmetry in the head region of *Limnea stagnalis. Acta Embryol. Exp.* **10**, 211–227.

Volm, M., Schwartz, V. and Wayss, K. (1970). Effect of lithium and thiocyanate on the nucleic acid-synthesis of *Tetrahymena*. *Naturwissenschaften* **57**, 250.

Von Ubisch, L. (1925). Uber die Entodermisierung ektodermaler Bezirke des Echinoiden Kermes und die Reversion dieses Verganges. *Verh. phys. med. Ges. Würzburg* **50**, 13–19.

Waisel, Y. (1962). The absorption of Li^+ and Ca^{2+} by barley roots. *Acta Botan. Neerlandica* **11**, 56–68.

Wallace, A., Ronney, E.M., Cha, J.W. and Chaudhry, M. (1977). Lithium toxicity in plants. *Comm. Soil Sci. Plant Anal.* **8**, 773–780.

Weinstein, M.R. and Goldfield, M.D. (1975). Cardiovascular malformations with lithium use during pregnancy. *Am. J. Psychol.* **132**, 529–531.

Wissocq, J.C., Heurteaux, C., Hennequin, E. and Thellier, M. (1985). Microlocating lithium in the mouse embryo by use of a $(n, \propto)$ nuclear reaction. *W. Roux's Arch. Dev. Biol.* **194**, 433–435.

Wolcott, D.L. (1981). Effect of potassium and lithium ions on protein synthesis in the sea urchin embryo. *Exp. Cell Res.* **132**, 464–468.

Wolniak, S.M. (1987). Lithium alters mitotic progression in stamen hair cells of *Tradescancia* in a time dependent and reversible fashion. *Eur. J. Cell Biol.* **44**, 286–293.

Wright, T.L., Hoffman, L.H. and Davies, J. (1971). Teratogenic effects of lithium in rats. *Teratology* **4**, 151–156.

Yasugi, S. (1974). Observations on supernumerary head formation induced by lithium chloride treatment in the regenerating hydra, *Pelmatohydra robusta*. *Devl. Growth Differ.* **16**, 171–180.

Ziaie, Z. and Kefalides, N.A. (1989). Lithium chloride restores host protein synthesis in herpes simplex virus-infected endothelial cells. *Biochem. Biophys. Res. Commun.* **160**, 1073–1078.

3 New Approaches to Identification of Metal–Ligand Interactions in Solution and Application to Some Lithium Complexes

DAVID R. CROW

School of Applied Sciences, Wolverhampton Polytechnic, Wolverhampton WV1 1LY, UK

Introduction

Hydrated, or otherwise solvated, metal ions are complexes in which coordination with solvent molecules produces a solute species whose effective size is significantly different from that of the free ion as would be understood in the crystallographic sense.

Replacement of such solvation molecules by other, competing, complexing species alters the size and/or shape of the complex. Such changes are in turn reflected in changes in the values of measurable physical parameters of either metal ion or ligand. Stepwise replacement of inner sphere solvation molecules by another complexing species may be identified by means of suitable treatment of measured complementary changes in a variety of convenient parameters. This leads to numerical values for formation constants of the various species formed; even though these may not have thermodynamic significance, due to the use of concentration rather than activity terms, values emerging for such functions which are significantly greater than unity provide strong evidence for the existence of the various complexes.

The general principles of methods developed by the author for the evaluation of such constants will be discussed and their application to a number of lithium–ligand systems considered in detail.

General Principles

The analysis used is based upon that devised by Leden (1941) combined with the polynomial function proposed by Fronaeus (1950), namely

$$F_0[X] \approx 1 + \beta_1[X] + \beta_2[X]^2 + \cdots + \beta_N[X]^N \qquad (1)$$

LITHIUM AND THE CELL
ISBN 0-12-099300-7

where $[X]$ stands for free ligand concentration, βs for overall formation constants, $\beta_j = [MX_j]/[M][X]^j$, and $F_0[X]$ functions are so labelled to emphasize that they are increasing functions of *free* ligand concentration.

There are various ways of calculating values of $F_0[X]$ and in this chapter we are concerned with a novel approach. Before dealing with this, however, it is constructive to consider the application of the graphical method of Leden.

Looking at Equation 1, it is clear that a plot of $F_0[X]$ against $[X]$ yields a curve with intercept $= 1$. A new function, $F_1[X]$, may be defined by

$$F_1[X] = \frac{F_0[X] - 1}{[X]} = \beta_1 + \beta_2[X] + \cdots + \beta_N[X]^{N-1} \quad (2)$$

A plot of $F_1[X]$ against $[X]$ yields a curve with an intercept of β_1. Similarly, a succession of functions may be defined up to the penultimate

$$F_{N-1}[X] = \beta_{N-1} + \beta_N[X] \quad (3)$$

and the final

$$F_N[X] = \beta_N \quad (4)$$

Means of determining values of $F_0[X]$ functions

Direct methods use well-characterized equations which relate $F_0[X]$ to the shifts in value of a parameter of the metal ion which is of *thermodynamic* significance. Thus, the method developed by DeFord and Hume (1951a,b) employs the shift in reversible polarographic half-wave potential, $\Delta E_{1/2}$, of a metal ion induced by a ligand. Under such circumstances, $\Delta E_{1/2}$ is a function of $F_0[X]$; in principle, a set of thermodynamic constants should emerge. The attraction of this method is that it may yield, via one set of experimental data, equilibrium information for a large number of consecutive interactions. Its early promise, however, has been marred by the difficulty of the measurements, their time-consuming collection and the more fundamental fact that only comparatively rarely are electrochemical parameters of strictly thermodynamic significance.

Indirect methods rely upon the use of the data $\bar{n}$, $\log [X]$, where $\bar{n}$ is the average number of ligands attached to each metal ion for a given value of $[X]$. Many approaches to such assessments have been used with varying degrees of success. Although details of the methods are not required for this discussion, a feature which they all have in common with the new methods to be considered here is the concept of the *formation curve*. Such a curve results from a plot of $\bar{n}$ versus $\log [X]$, its shape being typically sigmoid (Fig. 1). Its significance may be better understood by considering Equation 1 in a less usual form. If, for a given value of $[X]$, rather than considering a power

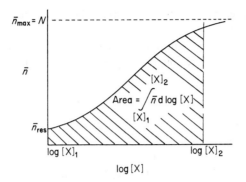

FIG. 1   Schematic formation curve.

series for $F_0[X]$ we examine its expression in terms of an overall *average* formation constant $\beta_{\bar{n}}$, corresponding to development of $\bar{n}$, we may write:

$$F_0[X] \sim \beta_{\bar{n}}[X]^{\bar{n}} \tag{5}$$

or

$$\log F_0[X] \sim \log \beta_{\bar{n}} + \bar{n} \log [X] \tag{6}$$

so that

$$\frac{d \log F_0[X]}{d \log [X]} \sim \bar{n} \tag{7}$$

Equation 7 is of great significance and central to later arguments. It follows from Equation 7 that

$$d \log F_0[X] = \bar{n}\, d \log [X]$$

and

$$\log F_0[X] = \int_0^{[X]} \bar{n}\, d \log [X] \tag{8}$$

The significance of Equation 8 is that it represents the area under the formation curve from $[X] = 0$ up to any value. It may be shown that the residual

$$(F_0[X])_{[X]\to 0} \sim \frac{1}{1 - \bar{n}_{res}}$$

if $\bar{n}_{res} \ll 1$, so that at $[X]_2$

$$\log F_0[X] = \log\left(\frac{1}{1 - \bar{n}_{res}}\right) + \int_{\log[X]_1}^{\log[X]_2} \bar{n}\, d \log [X] \tag{9}$$

Thus, in principle, values of $F_0[X]$ may be generated from graphical integration over a wide range of $[X]$.

If access may be had to a function, z, whose change in average value, $\Delta\bar{z}$, varies with ligand concentration in a way which parallels the formation curve, it may be postulated that

$$\Delta\bar{z} = k\bar{n}; \quad \Delta z_{max} = kN \tag{10}$$

A graph of $\Delta\bar{z}$ versus $\log[X]$ will then have essentially the same shape as a formation curve; although this of itself can give no direct indication of $\bar{n}$, its position relative to the $\log[X]$ axis will be the same as that of the (as yet unknown) formation curve. A plot of $\Delta\bar{z}$ versus $\log[X]$ is known as a pseudo-formation curve (Crow, 1982a,b, 1986, 1988a,b); such curves have been constructed and analysed which have been based upon diffusion coefficients, chemical shifts, molar conductivities, fluorescence intensities and circular dichroism signals.

Nature of pseudo-formation curves and generation of $F_0[X]$ data

By analogy with Equation 8, integration of a pseudo-formation curve should yield a function $F_0^1[X]$ defined by

$$\int_0^{[X]} \Delta\bar{z}\, d\log[X] = \log F_0'[X] \tag{11}$$

If the functions $F_0[X]$ and $F_0'[X]$ may be related, a route is open to $\beta_1 \ldots \beta_N$. Since

$$\frac{d\log F_0[X]}{d\log[X]} = \bar{n} = k'\Delta\bar{z} \tag{12}$$

(where $k' = 1/k$, see Equation 10) it follows that

$$\frac{d\log F_0[X]}{k'd\log[X]} = \Delta\bar{z} = \frac{d\log f_0^1[X]}{d\log[X]} = \frac{\bar{n}}{k'} \tag{13}$$

and that

$$\log F_0[X] = k'\log F_0'[X] \tag{14}$$

Identification of k' or k is crucial to the determination of $F_0[X]$; this represents a unique solution and the only one capable of producing realistic Leden analysis.

First, the relationship $\Delta\bar{z} = k\bar{n}$ will be *established*, after which the theoretical foundation for experimental determination of k' will be considered in some detail. Finally, a series of examples involving analysis of systems for different degrees of coordination involving lithium will be given.

Consider stepwise replacement of hydration water in the aqua complex $M(H_2O)_N$ by ligand species X:

$$M(H_2O)_N + xX + (w - N)H_2O \qquad \overset{k_1}{\rightleftharpoons}$$

$$M(H_2O)_{N-1}X + (x - 1)X + (w - (N - 1))H_2O \overset{k_2}{\rightleftharpoons}$$

$$M(H_2O)_{N-2}X_2 + (x - 2)X + (w - (N - 2))H_2O \ldots \qquad (15)$$

$$\overset{k_N}{\ldots \rightleftharpoons}$$

$$MX_N + (x - N)X + wH_2O$$

The various step constants may be expressed in terms of activities $\{\ \}$ by:

$$k_1 = \frac{\{M(H_2O)_{N-1}X\}\{H_2O\}}{\{M(H_2O)_N\}\{X\}}; \quad \{M(H_2O)_{N-1}X\} = k_1\{M(H_2O)_N\}\frac{\{X\}}{\{H_2O\}}$$

$$\therefore \quad m_{MX} \approx k_1 m_M \frac{\{X\}}{\{H_2O\}} \qquad (16a)$$

In a similar way it is easily shown that

$$m_{MX_2} \approx k_1 k_2 m_M \left(\frac{\{X\}}{\{H_2O\}}\right)^2 \qquad (16b)$$

and

$$m_{MX_N} \approx k_1 k_2 \ldots k_N m_M \left(\frac{\{X\}}{\{H_2O\}}\right)^N \qquad (16c)$$

The shift, $\Delta\bar{z}$, in the value of $\bar{z}$, induced by a given value of $\{X\}$ may be expressed in terms of *species* shifts, thus

$$\Delta\bar{z} = \sum_i x_i \Delta z_i \qquad (17)$$

where x_i is the mole fraction of species i and Δz_i is the *intrinsic species shift*.

$$\therefore \quad \Delta\bar{z} = x_{MX}\Delta z_{MX} + \cdots + x_{MX_N}\Delta z_{MX_N} \qquad (18)$$

If, further, it is assumed that *species shifts* are in proportion to $\{X\}$, then in practical terms this implies non-cooperativity,

$$\Delta z_{MX} = \frac{\Delta z_{max}}{N}; \quad \Delta z_{MX_2} = \frac{2\Delta z_{max}}{N}; \ldots \qquad (19)$$

Then

$$\Delta \bar{z} = x_{MX} \frac{\Delta z_{max}}{\mathcal{N}} + x_{MX_2} \frac{2\Delta z_{max}}{\mathcal{N}} + \cdots \qquad (20)$$

Therefore,

$$\frac{\Delta \bar{z}}{\Delta z_{max}} = \frac{1}{\mathcal{N}} \{ x_{MX} + 2x_{MX_2} + \cdots + \mathcal{N} x_{MX_{\mathcal{N}}} \} \qquad (21)$$

Now,

$$x_{MX} = \frac{m_{MX}}{\sum_i m_{MX}} \cdots \qquad (22)$$

This means that, in terms of Equations 16, Equation 21 becomes

$$\frac{\Delta \bar{z}}{\Delta z_{max}}$$

$$= \frac{1}{\mathcal{N}} \left\{ \frac{k_1 \left(\frac{\{X\}}{\{H_2O\}} \right) m_M + 2k_1 k_2 \left(\frac{\{X\}}{\{H_2O\}} \right)^2 m_M + \cdots + \mathcal{N} k_1 k_2 \ldots k_{\mathcal{N}} \left(\frac{\{X\}}{\{H_2O\}} \right)^{\mathcal{N}} m_M}{m_M + k_1 \left(\frac{\{X\}}{\{H_2O\}} \right) m_M + k_1 k_2 \left(\frac{\{X\}}{\{H_2O\}} \right)^2 m_M + \cdots + k_1 k_2 \ldots k_{\mathcal{N}} \left(\frac{\{X\}}{\{H_2O\}} \right)^{\mathcal{N}} m_M} \right\}$$

which, if $\{H_2O\} \sim 1$ and $\{X\} \sim [X]$ becomes

$$\frac{\Delta \bar{z}}{\Delta z_{max}} = \frac{1}{\mathcal{N}} \frac{k_1[X] + 2k_1 k_2[X]^2 + \cdots + \mathcal{N} k_1 k_2 \ldots k_{\mathcal{N}}[X]^{\mathcal{N}}}{(1 + k_1[X] + k_1 k_2[X]^2 + \cdots + k_1 k_2 \ldots k_{\mathcal{N}}[X]^{\mathcal{N}})} \qquad (23)$$

Therefore

$$\frac{\Delta \bar{z}}{\Delta z_{max}} = \frac{1}{\mathcal{N}} \frac{\beta_1[X] + 2\beta_2[X]^2 + \cdots + \mathcal{N} \beta_{\mathcal{N}}[X]^{\mathcal{N}}}{1 + \beta_1[X] + \beta_2[X]^2 + \cdots + \beta_{\mathcal{N}}[X]^{\mathcal{N}}} \qquad (24)$$

Or, in general,

$$\frac{\Delta \bar{z}}{\Delta z_{max}} = \frac{\bar{n}}{\mathcal{N}} \qquad (25a)$$

or

$$k' \Delta \bar{z} = \bar{n} \qquad (25b)$$

and

$$k' \Delta z_{max} = \mathcal{N} \qquad (25c)$$

Step formation constants and Scatchard slope K'

If the individual step constants are governed by statistical requirements only (i.e. no cooperativity is involved) they may be related via a common constant K' in the expressions

$$k_1 = NK'; \; k_2 = \frac{(N-1)K'}{2}; \; \dots \; k_j = \frac{[N-(j-1)]K'}{j}; \; \dots \; k_N = \frac{K'}{N} \quad (26)$$

i.e.

$$K' = (k_1 k_2 \dots k_N)^{1/N} = (\beta_N)^{1/N} \quad (27)$$

Substitution of Equations 26 and 27 into Equation 23 yields, after factorizing,

$$\frac{\Delta \bar{z}}{\Delta z_{max}} = \frac{K'[X]}{1 + K'[X]} = \frac{\beta_N^{1/N}[X]}{1 + \beta_N^{1/N}[X]} \quad (28)$$

Or, in terms of a more explicit expression of the left-hand side,

$$\frac{z_M - \bar{z}}{z_M - z_{MX_N}} = \frac{K'[X]}{1 + K'[X]} \quad (29)$$

Where, if z is a characteristic parameter of metal ion M, z_M corresponds to the 'free' (i.e. totally solvated) species, $\bar{z}$ to the average behaviour of all complex species for given $[X]$, while z_{MX_N} corresponds to the limiting value of $\bar{z}$ observed for the highest complex. Conversely, for a parameter characteristic of a ligand species interacting with a number of metal ions to form XM_N, we may write

$$\frac{\Delta \bar{z}}{\Delta z_{max}} = \frac{z_X - \bar{z}}{z_X - z_{XM_N}} = \frac{K'[M]}{1 + K'[M]} \quad (30)$$

Equation 29 may be rearranged into a number of linear forms.

(i) $$(z_M - \bar{z})(1 + K'[X]) = (z_M - z_{MX_N})K'[X]$$

therefore

$$\frac{(z_M - \bar{z})}{[X]} = K'(z_M - z_{MX_N}) - K'(z_M - \bar{z}) \quad (31)$$

This is the formulation usually attributed to Scatchard (1949) and is probably the most useful in practical terms.

(ii) If the reciprocal of Equation 29 is taken, namely

$$\frac{z_M - z_{MX_N}}{z_M - \bar{z}} = \frac{1 + K'[X]}{K'[X]}$$

then

$$\frac{1}{(z_M - \bar{z})} = \frac{1}{(z_M - z_{MX_N})} \cdot \frac{1}{K'[X]} + \frac{1}{(z_M - z_{MX_N})} \tag{32}$$

which is the Benesi–Hildebrand formulation (1949).

(iii) Slight rearrangement of Equation 32 gives the expression adopted by Scott (1956), namely

$$\frac{[X]}{(z_M - \bar{z})} = \frac{1}{K'(z_M - z_{MX_N})} + \frac{[X]}{(z_M - z_{MX_N})} \tag{33}$$

The complementary Equations 31, 32 and 33 all contain the terms K' and Δz_{max} and values of these two functions may, in principle, be obtained from graphical exploitation of the equations. Two important points should be emphasized at this juncture. First, an experimental value of K' may reflect a variety of complexing behaviour, ranging from simple 1:1 interactions to a sequence of multiple equilibria involving both mononuclear and polynuclear associations. There is no way of knowing at this stage which state of affairs exists. Second, accurate identification of Δz_{max} is fundamental to the method of analysis to be described and applied. While some workers have employed double reciprocal plots of $1/\Delta\bar{z}$ against $1/[X]$, as suggested by the Benesi–Hildebrand formulation, this can be of dubious reliability. A better approach is provided by the Scatchard method involving a plot of $\Delta\bar{z}/[X]$ against $\Delta\bar{z}$, which can usually serve to identify Δz_{max} reliably even when a significant degree of cooperativity is involved.

A variation on the Scatchard method makes use of the values of the function

$$\frac{\Delta\bar{z}}{(\Delta z_{max} - \Delta\bar{z})[X]} = K'$$

(Crow, 1988a,b), over the range of $[X]$ used to identify Δz_{max} and K'. Should it not prove possible to achieve constancy of this function, the reasons, apart from inadequate or poorly determined data, are interventions of other interactions such as self-association, polynuclear complex formation or extreme cooperativity. If Δz_{max} is known reliably, the constant k' in Equation 25c may be calculated so long as N is known: this is, of course, the problem which must be resolved if $\beta_1 \ldots \beta_N$ are to be disentangled from K'. It is rarely possible to determine k' or N by direct means and one must resort to testing a range of values. However, this range is usually small and the correct choice, once identified, will yield a value of k' which may be combined with $\log F_0^1[X]$ data to provide authentic $F_0[X]$ data. In terms of Equation 14 it is seen that

$$(F_0'[X])^{k'} = 1 + \beta_1[X] + \beta_2[X]^2 + \cdots + \beta_N[X]^N \tag{34}$$

There is clearly only one value of k' which, applied to experimentally derived $F_0'[X]$, can produce a sensible Leden analysis.

Use of average value, $\bar{z}$, when analysis of a pseudo-formation curve has established the value of $\mathcal{N}$

$\bar{z}$ may be expressed as

$$\bar{z} = \frac{\sum_0^{\mathcal{N}} z_j \beta_j [X]^j}{\sum_0^{\mathcal{N}} \beta_j [X]^j} = \frac{z_M + z_{MX}\beta_{MX}[X] + \cdots + z_{MX_{\mathcal{N}}}\beta_{MX_{\mathcal{N}}}[X]^{\mathcal{N}}}{1 + \beta_{MX}[X] + \cdots + \beta_{MX_{\mathcal{N}}}[X]^{\mathcal{N}}} \tag{35}$$

from which it is simple to show that

$$(z_M - \bar{z}) = (\bar{z} - z_{MX})\beta_{MX}[X] + \cdots + (\bar{z} - z_{MX_{\mathcal{N}}})\beta_{MX_{\mathcal{N}}}[X]^{\mathcal{N}} \tag{36}$$

where z_{MX_j} represent the *species* values of z.

A function, G_1, may now be defined (Crow, 1983):

$$G_1 = \frac{(z_M - \bar{z})}{[X]} = (\bar{z} - z_{MX})\beta_{MX} + \cdots + (\bar{z} - z_{MX_{\mathcal{N}}})\beta_{MX_{\mathcal{N}}}[X]^{\mathcal{N}-1} \tag{37}$$

A plot of G_1 versus $(\bar{z} - z_{MX})$ may be used to obtain β_{MX} as

$$\beta_{MX} = \left[\frac{G_1}{(\bar{z} - z_{MX})}\right]_{[X]=0}$$

Similarly, in terms of a new function G_2, given by

$$G_2 = \frac{G_1 - (\bar{z} - z_{MX})\beta_{MX}}{[X]} = (\bar{z} - z_{MX_2})\beta_{MX_2} + \cdots + (\bar{z} - z_{MX_{\mathcal{N}}})\beta_{MX_{\mathcal{N}}}[X]^{\mathcal{N}-1}$$

$$\tag{38}$$

the plot of G_2 versus $(\bar{z} - z_{MX_2})$ may be used to generate

$$\beta_{MX_2} = \left[\frac{G_2}{(\bar{z} - z_{MX_2})}\right]_{[X]=0}$$

Equation 36 may further be expressed in terms of Δz_{max} and the various *species shifts* $\Delta z_{MX} \ldots \Delta z_{MX_{\mathcal{N}}}$, i.e. $(z_M - z_{MX}) \ldots (z_M - z_{MX_{\mathcal{N}}})$. Since any $\Delta\bar{z}$ is given by $(z_M - \bar{z})$ we may write, assuming that the various species shifts are in proportion to $[X]$:

$$
\left.\begin{array}{l}
(z_M - z_{MX}) - (z_M - \bar{z}) = (\bar{z} - z_{MX}) = \Delta z_{MX} - \Delta\bar{z} = \left(\dfrac{1}{\mathcal{N}}\Delta z_{max} - \Delta\bar{z}\right) \\[2mm]
(z_M - z_{MX_2}) - (z_M - \bar{z}) = (\bar{z} - z_{MX_2}) = \Delta z_{MX_2} - \Delta\bar{z} = \left(\dfrac{2}{\mathcal{N}}\Delta z_{max} - \Delta\bar{z}\right) \\[2mm]
\qquad\vdots \qquad\qquad\qquad \vdots \qquad\qquad \vdots \qquad\qquad \vdots \\[2mm]
(z_M - z_{MX_N}) - (z_M - \bar{z}) = (\bar{z} - z_{MX_N}) = \Delta z_{MX_N} - \Delta\bar{z} = \quad (\Delta z_{max} - \Delta\bar{z})
\end{array}\right\}
\tag{39}
$$

Equation 36 now becomes

$$
\Delta\bar{z} = \left(\frac{1}{\mathcal{N}}\Delta z_{max} - \Delta\bar{z}\right)\beta_{MX}[X] + \cdots + \left(\frac{j}{\mathcal{N}}\Delta z_{max} - \Delta\bar{z}\right)\beta_{MX_j}[X]^j
$$
$$
+ \cdots + (\Delta z_{max} - \Delta\bar{z})\beta_{MX_N}[X]^{\mathcal{N}}
\tag{40}
$$

For the special case $\mathcal{N} = 2$, Equation 39 adopts the form:

$$
\frac{\Delta\bar{z}}{[X]} = (\tfrac{1}{2}\Delta z_{max} - \Delta\bar{z})\beta_{MX} + (\Delta z_{max} - \Delta\bar{z})\beta_{MX_2}[X]
$$

Thus,

$$
\frac{\Delta\bar{z}}{(\tfrac{1}{2}\Delta z_{max} - \Delta\bar{z})[X]} = \beta_{MX} + \beta_{MX_2}\left(\frac{\Delta z_{max} - \Delta\bar{z}}{\tfrac{1}{2}\Delta z_{max} - \Delta\bar{z}}\right)[X]
\tag{41}
$$

By subsitution of $k'\Delta z = \bar{n}$, $\Delta z_{max} = 2/k'$ and $\tfrac{1}{2}\Delta z_{max} = 1/k'$ and Equation 41 leads to

$$
\frac{\bar{n}}{(1 - \bar{n})[X]} = \beta_{MX} + \beta_{MX_2}\left(\frac{2 - \bar{n}}{1 - \bar{n}}\right)[X]
\tag{42}
$$

Equation 42 is an independently established expression whose form engages the symmetry of a formation curve for $\mathcal{N} = 2$.

Equations presented in this section, requiring as they do a knowledge of values of *species* functions, may only be effectively applied when $\mathcal{N}$ is known unambiguously. This means that preliminary analysis via an appropriate pseudo-formation curve is necessary.

In the sections which follow, the theory developed will, in its various forms, be applied to a number of lithium–ligand systems for which published data are available or for which data have recently been determined. In the former case the usefulness of the new methods is demonstrated by comparison of the data which they produce with those generated by alternative methods. In the latter case some previously unidentified interactions are given quantitative significance.

Lithium–Ligand Systems

Characterization of lithium–valinomycin and lithium–ionophore A23187 interactions

Sankaran and Easwaran (1982) studied the variation of the circular dichroism spectra of valinomycin in acetonitrile in the presence of increasing concentrations of Li^+ as perchlorate. The analysis described here derives from the relative molar ellipticity data reported by the above authors. Starting with valinomycin at a concentration of 4.87 mM, the lithium/ligand ratio was increased up to 100, i.e. the maximum value of $[Li^+] = 0.487$ M and is in sufficient excess. The Scatchard plot shows sensible linearity over the whole range of $[Li^+]$, and yields a value of 26.2 for K'. In this and the pseudo-formation curve shown in Fig. 2, the $\Delta[\theta]$ data are expressed in arbitrary units; one of the conveniences of the methods to be demonstrated is that this is feasible.

Attempted Leden analysis assuming that $\mathcal{N} = 1$, generated $F_0[Li^+]$ data which is shown plotted in Fig. 3. Clearly complexation does not extend beyond the 1 : 1 stage and $\beta_1 = 26$. This result confirms the very weak, but identifiable, interaction obtained by Sankaran and Easwaran using the essentially curve-fitting technique due to Reuben (1973). By this means a value of 48 was obtained. It would seem that the value of 26 is more reliable by virtue of its

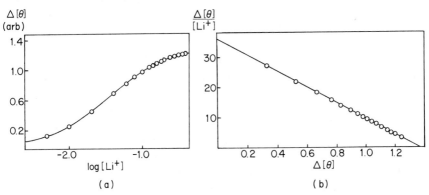

FIG. 2   (a) Pseudo-formation curve and (b) Scatchard plot for the system Li^+–valinomycin.

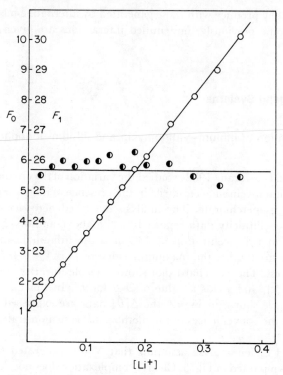

FIG. 3   Leden analysis for the Li^+–valinomycin system: demonstration of satisfactory nature for $N = 1$.

consistency with the Scatchard slope, about which there is little question of ambiguity.

Taylor *et al.* (1985) investigated the complexing reactions between Li^+ (and Na^+) and ionophore A23187 occurring in methanol–water and those occurring in the presence of phospholipid vesicles. The purpose was to establish the effects of membrane association of the ionophore on the stability of metal ion–ionophore complexes. Interactions were studied by observation of changes in the fluorescence emission spectra of the ionophore in the presence of increasing amounts of Li^+. Thus, the concentrations of phospholipid vesicles (DMPC) and ionophore A23187 were 2.5×10^{-3} M and 3.33×10^{-6} M, respectively, while the concentration of Li^+ ion was varied over the range 5×10^{-5} M to 4×10^{-3} M. The pseudo-formation curve and Scatchard plot are shown in Fig. 4; inspection of the position of the former with respect to the $\log [Li^+]$ scale with that for the corresponding curve of the previous system, reveals immediately the significantly higher stability of the interactions for this case. Generation of $F_0[Li^+]$ data on the assumption that $N = 1$,

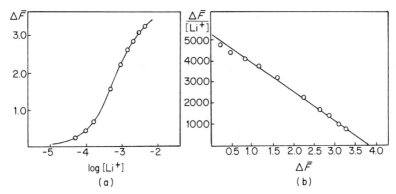

FIG. 4   (a) Pseudo-formation curve and (b) Scatchard plot expressed in terms of change in fluorescence emission spectral signals for ionophore A23187 in the presence of increasing concentration of lithium ion. A constant concentration of phospholipid vesicles was maintained since the study was concerned with comparative complexing tendency of membrane-associated ionophore.

followed by Leden analysis, gave unambiguous confirmation of 1 : 1 interaction with $\beta_1 = 1320$; this compares well with the published value of 1660.

Lithium–lasalocid A complexes

Shastri and Easwaran (1984) investigated these interactions by studying the variations of parameters characteristic of the ligand as Li^+ was added and those characteristic of lithium as ligand was added. For the former experiments circular dichroism and nuclear magnetic resonance (NMR) (^{13}C and ^1H) of lasalocid were used, in the latter ^7Li NMR. Ligand variables were expressed in terms of $[Li^+]/[LS]$, and those for the metal ion in terms of the ratio $[LS]/[Li^+]$. Although a knowledge of $[LS]$ in the former case and $[Li^+]$ in the latter yields $[Li^+]$ and $[LS]$, respectively, which are necessary for the analysis described here, there is a disadvantage in that the values are not sufficiently in excess of that of the interacting material over the whole range. Nevertheless, the formation constant data which emerge via the pseudo-formation curves show some interesting agreements with those of Shastri and Easwaran, who used algebraic methods. Some clear differences also emerge, however. More detailed experiments may be necessary to resolve these finally, but the preliminary results are presented here with some tentative explanations.

   Shastri and Easwaran interpret all their experimental data in terms of the species $LiLS$ and $Li(LS)_2$. The present treatment shows that the shifts of lasalocid function with respect to Li^+ concentration are apparently consistent with the polynomial

$$F_0'[Li] = 1 + \beta_1'[Li^+] + \beta_2'[Li^+]^2$$

while the shifts of Li$^+$ function with respect to [LS] apparently fit the relation

$$F_0''[\text{LS}] = 1 + \beta_1''[\text{LS}] + \beta_2''[\text{LS}]^2$$

Pseudo-formation curves for the two cases are compared in Fig. 5.

The Scatchard plot for the study with varying [Li$^+$] yielded a value of $\Delta[\theta]_{\text{max}}$ of 11.32 (in arbitrary units), which implies values of k' (Equation

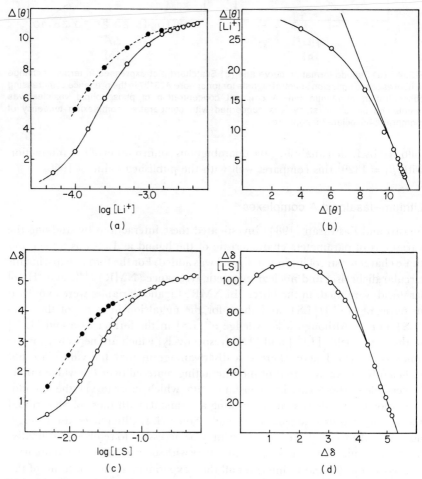

FIG. 5   (a) Pseudo-formation curve and (b) Scatchard plot based on variation of $[\theta]_{\text{max}}$ for lasalocid (0.0005 M) in the presence of increasing concentration of Li$^+$ (0–0.005M). ○, uncorrected data; ●, corrected data.

(c) Pseudo-formation curve and (d) Scatchard plot based on variation of $\Delta\delta$ for ^{7}Li$^+$ (0.054 M) in the presence of increasing concentration of lasalocid (0–0.55 M). ○, uncorrected data; ●, corrected data.

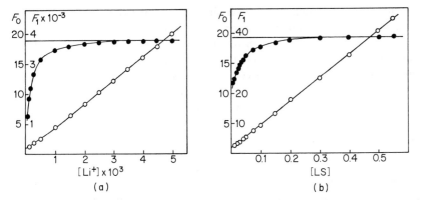

FIG. 6 (a) Leden analysis, assuming $N = 1$, of F_0' data based on constant concentration of lasalocid and varying concentration of lithium ion.

(b) Leden analysis, assuming $N = 1$, of F_0' data based on constant concentration of lithium ion and varying concentration of lasalocid.

13) of 0.088 if $N = 1$, or 0.177 if $N = 2$. Leden analysis at first sight would appear to be impossible for $N = 1$ (Fig. 6). However, it should be noted that Scatchard plots show significant deviations from linearity at low analytical concentrations of $[Li^+]$ where these values are no longer the fair approximation to $[Li^+]_{free}$ which the Leden treatment requires. Correction of $\Delta[\theta]$ values by graphical reappraisal as indicated in Fig. 5(b), allows a corrected pseudo-formation curve to be drawn. The value of the constant β_{LiLS} of 3.8×10^3 agrees favourably with the several values of different origin reported by Shastri and Easwaran, namely 1.4×10^3, 3.2×10^3 and 9×10^4. There is no indication of further complexation as suggested by these authors. There is, however, a point of some interest. If a system *is*, in fact, characterized by $N = 1$, then clearly $F_0[X] \sim 1 + \beta_1[X]$. If analysis is for such cases carried out as though $N = 2$, then the function generated may be readily shown to be given by

$$F_0^*[X] \sim 1 + 2\beta_1[X] + \beta_1^2[X]^2$$

and Leden analysis will apparently give $\beta_1^* = 2\beta_1$ and $\beta_2^* = \beta_1^2$. The rather unique relation between β_2^* and β_1^* combined with careful scrutiny of the analysis for $N = 1$ should avoid confusion. It should perhaps be noted that in the present case an apparently good analysis for $N = 2$ may be obtained if uncorrected values of $[Li^+]$ are used. Misinterpretation of what is *correct treatment* to *uncorrected* data can prove to be misleading. Similar behaviour is found for data deriving from progressive addition of ligand to metal ion: correction for $[LS]_{free}$ yields a single formation constant but of quite different value to that obtained by the circular dichroism, ^1H and ^{13}C NMR methods.

D.R. Crow

A value of 62 obtained from the limiting Scatchard slope and 39 from the Fronaeus–Leden treatment of a corrected pseudo-formation curve compares favourably with the value of 48 reported by Degani and Friedman (1973). These latter authors carried out fluorimetric titration of ligand with metal ions in methanol using alternative graphical methods. The singularly different values obtained for the experiments in which 'titrant' and 'titrand' are interchanged would suggest the participation of polynuclear species $Li_x(LS)_y$.

The lithium–imidazole system

Since imidazole is an essentially neutral ligand, it should contribute negligibly to the molar conductivity of an aqueous solution. If such species complex with metal ions (and imidazole has significant donor properties), the size of metal ions will change when hydration molecules are exchanged for other neutral ligand species. This will cause changes in the diffusion coefficient (D), which in turn is reflected in a change of Λ by virtue of the Nernst–Einstein equation:

$$\Lambda_i = \frac{D_i z_i^2 F^2}{RT}$$

where other terms have their usual significance. Variations, $\Delta\bar{\Lambda}$, of Li^+ with concentration of imidazole were found by the present author to give rise to pseudo-formation curves from which equilibrium data could be resolved. A concentration of 0.01 M of lithium was used with the concentration of ligand varying from 0.02 to 0.19 M. The Leden data are presented with the results of graphical analysis in Table 1. Analysis for four complexes of rather low stability was obtained.

TABLE 1   Leden analysis for the system lithium–imidazole ($k' = 2.35$).

$\Delta\bar{\Lambda}$ $(\Omega^{-1}\,cm^2\,mol^{-1})$	[Im] (M)	$\log F_0'$ [Im]	F_0 [Im]	F_1	F_2	F_3	F_4
0.00	0.00		1.000	—	—	—	—
0.60	0.01	0.0269	1.157	15.70	—	—	—
1.18	0.02	0.0537	1.337	16.85	118	—	—
2.35	0.04	0.1068	1.782	19.55	126	350	—
3.50	0.06	0.1583	2.355	22.58	135	383	2883
4.61	0.08	0.2090	3.098	26.23	147	438	2850
5.70	0.10	0.2590	4.061	30.61	161	490	2800
6.72	0.12	0.3080	5.294	35.78	177	542	2767
7.70	0.14	0.3562	6.876	41.97	196	600	2857
8.64	0.16	0.4037	8.886	49.29	217	656	2788
9.50	0.18	0.4500	11.416	57.87	241	716	2811
10.20	0.20	0.4950	14.587	67.94	267	775	2825

$\beta_1 = 14.5$; $\beta_2 = 112$; $\beta_3 = 210$; $\beta_4 = 2820$.

Lithium–Polyethyleneglycol 400 interactions

Some recent investigations by the author of polynuclear complex systems are typified by the behaviour observed for lithium-polyethyleneglycol (PEG) 400 species. Here observed behaviour and preliminary interpretation for Li^+–PEG systems contrast sharply with the resolved behaviour of the complexing system involving nickel and imidazole. The latter has been analysed in detail in terms of both diffusion coefficient and molar conductivity data and a sequence of *six* complexing reactions given quantitative significance. The values of formation constants deriving from principles outlined are comparable to those reported previously in terms of pH and voltammetric methods. Such correspondence may be taken as adequate verification for the theoretical principles outlined earlier in this chapter. In Fig. 7 are shown the variations of a chosen parameter for Ni^+ and Li^+ with concentration of imidazole and PEG 400,

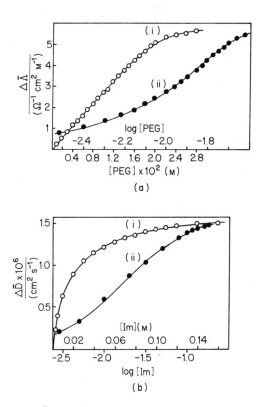

FIG. 7 (a) Variation of $\Delta\bar{\Lambda}$ for Li^+ with (i) [PEG] and (ii) log [PEG]. Extended, somewhat ill-defined curves are characteristic of systems involving significant cooperativity.

(b) Variation of $\Delta\bar{D}$ for Ni^{2+} with (i) [Im] and (ii) log [Im]. Clear, sharp variation of signal for metal ion with increasing ligand concentration and more symmetrical formation curve are characteristic of systems involving negligible cooperativity.

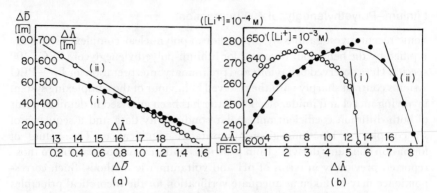

FIG. 8   (a) Linear Scatchard plots for Ni^{2+} –imidazole system based on shifts in (i) diffusion coefficient, $\Delta \bar{D}$, and (ii) molar conductivity, $\Delta \bar{\Lambda}$, clearly indicating the lack of cooperativity.

(b) Scatchard plots for Li^+–PEG 400 system. Extreme departures from linearity are characteristic of significant positive cooperativity.

respectively. The rather different shapes of the curves are further enhanced in the pseudo-formation curves (Fig. 7). Scatchard plots (Fig. 8) for the two systems clearly emphasize the distinct differences between the two types of system. For nickel–imidazole no cooperativity is involved and this is evidently a feature prevailing over the range of six interactions which are now well characterized. For lithium–PEG 400, the extent of cooperativity is very significant—the small metal ion concentrations relative to that of ligand has ensured in this case that the analytical concentrations of ligand may be reasonably approximated to the free, unbound, concentrations.

Analysis of the pseudo-formation curve for Li–PEG 400 reveals significant interaction—apparently extending to the formation of three complexes. Leden analysis is particularly clean at metal ion concentrations of both 10^{-4} M and 10^{-3} M, although the numerical values of the formation constants differ for the two cases.

Such variation is characteristic of polynuclear complex formation. It is not immediately clear what the progressive complexation process produces: what is interesting is that the concept of pseudo-formation curve and its analysis would seem to remain valid with systems showing both cooperativity and polynuclear involvement. These induce distortions to the pseudo-formation curve which reflect a variation of the value of k' in Equation 34 for the overlapping interactions for which it is constituted. That an averaged value of k', derived by identification of $\Delta \Lambda_{max}$ as described earlier, actually works is clearly useful in distinguishing the various species, although the values of formation constants which emerge must be regarded somewhat conservatively. The immediate practical effect for the present case is rather curious. The value of k' in terms of $N = k' \Delta \Lambda_{max}$ for N taken as 2 provides a perfect Leden analysis for *three* complexes!

The values of $\bar{\Lambda}$ for Li–PEG solutions are quite high relative to values observed for uncomplexed metal ions, i.e. the change in $\Delta\bar{\Lambda}$ is significant but not as large as might be expected for species of the size of PEG 400. On the other hand, it would be naive to treat the interactions as simply Li(PEG), Li(PEG)$_2$ and Li(PEG)$_3$, even though the heavy hydration of both species would tend to smear out changes in volume on complexation and reduce the value of $\Delta\bar{\Lambda}$. There are 7 to 8 binding sites at 'ether oxygens' down the length of the molecular chain

$$HO.(CH_2CH_2O)_nCH_2CH_2OH$$

One model of the complexing process which is in accord with observed behaviour involves effective formation of $(Li^+)_8$ PEG in which the successive formation (Li^+) PEG, $(Li^+)_2$ PEG, and so on, is characterized by an approximately constant step constant $k_1 \sim k_2 \sim k_3 \ldots k_8$. Thus

$$(k_1)^8 \sim (k_2)^8 \ldots (k_8)^8 \sim (k_1k_2 \ldots k_8) \sim \beta_1.$$

A second polymer chain can then be imagined as attaching to the first, followed by a third.

$$[Li(PEG)]^+ \rightleftharpoons [Li(PEG)_2]^+ \rightleftharpoons [Li)PEG)_3]^+$$

$$\vdots \qquad\qquad \vdots \qquad\qquad \vdots$$

$$[Li_8(PEG)]^+ \rightleftharpoons [Li_8(PEG)_2]^+ \rightleftharpoons [Li_8(PEG)_3]^+$$

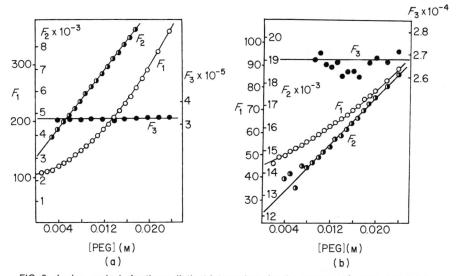

FIG. 9 Leden analysis for three distinct interactions for the system Li$^+$–PEG 400: (a) for [Li$^+$] = 10^{-4} M; (b) for [Li$^+$] = 10^{-3} M. Clear resolution of the interactions is in no way distorted by the cooperativity operating in the stepwise complex formation.

TABLE 2 Selected derived values of G_j and of complementary $(\bar{\Lambda} - \Lambda_{MX_j})$ functions for identification and characterization of complexes with the Li$^+$–polyethyleneglycol 400 system. $[Li^+] = 10^{-4}$ M. Species values taken as

$$\Lambda_{MX} = 217, \; \Lambda_{MX_2} = 203, \; \Lambda_{MX_3} = 189 \; \Omega^{-1} \, cm^2 \, mol^{-1}.$$

[PEG]	$\bar{\Lambda}$	G_1	$(\bar{\Lambda} - \Lambda_{MX})$	$\beta_{MX}(\bar{\Lambda} - \Lambda_{MX})$	G_2	$(\bar{\Lambda} - \Lambda_{MX_2})$	$\beta_{MX_2}(\bar{\Lambda} - \Lambda_{MX_2})$	G_3	$(\bar{\Lambda} - \Lambda_{MX_3})$	$\dfrac{G_3}{(\bar{\Lambda} - \Lambda_{MX_3})}$
0.000	245		28.0			42.0			56.0	
0.003	237	2667	20.0	1740	3.09×10^5	34.0	2.34×10^5	2.48×10^7	48.0	5.2×10^5
0.005	231.6	2680	14.6	1270	2.82×10^5	28.6	1.97×10^5	1.70×10^7	42.6	4.0×10^5
0.007	226.0	2714	9.0	783	2.76×10^5	23.0	1.59×10^5	1.67×10^7	37.0	4.5×10^5
0.009	220.6	2711	3.6	313	2.66×10^5	17.6	1.21×10^5	1.61×10^7	31.6	5.1×10^5
0.010	217.8	2720	0.8	70	2.65×10^5	14.8	1.02×10^5	1.63×10^7	28.8	5.6×10^5
0.011	215.1	2709	-1.8	-157	2.61×10^5	12.2	0.84×10^5	1.60×10^7	26.2	6.1×10^5
0.012	212.5	2708	-4.5	-392	2.58×10^5	9.5	0.66×10^5	1.60×10^7	23.5	6.8×10^5
0.013	210.0	2692	-7.0	-609	2.54×10^5	7.0	0.48×10^5	1.58×10^7	21.0	7.5×10^5

The clear identification of three major interactions would suggest that at all PEG concentrations used in construction of the pseudo-formation curve, virtual saturation with Li^+ occurs. Figures 9(a) and 9(b) show the plots of F_1, F_2 and F_3 for two concentrations of Li^+. The two sets of formation constants have the values $\beta_1 = 102$; $\beta_2 = 3000$; $\beta_3 = 3 \times 10^5$ for $[Li^+] = 10^{-4}$ M and $\beta_1 = 44$; $\beta_2 = 1200$; $\beta_3 = 2.7 \times 10^4$ for $[Li^+] = 10^{-3}$ M. The differences are consistent with the suspected polynuclear character of the complexing process.

The system in which $[Li^+] = 10^{-4}$ M has also been successfully treated in terms of the functions G_1, $G_2 \cdots$ (Equations 37, 38) involving the species values Λ_{MX_j}. There is some difficulty in deciding upon the latter, since species shifts are unlikely to vary in proportion to $[PEG\ 400]$ for a system displaying

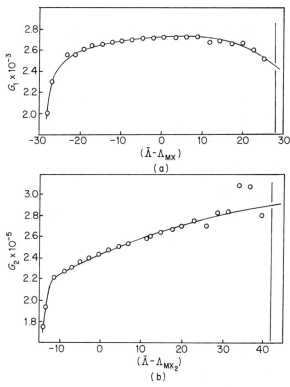

FIG. 10   Plots based on Equations 37 and 38 for the lithium–polyethyleneglycol 400 system.

(a) G_1 versus $(\bar{\Lambda} - \Lambda_{MX})$;

$$\beta_1 = \left[\frac{G_1}{(\bar{\Lambda} - \Lambda_{MX})} \right]_{[X]=0} = \frac{2440}{28} = 87$$

(b) G_2 versus $(\bar{\Lambda} - \Lambda_{MX_2})$;

$$\beta_2 = \left[\frac{G_2}{(\bar{\Lambda} - \Lambda_{MX_2})} \right]_{[X]=0} = \frac{290\,000}{42} = 6900$$

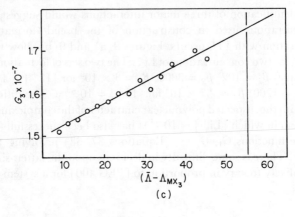

(c) G_3 versus $(\bar{\Lambda} - \Lambda_{MX_3})$;

$$\beta_3 = \left[\frac{G_3}{(\bar{\Lambda} - \Lambda_{MX_3})}\right]_{[X]=0} = \frac{1.76 \times 10^7}{56} = 3.14 \times 10^5$$

such a marked degree of cooperativity. In fact, treatment in terms of equal species shifts produces a less satisfactory analysis than one in which such shifts are graded. Selected data deriving from such analysis, with associated graphs which demonstrate the applicability of the technique, are given in Table 2 and Fig. 10. The apparent initial ambiguity over the number of species and the assessment of k' would appear to be largely resolved: it is not possible to 'force' an incorrect value of N or k' on the system and the analysis shown in Table 2 would not be possible if this were to be attempted.

Conclusions

Interactions between metal ions and complexing ligands are reflected in the variation of a variety of physical parameters of either species in the presence of increasing concentration of the other.

Careful measurement of the changes in the values of such parameters, as a function of the relative concentrations of complexing agent and metal ion, allows unambiguous assessment of both the extent of complexation and the relative stabilities of the species formed.

In principle, the methods developed may be applied to systems of a significant number of step equilibria, to those involving characteristic cooperativity, and evidently to those including polynuclear interactions.

References

Benesi, H.A. and Hildebrand, J.H. (1949). A spectrophotometric investigation of the interaction of iodine with aromatic hydrocarbons. *J. Amer. Chem. Soc.* **71**, 2703–2707.

Crow, D.R. (1982a). Diffusion coefficients and complex equilibria in solution. Part 1. *Talanta* **29**, 733–737.

Crow, D.R. (1982b). Diffusion coefficients and complex equilibria in solution. Part 2. *Talanta* **29**, 737–742.

Crow, D.R. (1983). Diffusion coefficients and complex equilibria in solution. Part 3. *Talanta* **30**, 659–664.

Crow, D.R. (1986). Diffusion phenomena and metal complex formation equilibria. Part 1. *J. Chem. Soc., Faraday Trans. 1* **82**, 3415–3430.

Crow, D.R. (1988a). Diffusion phenomena and metal complex formation equilibria. Part 2. *J. Chem. Soc., Faraday Trans. 1* **84**, 4285–4297.

Crow, D.R. (1988b). New methods for the determination of the extent and stability of interaction of Li^+ with further solute species. *In* "Lithium: Inorganic Pharmacology and Psychiatric Use" (ed. N.J. Birch), pp. 305–307. IRL Press Ltd, Oxford.

DeFord, D.D. and Hume, D.N. (1951a). The determination of consecutive formation constants of complex ions from polarographic data. *J. Amer. Chem. Soc.* **73**, 5321–5322.

DeFord, D.D. and Hume, D.N. (1951b). A polarographic study of the cadmium thiocyanate complexes. *J. Amer. Chem. Soc.* **73**, 5323–5325.

Degani, H. and Friedman, H.L. (1973). Fluorimetric complexing constants and circular dichroism measurements for antibiotic X-537 A with univalent and bivalent cations. *J. Chem. Soc. Chem. Commun.* 431–432.

Fronaeus, S. (1950). A new principle for the investigation of complex equilibria and the determination of complexity constants. *Acta. Chem. Scand.* **4**, 72–87.

Leden, I. (1941). Potentiometric measurements for the determination of complex ions in cadmium salt solutions. *Z. Phys. Chem.* **188**, 160–181.

Reuben, J. (1973). Complex formation between $Eu(fod)_3$, a lanthanide shift reagent, and organic molecules. *J. Amer. Chem. Soc.* **95**, 3534–3540.

Sankaran, M.B. and Easwaran, K.R.K. (1982). CD and NMR studies on the interaction of lithium ion with valinomycin and gramicidin-S. *Biopolymers* **21**, 1557–1567.

Scatchard, G. (1949). The attraction of proteins for small molecules and ions. *Ann. N.Y. Acad. Sci.* **51**, 660–672.

Scott, R.L. (1956). Some comments on the Benesi–Hildebrand equation. *Rec. Trav. Chim.* **75**, 787–789.

Shastri, B.P. and Easwaran, K.R.K. (1984). Conformations of lasalocid A–lithium complexes in acetonitrile. *Int. J. Biol. Macromol.* **6**, 219–223.

Taylor, R.W., Chapman, C.J. and Pfeiffer, D.R. (1985). Effect of membrane association on the stability of complexes between ionophore A23187 and monovalent cations. *Biochemistry* **24**, 4852–4859.

4 Methods for the Determination of the Distribution of Lithium at the Histological and Cytological Levels

MICHEL THELLIER[1] and JEAN-CLAUDE WISSOCQ[2]

[1] *Echanges Cellulaires, Faculté des Sciences et Techniques de l'Université de Rouen-Haute Normandie, BP 118, F 76134 Mont-Saint-Aignan Cedex, France*

[2] *Biologie Animale et Cellulaire, Faculté des Sciences de l'Université de Picardie, F 80039 Amiens Cedex, France*

Introduction

In less than 20 years, the number of methods available for the determination of lithium in biology has increased significantly. Most of these methods were developed initially by physicists specifically for solid-state physics, but they are being progressively extended to the study of biological objects. Some of them may be used *in vivo*, but most are used to study the distribution of lithium in fixed histological preparations. With some methods, only the element lithium is detected and possibly imaged, whereas other methods can discriminate the stable isotopes, ^6Li and ^7Li. In some cases, the measurements are independent of the microenvironment of lithium; in other cases they are not, and may thus provide some information about possible chemical binding of lithium. The present contribution is an attempt to summarize the available methods for lithium study and to give their main characteristics.

Non-imaging Studies of Local Lithium Contents

Measurements of mean Li^+ concentrations in body fluids or tissue fragments

Principle of the method

The simplest and oldest method for studying the distribution of lithium in Li-treated organisms has consisted of dissecting the specimen into pieces and measuring the mean Li^+ concentration in each fragment. The solid tissues generally have to be transformed to a liquid form before the measurements.

The spectrophotometric methods that can be used to determine lithium, as reviewed by Matusiewicz (1982), are flame atomic emission and absorption spectrometry, atomic absorption with a graphite furnace or carbon rod emission spectrography, inductively coupled plasma emission spectrometry, and direct-current plasma echelle spectrometry. However, flame atomic emission and absorption spectrometry have been used mainly for biological and medical lithium studies. In the absence of conventional colorimetric or fluorometric methods for lithium detection (Pacey *et al.*, 1987), one may use Li^+ extraction by crown ethers or cryptates, and subsequent measurement of the changes produced in the UV–visible absorption spectrum, or in the fluorescence spectrum of these complexing molecules (Pacey *et al.*, 1987). The conventional methods of sampling (which require a minimum of approximately 1 ml of sample solution) have been replaced in some cases by discrete injection/single-pulse nebulization (Matusiewicz, 1982), or flow injection (Rocks *et al.*, 1982; Pacey *et al.*, 1987), requiring much smaller solution samples (as little as 10 μl).

The drawbacks of the fragmentation method are that (i) it does not portray the distribution of lithium, but can give only a set of mean values in each tissue fragment, (ii) the resolving power (which depends on the sharpness of the dissection) is often poor, and (iii) conventional spectrophotometric techniques, and the techniques with the crown ethers and ligands, do not discriminate isotopes, which makes it impossible to perform tracer experiments. Despite these difficulties, the method is still frequently used, especially for a preliminary study of the lithium distribution in a biological specimen. Moreover, a modification of the atomic absorption spectrophotometry (using the isotopic shift $^6Li/^7Li$) makes it possible to discriminate the stable isotopes of lithium.

Flame emission spectrophotometry

The technique is applicable to measuring Li concentrations in solutions, such as body fluids (urine, serum, cerebro-spinal fluid, etc.) or solubilized solid tissues. The solution under study is nebulized in a flame which is maintained as constant as possible in form and temperature. In the flame, some of the lithium ions are excited from their ground state energy level to a higher level of energy; then these ions return rapidly to the ground state by emission of photons with a maximum at a wavelength close to 670.8 nm. A calibration curve is constructed using standard Li^+ solutions. If the nebulized solutions were pure solutions of a lithium salt, the light intensity at the wavelength of 670.8 nm would be directly proportional to the concentration of the lithium ions. However, the presence of other cations (such as Na^+, K^+ or Ca^{2+}) causes interference effects, changing the apparent concentration of lithium. This is due to energy transfers between the excited and ground state energy levels of the different ions. To avoid such interference, the calibration solutions may be produced by dissolving known quantities of a lithium salt in a solution identical to the experimental ones (blood serum for instance). Another possibility consists of adding a massive proportion of one interfering ion, both in the calibration and in the experimental solutions, in order to saturate the interference effects in all these solutions.

The first determinations of lithium in biological samples, using flame spectrophotometry, were produced by Trautner *et al.* (1955) in urine samples, and by Schou (1956) in human serum. These studies were followed by a great number of others. For instance, Morrison *et al.* (1971) have measured the differential absorption of Li^+ by the rat gut following intraperitoneal or oral administration of either LiCl or Li_2CO_3. Lehmann (1975) has studied the kinetics of lithium in the serum, liver and brain of the rat. Siebers and Maling (1988) have evaluated the sodium/lithium countertransport rate by direct measurement of erythrocyte lithium concentrations.

Flame atomic absorption spectrophotometry

In this technique, the solution samples (body fluids or solubilized tissues) are nebulized in a flame, while light from a hollow cathode lamp containing Li, with the characteristic wavelength of lithium, is directed through the flame. In the flame, those Li atoms which are in the ground state tend to absorb the photons emitted by the lamp. The amount of light thus absorbed by the flame at the wavelength which is characteristic of lithium is a measurement of the Li concentration in the nebulized solution. A calibration curve is again constructed from standard Li solutions. According to Amdisen (1975), atomic absorption spectrophotometry is preferable to flame photometry because of the simplicity of the first procedure and of its higher accuracy at low Li concentrations. To overcome interferences, see, for instance, Birch and Jenner (1973).

Among the many Li studies based upon atomic absorption spectrophotometry measurements, we may quote a few representative examples. Notable interindividual differences in the Na^+-dependent Li^+ countertransport system, and in the Li^+ distribution ratio across the red cell membrane, have been found to exist among Li^+-treated patients by Greil *et al.* (1977), while Trevisan *et al.* (1981) have found that the ratio of intra- to inter-individual variances was small in healthy humans, for the Na^+ stimulated Li^+ efflux. Ebara and Smith (1979) have studied the Li^+ level in blood platelets, serum, red blood cells and brain regions in rats given acute or chronic lithium salt treatments.

Plenge *et al.* (1981) have shown that the serum lithium was relatively constant during the day in perorally Li^+-treated rats, whereas, in Li^+-injected rats, lithium reached a high peak value just after the injection followed by a decrease to very low values. Selinger *et al.* (1982) have used salivary lithium concentrations as an indicator of plasma lithium levels.

Use of the isotopic shift in lithium spectrophotometry

The principle behind isotopic analysis by absorption spectrometry (Wheat, 1971) is as follows. In its detailed structure, the resonance line of lithium at 670.8 nm undergoes an isotopic shift of 0.015 nm; moreover it is a doublet for each of the two stable isotopes, ^6Li and ^7Li, with a doublet spacing of 0.015 nm. For a mixture of both Li isotopes, the 670.8 line thus appears as a triplet. The first peak of the triplet corresponds to the short-wavelength peak of ^7Li. The second peak, 0.015 nm from the first, is the superimposition of the long-wavelength peak of ^7Li and the short-wavelength peak of ^6Li. The last peak, 0.015 nm from the second, is the long-wavelength peak of ^6Li. One might imagine one could perform direct measurements of the ^6Li and ^7Li concentrations in a mixture, by flame photometry, using only the extreme peaks of the triplet. However, because these peaks are so close to each other, this would require a spectrophotometer with an extremely good energy resolution: we have not found any mention of such measurements. However, atomic-absorption spectrometry measurements have revealed themselves to be possible. These are based on the fact that ^6Li atoms attenuate ^6Li emissions more than they attenuate ^7Li-emission, and vice versa. Applications of this theoretical possibility have been published (Zaidel and Korennoi, 1961; Manning and Slavin, 1962; Wheat, 1971; Chapman and Dale, 1976; Birch *et al.*, 1978a,b).

The advantage of this method, compared with the two methods described earlier, is that it can measure isotopic ratios, $^6Li/^7Li$. This makes it possible to label lithium by use of highly purified isotopes, ^6Li and ^7Li, and possibly to measure unidirectional Li fluxes and perform Li^+ pharmacokinetic studies.

Extraction of lithium by specific Li^+ ionophores such as crown ethers or cryptates

The crown ethers were introduced by Pedersen (1967). The crowns are macroheterocycles containing the repeating unit ($-Y-CH_2-CH_2-$) where Y is a heteroatom like O, S, N or P; they are usually monocyclic and essentially two-dimensional; the name 'crown' arises from the appearance of the molecular model of these substances, which may be very similar to the structure of a regal crown (Gokel and Durst, 1976). The representation for a crown is x-crown-y, where x is the total number of atoms in the ring and y is the number of heteroatoms in this ring (for instance 12-crown-4). The most

 12-crown-4

important class of rings of crown ethers consists of molecules of the type ($-CH_2-CH_2-O-$)$_n$, with $n > 3$. Crown ethers may complex cations very efficiently and specifically. For instance, 18-crown-6 is specific for K^+, 15-crown-5 for Na^+, and 12-crown-4 for Li^+ (Cook *et al.*, 1974). By adding a third strand to a crown, Lehn and collaborators have produced three-dimensional polyheteroatom macrocycles which they have termed 'cryptates'. For instance the (1.1.1 cryptate) possesses one oxygen atom on each of its three strands, while the (2.2.2 cryptate) has two oxygen atoms on each strand. Such compounds may serve as very specific 'cages' for alkali metal ions. For a review of the cryptates, see Lehn (1978).

Pacey *et al.* (1987) have developed several chromogenic and fluorogenic crown ether or cryptate compounds which may be utilized in the selective determination of alkali metal ions. Two such ionophores have been particularly studied: (i) (2.1.1 cryptate) with resazurin, and (ii) the chromogenic aza-12-crown-4 crown ether (1-(2'-hydroxy-5'-nitrobenzyl)-1-aza-4,7,10-trioxa-cyclododecane). When such ionophores are present in a lithium-containing mixture, they complex the alkali metal ion, thus extracting it from the solution. This changes the absorbance of the ionophores in the UV–visible region (or their fluorescence). The (2.1.1 cryptate) with resazurin exhibited a linear range for Li^+ of 70 ppb to 2.1 ppm, and the method could tolerate up to 1000 ppm sodium ions; the chromogenic aza-12-crown-4 exhibited a linear range for Li^+ of 0.3 to 2 ppm, while a sodium ion concentration of 230 ppm was tolerated (Pacey *et al.*, 1987).

Specific lithium electrodes and microelectrodes

Details about the principle and the construction of ion-sensitive microelectrodes may be found in R.C. Thomas (1978). Briefly summarized, a specific ion-sensitive microelectrode consists of a micropipette with a sharp-pointed tip. The pipette is filled with a concentrated KCl solution, while the tip contains a liquid 'ion-sensitive membrane' or 'liquid ion-exchanger' with the best possible specificity for the ion under study. The pipette is connected to the electric circuit of measurement via an Ag/AgCl wire. The liquid ion-exchanger may be composed of a crown ether or of a cryptate (see previous section), or of some other specific ionophore, dissolved in a water-immiscible solvent. For a perfectly specific microelectrode, the liquid ion-exchanger should be sensitive to only one ion. In fact, the existing liquid ion-exchangers are never perfectly specific; therefore interferences from other ions are likely to be encountered. Apart from the pH microelectrodes, the ion for which the best specificity is available is K^+, and usable liquid ion-exchangers exist for Cl^-, Ca^{2+} and Na^+. In order to prevent the displacement by water of the water-immiscible liquid, the glass at the micropipette tip is made hydrophobic by treatment with organic silicon compounds. The microelectrodes can have very small tips, allowing them to be inserted into living cells. However, the smaller the tip, the higher is the resistance (approximately $10^8 \Omega$, for a tip diameter in the range $0.1-1 \mu m$). The response times are of the order of milliseconds (Thomas, 1978). In contrast with most of the other methods of analysis, the ion electrodes measure ionic *activities* rather than *concentrations*.

For lithium, the lipophilic neutral carrier ETH 1810 (N,N-dicyclohexyl-N'-N'-diisobutyl-*cis*-cyclohexane-1,2-dicarboxamide) exhibits attractive properties with regard to the Li^+/Na^+ selectivity (Metzger *et al.*, 1986). Apart from ETH 1810, UWXC 10 (N,N-diethyl-5,5-dimethyl-N,N-bis(3-oxapentyl)-3,7-dioxanonanediamine) and 14-crown-4 (3-dodecyl-3-methyl-1,5,8,12-tetra-oxacyclotetradecane) were also assayed for their selectivity to lithium (Xie and Christian, 1987). The addition of TOPO (trioctylphosphine oxide) had a negative effect on the selectivities of ETH 1810 and of UWXC 10. However, when TOPO was added to 14-crown-4, the sensitivity to Li^+ was increased by a factor of 3, while that to Na^+ was practically unchanged; therefore, the selectivity Li^+/Na^+ was also increased by a factor of 3 (Xie and Christian, 1987). In the cases which we have been discussing (Metzger *et al.*, 1986; Xie and Christian, 1987), the lithium ionophores (ETH 1810, UWXC 10, 14-crown-4 and TOPO) in fact were not incorporated to microelectrodes but only to ordinary electrodes meant for measuring lithium in body fluids such as serum. The performance of the lithium ion-selective electrode of the du Pont $Na^+/K^+/Li^+$ analyser (E.I. du Pont de Nemours & Co, Wilmington, DE 19810) was also evaluated for lithium measurements in serum: the lithium ion-selective electrode appeared to be relatively free from interferences from Na^+, Mg^{2+} and Ca^{2+} concentrations ranging from normal or less to as much as 49% (for Ca^{2+}) above the upper limit of the normal reference interval (Bertholf *et al.*, 1988). With a lithium ion-selective solvent polymeric membrane, based on the lipophilic diamide N,N,N',N'-tetraisobutylcyclohexane-*cis*-1,2-dicarboxylic diamide, the selectivity coefficients for lithium ions over other alkali metal ions were greater than 100, while over alkaline earth metal cations they were about 1000 (Zhukov *et al.*, 1981). Electrodes with a very good Li^+ specificity have been produced by Kimura *et al.* (1987).

A proper Li^+-sensitive microelectrode was constructed by R.C. Thomas and co-workers in 1975. It was made from borosilicate glass micropipettes, which were siliconized by dipping into a solution of tri-*n*-butyl chlorosilane in chloronaphthalene. The tip of the micropipette was filled with a Li^+ ligand solution whose composition was (by weight) 9.7% Li^+ carrier, 85.5% TEHP

(Tris-2(2 ethylhexyl)-phosphate) and 4.8% sodium tetraphenyl borate. The Li^+ carrier was that proposed by Güggi *et al.* (1975), i.e. N,N'-diheptyl-$N,N',5,5$-tetramethyl-3,7-dioxanonane diamide. Measurements of the Li^+ content of Li^+-injected snail neurones were performed using this microelectrode. It appeared that Li^+ was being actively transported out of the cell against both an electrical and a concentration gradient (R.C. Thomas *et al.*, 1975). Liquid-ion exchanger microelectrodes were used to monitor the extracellular activity of K^+ and Li^+ in the cerebellar cortex of rats receiving acute or chronic lithium treatment; the conclusion was that there was an inhibitory action of Li^+ upon the activity of the Na^+/K^+ pump (Ullrich *et al.*, 1980). Lithium-sensitive microelectrodes were used to investigate the transmembrane distribution of Li^+ in motoneurons of the isolated frog spinal cord (Grafe *et al.*, 1982). Shanzer *et al.* (1983) have described a new series of lipophilic Li^+ ionophores whose structure is based on an acyclic system in which a hexafunctional lipophilic envelope is formed around the metal ion in an octahedral arrangement, and they have stated that this Li^+ ionophore was well suited for the manufacture of a specific electrode. However, in recent years, we have found no mention of Li^+ studies using microelectrodes, except for a contribution (in Chinese) by Zhuang and Qi (1989).

Ion-exchanging resins and high performance liquid chromatography

A protonated anionic resin, RH, will fix Li^+ according to

$$RH^+ + Li^+ \rightleftarrows RLi^+ + H^+$$

if its affinity is greater for Li^+ than it is for H^+; then one can displace Li^+ by another cation whose affinity for the resin is greater than that for Li^+. This conventional ion exchange of lithium has been refined in two ways: by the use of ionic HPLC (high performance liquid chromatography), and by enrichment of stable isotopes, ^6Li and ^7Li, by ion exchange.

In the Waters ILC Series Ion/Liquid Chromatographs, an eluant solution perfuses a resin column at a rate of 0.1 to 9.9 ml min^{-1}. A given volume, $v(v \leqslant 100 \,\mu l)$, of the solution whose lithium content has to be determined is injected. The various ions present in this solution are carried along the column, and reach the base of the column, one after the other, according to their ionic mobility. The quantitative detection is performed using conductivity measurement. For monovalent cations, the eluant may be a nitric acid solution, the time taken per run is approximately 15 min, and one can achieve parts-per-trillion sensitivity. The calibration is performed using simple salt solutions. The method is well suited to the analysis of pure salt solutions; it may also be used for more complex solutions (for instance containing organic solutes), however with a risk of clogging the resin.

The separation of the stable isotopes, $^6Li^+$ and $^7Li^+$, has been performed using displacement chromatography (see, for instance, Hagiwara and Takakura, 1969; Fujine *et al.*, 1982). The principle is as follows. Water molality is reduced in the resin phase, and the water structure is changed close to the resin matrix: lithium ions are thus less strongly hydrated and less strongly bonded in the resin than they are in the surrounding solution. Since $^7Li^+$ tends to concentrate preferentially in the more strongly bonded species, it tends to accumulate in the aqueous phase, while $^6Li^+$ tends to favour the resin. In displacement chromatography, the lithium solution is poured on top of an ion-exchange resin bed (in the H-form), where it forms a lithium adsorption band. A displacement reagent such as Na, Ca or Ba acetate, is introduced into the top of the band. While the band moves in the resin column, lithium isotopes are exchanged between the resin and solution phases; lithium isotope concentration profiles are thus formed in the band, with the isotopes ^7Li and ^6Li being concentrated, respectively, in the front and rear edges of the band. With the isotopic composition of natural lithium being of 92.6% ^7Li and 7.4% ^6Li, Hagiwara and Takakura (1969) have reported that, in one run, the same isotopes were enriched to 97.09% and 14.64%, respectively, in the band edges. Moreover, since there is no increase of elution volume between two operations, the operation may be repeated, and the isotopic separation can thus become quite good (Tremillon, 1965).

Nuclear Magnetic Resonance and Lithium Imaging *in vivo*

During the present decade lithium nuclear magnetic resonance (NMR) has developed significantly, using different approaches (chemical shift, relaxation, Fourier spectrometry), and it is beginning to become possible to use it for lithium imaging *in vivo*. For more detail about lithium NMR, see F.G. Riddel, Chapter 5, and M. S. Hughes, Chapter 10, this volume.

Principle of the method

For a simple introduction to the use of NMR methods in biology, see, for instance, Lenk *et al.* (1979) or Janin (1985). Briefly summarizing, the intrinsic angular momentum, or spin (which has no equivalent in classical mechanics), is an important quantum mechanical characteristic of elementary particles. For electrons and nucleons (i.e. protons and neutrons), the spin may take only two values, $-1/2$ and $+1/2$. The spin, I, of an atomic nucleus is obtained by the combination of the proton and neutron spins. For atomic nuclei with even numbers of protons and neutrons (for instance ^{12}C or ^{16}O), $I = 0$; for atomic nuclei with an odd number of nucleons, I is half of an odd whole-number (for example, $I = 1/2$ for ^{1}H or ^{31}P, while $I = 3/2$ for ^{7}Li, ^{23}Na or ^{35}Cl); for the atomic nuclei with odd numbers of both protons and neutrons, the spin is a whole number (for example $I = 1$ for ^{2}H, ^{6}Li and ^{14}N). Atomic nuclei with I-values larger than $1/2$ are also characterized by an electric quadrupole moment, Q, different from zero; Q can be either positive or negative. The presence of an external magnetic field B_0 increases the energy of atomic nuclei with spin parallel to B_0, whereas it decreases the energy of those with spin antiparallel to B_0. The energy difference ΔE_z, between the two energy levels thus created, is given by

$$\Delta E_z = \gamma \frac{h}{2\pi} B_0$$

where γ is the gyromagnetic ratio of the atomic nuclei under consideration and h is Planck's constant. In fact with a given sample under study, the spins of the atomic nuclei are statistically distributed with only a small proportion, P (termed polarisation), of these atomic nuclei being actually parallel to B_0: the P values are usually in the range of 10^{-5}–10^{-6}.

For taking measurements, a small magnetic field, b (approximately 10^4 times smaller than B_0), is imposed at right-angles to B_0, and rotating with an imposed frequency, ω. This is equivalent to a radio frequency wave delivering energy quanta equal to $(h/2\pi)\omega$, and the rotating magnetic field is in fact imposed by a radio frequency source. When ω is equal to a particular value, ω_0, such that

$$\frac{h}{2\pi} \omega_0 = \Delta E_z$$

and

$$\omega_0 = \gamma B_0 \quad \text{(Larmor frequency)}$$

then the spin of the atomic nuclei under consideration can change from being parallel to B_0 to being antiparallel. Due to this resonance effect, there is a peak of absorption of energy when the frequency, ω, is just equal to ω_0. The NMR measurement records these peaks of absorption of energy. The magnetic field B_0 is usually of the order of magnitude of 1 tesla (10^4 gauss), and the frequency, ω, in the range 10^7–10^8 Hz (s^{-1}). Instead of changing ω at constant B_0, it is often preferable to keep a constant value of ω and vary B_0. Due to the low value of the polarization, the sensitivity of the NMR measurements is not very good, which is the main difficulty with this method. Since the magnetic field B_0 is never perfectly homogeneous, the measured peaks always have a non-zero width, which limits the resolving power of the method. Therefore, for good resolving power, it is necessary to improve the homogeneity of B_0 (measured by the ratio of the fluctuations to the mean value). The best

NMR devices presently available have an homogeneity up to 10^{-8}. Clearly, NMR measurements are possible only with atomic nuclei with spin non-zero.

Miscellaneous types of NMR measurements

By determining the Larmor frequencies of a sample for a given value of B_0, one may identify the types of atomic nuclei existing in the sample.

Of greater interest for us is the chemical shift. When an atomic nucleus is inserted into a molecule, the neighbouring electrons act as a magnetic shield in such a way that the magnetic field at the level of the nucleus under consideration is slightly different from the imposed field B_0 (diamagnetic effect). Therefore, there is a slight shift, $\Delta\omega_0$, of the Larmor frequency compared with the reference value, ω_0. The relative importance, σ, of this shift is usually expressed in ppm (part per million), by

$$\sigma = 10^6 \left(\frac{\Delta\omega}{\omega_0} \right)$$

The coupling of the spins of neighbouring atomic nuclei produces a fine structure of the peaks, which one may observe when using NMR devices with a very good homogeneity (high resolution).

Following an excitation pulse from the radio frequency source, the system of spins of atomic nuclei in the sample will progressively relax back to its original state. The first type of relaxation, termed 'spin-lattice' or 'longitudinal', corresponds to the relative proportions of spins parallel and antiparallel to B_0 evolving towards the value characteristic of the thermal equilibrium. The second type of relaxation, termed 'spin–spin' or 'transverse' relaxation, characterizes the increasing disorder of the system of spin. For atomic nuclei with spin values equal to $1/2$, each of these relaxations may be characterized by a single exponential term, each with a characteristic time (T_1 and T_2, respectively). For atomic nuclei like ^7Li, whose spin value is $3/2$, the relaxations are biexponential.

For NMR imaging, a small magnetic field, b_0, whose intensity is a function of the space coordinates, is superimposed on B_0. Each space point is then characterized by a specific resonance frequency. From the distribution of the radio frequency spectrum, a computer reconstitutes the distribution of the atomic nuclei under observation. Using relaxation methods, the imaging contrast is much improved. An important application is that of the medical imaging of protons: the soft tissues (which are particularly rich in hydrogen) are imaged with practically no disturbance from the bones, and the contrast between neighbouring tissues (for instance white and grey matter in the brain) is considerably better than that of X-radiography.

In practice the radio frequency field may either be permanent, or may be delivered in the form of very intense and very short rectangular impulses (a few μs in duration). The second method makes it possible to measure the relaxation times T_1 and T_2. Moreover, several such transient responses may be stored in a computer memory, thus improving the sensitivity. It is possible to improve the impulse methods further by calculating the Fourier transform of the transient response of the spins of atomic nuclei. The latter method, termed Fourier spectrometry, is 100 times as sensitive as the method using a permanent radio frequency field; moreover, it is possible to evaluate the relaxation times, T_1, of the individual peaks in complex systems.

Applications to lithium problems in biology

Using spin-lattice relaxation and Fourier transforms, Hutton *et al.* (1977) have studied the structure of complexes of pyruvate kinase with Li^+ and other cations, in connection with the activation of this enzyme by Li^+, while Grisham and Hutton (1978) have used ^7Li-NMR as a probe of monovalent cation sites at the active site of (Na^+-K^+) ATPase from kidney. Ion transport across the membrane of phosphatidylcholine vesicles was measured using ^{23}Na and ^7Li NMR (Degani and Elgavish, 1978). The interaction of lithium with cryptates was studied using ^7Li-NMR (Cambillau and Ourevitch, 1981); ^7Li-NMR was applied to study the cellular transport of lithium, especially in rat hepatocytes (G.M.H. Thomas and Olufunwa, 1988; G.M.H. Thomas *et al.*, 1988), and in human erythrocyte (Partridge *et al.*, 1988; Hughes *et al.*, 1988; Espanol *et al.*, 1989). From multinuclear NMR studies in lithium pharmacology, it was suggested that free lithium concentration in cells was significantly lower than hitherto had been suspected (Hughes, 1988). *In vivo* ^7Li-NMR spectroscopy was used to measure mean lithium levels in muscle and brain following lithium treatment of humans, thus revealing a relatively slow accumulation of lithium in the brain (Renshaw and Wicklund, 1988).

Renshaw *et al.* (1985, 1986) have begun to apply NMR imaging techniques to ^7Li, with a view to assessing the distribution of lithium in human beings who receive lithium therapeutically. They have already obtained lithium images of aqueous phantoms, and of a rat abdomen *in vivo*. At 2 mM Li^+, 4 h of signal averaging produced an image with resolution of the order of several mm.

Methods for Imaging Lithium, or Li Isotopes, in Histological Sections

Methods derived from electron microscopy

The electron microscopy images are obtained via an interaction between a beam of accelerated, monokinetic electrons, and the sample under study. In this interaction, the incident electrons lose a part of their energy, which they give to the electronic clouds of the sample molecules, and in doing so excite these molecules, which subsequently will lose their excitation energy in the form of emitted photons. When the excitation/de-excitation process affects the deep (or 'heart') electrons, the energy loss of the incident electrons, and the photon emission of the excited molecules, is characteristic of the individual atoms present in the sample. This is the principle of two major methods for element imaging, termed respectively electron energy loss analysis and electron microprobe analysis.

The electron microprobe analyses the X-rays which are emitted in the de-excitation of the excited sample atoms. The sensitivity is highest for elements with atomic numbers ranging from 13 to 30, but it is inaccurate for the lightest elements and, unfortunately, not applicable to lithium. Related methods exist, such as the photon induced X-ray emission (PIXE) and the X-ray fluorescence microprobe, where the sample molecules are excited using a beam of accelerated photons or a beam of X-rays, respectively, instead of the beam of accelerated electrons of an electron microscope. However, again because there is no emission of suitable X-rays in the de-excitation of the lightest atoms, these methods also are not useful for lithium imaging.

In electron microscopy, the energy losses of the incident monokinetic electrons are characteristics of the interatomic bonds (in the range of a few eV) or of the atoms (of the order of 100 eV, or even more) present in the sample. Therefore, the electrons emerging from the sample can be filtered, in order to keep only those with the value of energy loss characteristic of the atom under study, and then these filtered electrons can be used to build an electron microscopic image of the distribution of this particular atom. Spatial resolution is considered to be about 0.5 nm. However, quantitative analysis remains difficult, and the method does not discriminate between isotopes. Up to now, electron energy loss analysis has been used mainly by solid state physicists, and only rarely by biologists (for details, see Henry and Duval, 1975, and Galle *et al.*, 1979). The mapping of boron-labelled proteins on histological sections of the rat has been published recently (Bendayan *et al.*, 1989), but, to our knowledge, the method has not yet been applied to lithium imaging in biological samples.

Nuclear methods

Lithium has no radioisotope with a suitable half-life, and so lithium autoradiography is impossible. However, one may image the distribution of lithium in a histological preparation using specific nuclear reactions, either with neutrons or with charged particles (especially protons or deuterons). Let us recall that a nuclear reaction is symbolized by $X(x, y)Y$, meaning that bombardment of a target nuclide X by a radiation x gives another nuclide Y and a radiation y. The various isotopes of an element are not subjected to identical nuclear reactions, which makes it possible to discriminate isotopes from one another.

Nuclear reactions with neutrons

The two stable isotopes of lithium, ^{6}Li and ^{7}Li, may be engaged in nuclear reactions with thermal neutrons, i.e.

$$^6\text{Li}(n, \alpha)^3\text{H} \quad \text{and} \quad ^7\text{Li}(n, \gamma)^8\text{Li}$$

Therefore, the detection of ^{6}Li and ^{7}Li can be achieved by detecting the induced radioactivity. For instance (Wiernik and Amiel, 1969; Heydorn *et al.*, 1977), ^{7}Li was detected by measuring the induced ^{8}Li activity (β^-, with a half-life of 0.88 s), and ^{6}Li by measuring the radioactivity induced in the secondary reactions, $^{16}\text{O}(^3\text{H}, n)^{18}\text{F}$ and $^{18}\text{O}(^3\text{H}, \alpha)^{17}\text{N}$, produced by the tritons, ^{3}H, created by the $^6\text{Li}(n, \alpha)^3\text{H}$ nuclear reaction (Wiernik and Amiel, 1969). Li was also measured *in vivo* (Vartsky *et al.*, 1985) by collecting the ^{3}H gas exhaled by lithium-treated subjects following neutron irradiation of the organ of interest. A new technique for analysis of ultratrace lithium ($< 10^{-8} \text{gg}^{-1}$) in biological samples consists of mass spectrometric assay of ^{3}He from decay of ^{3}H produced by neutron reaction on ^{6}Li (Clarke *et al.*, 1987).

However, the radioactivities induced in the nuclear reactions (^{3}H and consequently ^{18}F or ^{17}N, or ^{8}Li) are so small compared with those of the many other radionuclides produced by neutron irradiation of a biological specimen, that the mapping of ^{6}Li or ^{7}Li by the autoradiography of their neutron-induced radioactivities is not feasible in practice.

Nevertheless, the nuclear reaction $^6\text{Li}(n, \alpha)^3\text{H}$ may be used to image the distribution of ^{6}Li in a histological section if the section is pressed against a dielectric film and then irradiated with neutrons. When the α and ^{3}H particles produced by the reaction hit the dielectric film, they leave latent tracks which can be enhanced by etching in NaOH (Fleischer *et al.*, 1972). The distribution of these $\alpha/^3$H tracks in the detecting film reveals the distribution of lithium in the histological section. This is termed neutron capture radiography (NCR). The method may be applied to the detection of ^{6}Li-enriched lithium or to that of natural lithium (containing 7.4% ^{6}Li). Due to the short range of the α and ^{3}H particles, the spatial resolution is better than that of conventional autoradiography using β-emitters. Interferences exist, especially with boron, due to the existence of the $^{10}\text{B}(n, \alpha)^7\text{Li}$ nuclear reaction; but the latter interference may be avoided by placing a shield, a few μm thick, between the sample and the detecting film, in order to absorb the emitted α and ^{7}Li particles: only the ^{3}H tracks issued from ^{6}Li will then be recorded in the dielectric film. The detecting films most frequently employed are made either of cellulose nitrate (such as the LR 115 and CN 85 types manufactured by Kodak–Pathé) or of polycarbonates (such as the CR 39 manufactured by Pershore Mouldings). The thickness of such films is in the range of a few μm to a fraction of a mm. Zamenhof *et al.* (1989) have proposed the use of a 0.8 μm thick lexan detector which (a) exhibits higher resolution than the thicker detectors, and (b) permits etching of the detector without destruction of the tissue section, thereby ensuring precise spatial locating of the tracks with regard to the corresponding histological structures. Among many other possibilities, the NCR method has been applied to image the distribution of Li^+ in various animal organs (Nelson *et al.*, 1976; Thellier *et al.*, 1976a), in particular the brains of normal adults (Carpenter *et al.*, 1977; Thellier *et al.*, 1976b; Wissocq *et al.*, 1979; Heurteaux *et al.*, 1980; Nelson *et al.*, 1980), of normal embryos (Wissocq *et al.*, 1985) and of dysmyelinating mutants (Heurteaux *et al.*, 1986).

Nuclear reactions with charged particles

The lithium isotopes, ^6Li and ^7Li, undergo nuclear reactions with accelerated charged particles (such as protons or deuterons), for example $^6Li(d, 2\alpha)$, $^7Li(p, 2\alpha)$, $^7Li(p, \gamma)^8Be$, $^7Li(p, n)^7Be$, $^7Li(p, p'\gamma)^7Li$, $^7Li(d, \alpha)^5He$. The type of nuclear reaction produced depends on the energy of the bombarding particle; for instance, nuclear reaction $^7Li(p, 2\alpha)$ is favoured when the energy of the protons is below 1.88 MeV, whereas $^7Li(p, n)^7Be$ is favoured when it is above 1.88 MeV (Pannetier, 1965). The energies of the particles emitted in a nuclear reaction are also characteristic of this reaction; for instance, when a target containing a mixture of ^6Li and ^7Li is irradiated with 4.0 MeV deuterons, the α-particles emitted by nuclear reactions $^6Li(d, 2\alpha)$ and $^7Li(d, \alpha)^5He$ have respective energies 12.18 and 9.21 MeV, which makes it possible to distinguish the two lithium isotopes from each other. The method is well suited to determining the isotopic ratios, $^6Li/^7Li$, in a specimen. The analysis time per sample is only a few minutes. For ^6Li concentrations ranging from 7.4% (natural lithium) to 30%, the relative standard deviation was $\pm 4\%$ (Pretorius and Coetzee, 1972).

Räisänen and Lappalainen (1986) have used proton beams originating from a low-energy accelerator ($E_p \leqslant 2.7$ MeV) for the analysis of ^7Li via the $^7Li(p, 2\alpha)$ and $^7Li(p, p'\gamma)^7Li$ nuclear reactions. Isotope ^6Li is unaffected under these conditions. ^7Li is determined by detecting the α-particles of the first reaction or the γ-rays originating from the second one. Fluorine is the only element that can cause possible overlap in the dominant part of the α-spectrum of the $^7Li(p, 2\alpha)$ reaction. In the case of organic samples (orchard leaves, human blood serum, human bone), the optimal value of the proton energy to minimize the fluorine interference was 2.4 MeV; the measuring time was 5–10 min per sample, and the sensitivity better than 0.1 ppm by dry weight. For the $^7\text{Li}(\text{p, p}'\gamma)^7\text{Li}$ nuclear reaction, the optimum energy of the incident protons was in the range 1.7–2.0 MeV; the sensitivity was about 0.15 ppm by dry weight with the organic samples, but it was dependent on the relative concentrations of Li, B, Na and F in the sample; the measuring time was approximately 30 min.

In the nuclear microprobe (Pierce, 1975; Engelmann, 1978), a thin beam of bombarding particles (usually protons or deuterons) scans the specimen. At the point of impact of the beam, nuclear reactions are produced, and the excitation/de-excitation processes occurring in the electronic shells result in the emission of X-rays. Among all the emitted radiations, those characteristic of the nuclide under consideration are selected and counted, then a computer reconstitutes an image of the distribution of this nuclide. The nuclear microprobe can detect any isotope of any chemical element, with practically no risk of interferences. The measuring time is very short. The sensitivity is variable according to the nature of the nuclide under consideration; however, sensitivities of 10^{-12} g, or even better, have been reported. The spatial resolution is of the order of 1 μm, but the resolution in depth is much better. Although the method is not too destructive, it is possible to image a nuclide at any desired depth in the sample, or to draw a profile in depth: to do this, one has to adjust the initial energy of the bombarding protons or deuterons to a value such that their remaining energy at the desired depth is exactly equal to a resonance energy of the physical event under consideration.

To our knowledge the nuclear microprobe has not yet been applied to imaging the stable isotopes of lithium, ^6Li and ^7Li, in biological samples, although there is no reason why it could not be, using the proton or deuteron reactions described above.

Methods based on mass spectrometry

Mass spectrometry can detect lithium ions and discriminate between the stable isotopes, ^6Li and ^7Li (Omura and Morito, 1958; Chow and Goldberg, 1962; Lloyd and Field, 1981), as no other element has a naturally occurring isotope of mass number 6 or 7. Hence, there is practically no risk of interference, except perhaps with usually very rare ions such as $^{12}\text{C}^{2+}$ with $^6\text{Li}^+$. Moreover two refinements of the method, secondary ion emission microscopy and the use of the laser microprobe mass analyser (LAMMA), make it possible to image the distribution of ^6Li or ^7Li in a specimen, including biological specimens.

In secondary ion emission microscopy (for application to biology, see Lefèvre, 1975; Galle, 1982, 1984; Galle *et al.*, 1982) a beam of accelerated ions (such as ionized oxygen, argon or caesium), termed the primary ions, is projected onto the sample, where it erodes the superficial layers. Part of the matter thus ejected comes out in an ionized form (secondary ions); these secondary ions are accelerated and selected in charge and mass by mass spectrometry. Two different modes of functioning exist for imaging an ion. In the stigmatic-image mode, the beam of primary ions is large (diameter $\sim 250\,\mu m$), and the selected ions are focused by electromagnetic lenses into a stigmatic image. In the scanning mode, the beam of primary ions is extremely thin, and only the mean current of the selected ions is measured at each point hit by the beam; a computer then reconstitutes an image of the distribution of this particular type of ions, after scanning the preparation with the primary ion beam. The spatial resolution on the sample surface is better than $1\,\mu m$, and the resolution in depth is on the order of a few nm. The sensitivity is usually excellent. For absolute quantitative measurements, the preparation of reliable standards is often a problem, but the isotopic ratios are determined with accuracy. The progressive erosion of the specimen by the primary ions makes it possible to draw the profile, in the depth of the sample, of any ion under consideration. The method is well suited to imaging lithium, and the lithium isotopes, in biological samples, although this has not been achieved very often up to now. However, the method has already been applied to:

(1)   the measurement of the unidirectional influx and efflux of Li^+ through an epithelium (Thellier *et al.*, 1976a);

(2) Li^+/H^+ interactions in transmembrane transport (Hartmann *et al.*, 1983);

(3)   lithium absorption by developing eggs (Elbers, 1983);

(4)   the natural distribution of lithium in marine organisms (Chassard Bouchaud *et al.*, 1984);

(5)   the compartmental analysis of lithium in plants (Bielenski *et al.*, 1984) and in animals (Heurteaux *et al.*, 1985); and

(6)   the analysis of lithium in microdrops using isotopic dilution (Thellier *et al.*, 1989).

The principle of LAMMA is based on the excitation of a micro-volume of the sample by a focused laser beam. Then the ions produced in this interaction are collected and selected by mass spectrometry, in a way somewhat similar to that described in secondary ion emission microscopy. LAMMA can discriminate isotopes. The quantitative estimations require the preparation of adequate standards, but the isotopic ratios can be determined with accuracy. Although imaging is theoretically possible, especially by scanning the preparation with the laser beam, it seems that this possibility has still not been used very much. For further details about laser microprobe mass spectrometry and its biological and medical applications, see Kaufmann *et al.* (1975, 1979), Edelmann (1981), Schröder (1981), Vogt *et al.* (1981), Chamel and Eloy (1983), and Clarke (1989).

Main Technical Problems Encountered

There are specific problems associated with the application of each of the methods given above, described in detail in the original publications. In the present contribution, we are going to concentrate on two general types of difficulties, namely the preparation of histological sections with the aim of studying the distribution of mobile substances, and the possible occurrence of isotopic effects.

Histological sections

Lithium ions are very mobile. Therefore, when preparing histological sections in order to study the distribution of Li^+ *in vivo*, it is necessary to prepare the sections without disturbing the natural distribution of lithium.

The usual method for fixing the tissues consists of freezing them rapidly. This cryofixation has to be performed at a very low temperature in order to avoid the formation of ice crystals in the specimen. The temperature of boiling N_2 ($-195.8°C$) would be adequate. However, due to the 'calefaction' phenomenon, it is not sufficient merely to dip the specimen into boiling nitrogen. Rather, the specimen should be dipped into a liquid such as melting propane ($-189.7°C$), obtained by cooling in boiling nitrogen. An even better, but more expensive, possibility is to dip the specimen into melting nitrogen ($-209.9°C$). In these operations, it is clear that only minute tissue fragments may be used in order that the thermal equilibration of the specimen with its surroundings is practically instantaneous. The usual freezing methods give adequate freezing in the range of $10-20\,\mu m$ depth; however, high-pressure freezing may allow ultrastructural preservation up to $0.1-0.5\,mm$ depth. Moreover, if the frozen tissues have not been dehydrated, it is advisable to keep them at a low temperature. Rewarming to a temperature of about $-30°C$ (or above) is enough to cause a rapid phase change of the amorphous ice to ice crystals. For further details of cryofixation techniques, see the comprehensive reviews by Plattner and Bachmann (1982), Robards and Sleytr (1985), Gilkey and Staehelin (1986), Menco (1986) and Sitte *et al.* (1987).

The frozen tissues may be cryosectioned and freeze-dried for subsequent analysis of the lithium distribution. The method has the advantage of being fairly simple and rapid. It has been revealed to be satisfactory for the quantitative imaging of lithium at the histological level (Carpenter *et al.*,

1977; Nelson *et al.*, 1976, 1980; Thellier *et al.*, 1980a,b; Wissocq *et al.*, 1979, 1985; Heurteaux *et al.*, 1980, 1986). However, cryosectioning and freeze-drying imply significant rewarming, with the problems which have been described above. Moreover, the freeze-dried sections are fairly heterogeneous in their local composition and density, which is unfavourable for accurate quantitative determinations using most of the analytical methods described in the previous sections.

Frozen samples may also be freeze-substituted. For that purpose, they are bathed in solvents (such as methanol, ethanol or acetone) at low temperature (for example $-85°C$) for up to several weeks, until complete exchange has occurred of their internal water for the solvents. They are then infiltrated with a polymerizable resin. After polymerization the samples are sectioned using the usual histological methods. Freeze-substitution does not cause significant rewarming of the frozen tissues. Moreover, since the sample-water has been replaced by the polymerized resin, most of the mass of the freeze-substituted specimen corresponds to the resin: this means that this freeze-substituted specimen may be considered as practically homogeneous, as a first approximation, which makes it much simpler for the quantitative estimations. When the method is applied to the study of the distribution of Li^+ (or of other mobile substances), it is clear that the solvents to be employed for the substitution have to be chosen such that lithium salts are not soluble in these solvents. For details about freeze-substitution see Harvey (1982), Humbel and Müller (1984) and Steinbrecht and Müller (1987).

An alternative to freeze-substitution (i.e. to substituting the tissue-water with organic solvents) consists of chemically transforming this tissue-water into other substances, for instance solvents, *in situ*. Müller and Jacks (1975) have used 2,2-dimethoxypropane to chemically dehydrate a tissue by forming methanol and acetone, according to the endothermic reaction

$$CH_3-\underset{\underset{OCH_3}{|}}{\overset{\overset{OCH_3}{|}}{C}}-CH_3 + H_2O \rightarrow 2\,CH_3OH + CH_3-\overset{\overset{O}{\|}}{C}-CH_3$$

In this case, the reaction was accomplished at room temperature. It was therefore not suited to studying the distribution of mobile substances. More recent work has shown that the same principle was applicable at low temperature (Kaeser, 1989; Kaeser *et al.*, 1989). However, as has already been explained for freeze-substitution, such methods can be suitable for studying lithium distribution only if the solvents produced are such that the lithium salts are not soluble in them.

However ingenious they may be, the methods of sample dehydration which we have just described are generally complicated and time consuming, and none of them is perfectly satisfactory for preparing specimens with the aim of studying the distribution of mobile substances like Li^+. A better solution would be to freeze the tissues at a very low temperature, and analyse directly the distribution of Li^+ (or of any other mobile substance) on the frozen specimen. It is theoretically possible to perform secondary ion emission microscopy with frozen hydrated samples. Although this has not often been attempted (Bernius *et al.*, 1985), it is likely to bring the most significant progress in future investigations of mobile substances.

Isotopic effects

The basic postulate when using isotopes (whether stable or radioactive) for labelling experiments is that, apart from the nuclear properties, the various isotopes of a given element behave alike from the physical, chemical and biochemical points of view. In fact, this is never perfectly true, since isotopes differ from one another in at least one of their characteristics, i.e. their mass. For heavy elements this is not too important, since the difference in the mass of the isotopes relative to their mean mass is usually quite small. However, with the lithium isotopes, ^6Li and ^7Li, the relative difference of mass is over 15%, which cannot be taken as totally negligible. Therefore one may expect the occurrence of isotopic effects with lithium.

Several authors have reported that the biological effects of ^6Li and ^7Li are not always identical to each other (Sherman *et al.*, 1984; Alexander *et al.*, 1988; Stokes *et al.*, 1988).

There is also the possibility that isotopic effects occur in the analysis of the distribution of lithium in a specimen. This might particularly be the case with methods such as secondary ion emission microscopy, since the efficiency of extracting the secondary ions is likely to depend on the mass of these ions. In such a case, it is advisable to calibrate the method separately for each isotope, ^6Li and ^7Li, to be taken into consideration.

Conclusion

Up to now, most biologists have been interested in studying the cellular behaviour of heavy molecules, rather than that of small solutes, especially mineral ions, although these ions definitely play an important role in the functioning of living cells. The reason for this probably stems from the perceived methodological difficulties in determining the detailed location and movements of these mobile substances in biological systems. The increasing application of powerful physical methods to address biological problems means that the goal of studying the behaviour of mineral ions at a detailed subcellular level is no longer totally out of reach. These methods thus give access to an almost entirely new field of research in cellular biology. This is particularly interesting for lithium since this cation exhibits very specific biological effects, at relatively low concentrations and with practically no general toxicity, not only for the medical treatment of manic-depressive psychosis (Cade, 1949; Schou, 1957), but also for a number of physiological processes in animals and plants (see Wissocq *et al.*, Chapter 2, this volume).

References

Alexander, G.J., Lieberman, K.W. and Sechzer, J.A. (1988). Lithium-6 isotope toxicological, physiological and behavioural effects. *In* "Lithium Inorganic Pharmacology and Psychiatric Use" (ed. N.J. Birch), pp. 299–302. IRL Press, Oxford.

Amdisen, A. (1975). Monitoring of lithium treatment through determination of lithium concentration. *Dan. Med. Bull.* **22**, 277–291.

Bendayan, M., Barth, R.F., Gingras, D., Londono, I., Robinson, P.T., Alam, F., Adams, D.M. and Mattiazzi, L. (1989). Electron spectroscopic imaging for high-resolution immunocytochemistry: use of boronated protein A. *J. Histochem. Cytochem.* **37**, 573–580.

Bernius, M.T., Chandra, S. and Morrison, G.H. (1985). Cryogenic sample stage for the Cameca IMS 3F ion microscope. *Rev. Sci. Instrum.* **56**, 1347–1351.

Bertholf, R.L., Savory, M.G., Winborne, K.H., Hundley, J.C., Plummer, G.M. and Savory, J. (1988). Lithium determined in serum with an ion-selective electrode. *Clin. Chem.* **34**, 1500–1502.

Bielenski, U., Garrec, J.P., Demarty, M., Ripoll, C. and Thellier, M. (1984). Kinetic parameters in the compartmental analysis of Li^+ transport in *Lemna gibba* using the stable isotopes, ^6Li and ^7Li, as tracers. *Physiol. Plant.* **61**, 236–242.

Birch, N.J. and Jenner, F.A. (1973). The distribution of lithium and its effects on the distribution and excretion of other ions in the rat. *Br. J. Pharmacol.* **47**, 586–594.

Birch, N.J., Hullin, R.P., Inie, R.A. and Robinson, D. (1978a). Use of the stable isotope ^6Li in human pharmacokinetic studies. *Br. J. Clin. Pharmacol.* **5**, 351–352.

Birch, N.J., Robinson, D., Inie, R.A. and Hullin, R.P. (1978b). ^6Li a stable isotope of lithium determined by atomic absorption spectroscopy and use in human pharmacokinetic studies. *J. Pharm. Pharmacol.* **30**, 683–685.

Cade, J.F. (1949). Lithium salts in the treatment of psychotic excitement. *Med. J. Aust.* **36**, 349–352.

Cambillau, C. and Ourevitch, M. (1981). ^7Li NMR studies of an Li^+-ethyl acetoacetate enolate triple ion-Li^+ cryptate complex. *J. Chem. Soc. Chem. Comm.* **716**, 996–997.

Carpenter, B.S., Samuel, D., Wassermann, I. and Yuwiller, A. (1977). A study of lithium uptake and location in the brain using the nuclear track technique. *J. Radioanal. Chem.* **37**, 523–528.

Chamel, A. and Eloy, J.F. (1983). Some applications of the laser probe mass spectrograph in plant biology. *In* "Scanning Electron Microscopy" II, pp. 841–851. SEM Inc. A.M.F. O'Hare, USA.

Chapman, J.F. and Dale, L.S. (1976). The determination of lithium isotope abundances with a dual-beam atomic absorption spectrometer. *Anal. Chim. Acta* **87**, 91–95.

Chassard-Bouchaud, C., Galle, P., Escaig, F. and Miyakawi, M. (1984). Bioaccumulation du lithium par les organismes marins des zones côtières européennes, américaines et asiatiques: étude microanalytique par émission ionique secondaire. *C. R. Acad. Sci. Paris III* **299**, 719–724.

Chow, T.J. and Goldberg, E.D. (1962). Mass spectrometric determination of lithium in sea water. *J. Marine Res.* **20**, 163–167.

Clarke, N.S. (1989). Laser microprobe mass spectrometry. *Eur. Microsc. Anal.* **13**, 21–24.

Clarke, W.B., Koekebakker, M., Barr, R.D., Downing, R.G. and Fleming, R.F. (1987). Analysis of ultratrace lithium and boron by neutron activation and mass-spectrometric measurements of ^{3}He and ^{4}He. *Int. J. Appl. Radiat. Isot.* **38**, 735–743.

Cook, F.L., Caruso, T.C., Byrne, M.P., Bowers, C.W., Speck, D.H. and Liotta, C.L. (1974). Facile synthesis of 12-crown-4 and 15-crown-5. *Tetrahedron Lett.* **46**, 4029–4032.

Degani, H. and Elgavish, G.A. (1978). Ionic permeabilities of membranes. ^{23}Na and ^{7}Li NMR studies of ion transport across the membrane of phosphatidylcholine vesicles. *FEBS Lett.* **90**, 357–360.

Ebara, T. and Smith, D.F. (1979). Lithium levels in blood platelets, serum, red blood cells and brain regions in rats given acute or chronic lithium salt treatments. *J. Psychiat. Res.* **15**, 183–188.

Edelmann, L. (1981). Selective accumulation of Li$^+$, Na$^+$, K$^+$, Rb$^+$ and Cs$^+$ at protein sites of freeze-dried embedded muscle detected by LAMMA. *Fresenius Z. Anal. Chem.* **308**, 218–220.

Elbers, P.F. (1983). The site of action of lithium ions in morphogenesis of *Lymnaea stagnalis* analyzed by secondary ion mass spectroscopy. *Differentiation* **24**, 220–225.

Engelmann, C. (1978). La microsonde nucléaire, principe, performances et exemples d'applications. *In* "Microanalyse et Microscopie Electronique" (ed. F. Maurice *et al.*), pp. 497–501. Edition Physique d'Orsay, France.

Espanol, M.T., Ramasamy, R. and de Freitas, D.M. (1989). Measurements of lithium transport across human erythrocyte membranes by ^{7}Li NMR spectroscopy. *In* "Biological and Synthetic Membranes", pp. 33–43. Alan R. Liss Inc., New York.

Fleischer, R.L., Alter, H.W., Furman, S.C., Price, P.B. and Walker, R.M. (1972). Particle track etching. *Science* **178**, 255–263.

Fujine, S., Saito, K. and Shiba, K. (1982). Transient behavior of lithium isotope separation by displacement chromatography. *Sep. Sci. Technol.* **17**, 1309–1325.

Galle, P. (1982). Tissue localization of stable and radioactive nucleides by secondary-ion microscopy. *J. Nucl. Med.* **23**, 52–57.

Galle, P. (1984). Tissue microlocalization of isotopes by ion microscopy and by microautoradiography. *In* "Secondary Ion Mass Spectrometry" (ed. A. Benninghoven *et al.*), pp. 495–497. Springer-Verlag, Berlin.

Galle, P., Berry, J.P. and Lefevre, R. (1979). Microanalysis in biology and medicine. *In* "Scanning Electron Microscopy" II, pp. 703–710. SEM Inc., AMF O'Hare, USA.

Galle, P., Berry, J.P., Meignan, M. and Duckett, S. (1982). Analytical ion microscopy of biological tissue. *In* "Proceedings International Workshop on Physics and Engineering in Medical Imaging", pp. 45–48. IEEE Computer Soc., Los Angeles.

Gilkey, J.C. and Staehelin, L.A. (1986). Advances in ultra-rapid freezing for the preservation of cellular ultrastructure. *J. Electron Microsc. Techn.* **3**, 177–210.

Gokel, G.W. and Durst, H.D. (1976). Principles and synthetic applications in crown ether chemistry. *Synthesis* **3**, 168–184.

Grafe, P., Rimpel, J., Reddy, M.M. and ten Bruggencate, G. (1982). Lithium distribution across the membrane of motoneurons in the isolated frog spinal cord. *Pflügers Arch.* **393**, 297–301.

Greil, W., Eisenried, F., Becker, B.F. and Duhm, J. (1977). Interindividual differences in the Na^+-dependent Li^+ counter-transport system and the Li^+ distribution ratio across the red cell membrane among Li^+ treated patients. *Psychopharmacology* **53**, 19–26.

Grisham, C.M. and Hutton, W.C. (1978). Lithium-7 NMR as a probe of monovalent cation sites at the active site of $(Na^+ + K^+)$-ATPase from kidney. *Biochem. Biophys. Res. Commun.* **81**, 1406–1411.

Güggi, M., Fiedler, U., Pretsch, E. and Simon, W. (1975). A lithium ion-selective electrode based on a neutral carrier. *Anal. Lett.* **8**, 857–866.

Hagiwara, Z. and Takakura, Y. (1969). Enrichment of stable isotopes. III Enrichment of ^6Li and ^7Li by ion exchange column. *J. Nucl. Sci. Technol.* **6**, 326–332.

Hartmann, A., Bielenski, U., Lüttge, U., Garrec, J.P., Thoiron, A. and Thellier, M. (1983). Measurements of unidirectional fluxes of lithium, application to the study of Li^+/H^+ interactions with the transmembrane exchanges of *Lemna gibba* G1. *In* "Physical Chemistry of Transmembrane Ion Motions" (ed. G. Spach), pp. 591–597. Elsevier, Amsterdam.

Harvey, D.M.R. (1982). Freeze substitution. *J. Microsc.* **127**, 209–221.

Henry, L. and Duval, P. (1975). Analyse qualitative par perte d'énergie des électrons à travers la matière. *J. Microsc. Biol. Cell.* **22**, 381–388.

Heurteaux, C., Wissocq, J.C. and Thellier, M. (1980). Nuclear track-etch method: quantitative location of lithium in brain. *Trans. Amer. Nucl. Soc.* **34**, 150–151.

Heurteaux, C., Garrec, J.P., Ripoll, C., Wissocq, J.C. and Thellier, M. (1985). Analyse compartimentale de l'échange isotopique $^6Li/^7Li$ en conditions stationnaires dans le plasma de la souris. *C. R. Acad. Sci. Paris III* **300**, 529–534.

Heurteaux, C., Baumann, N., Lachapelle, F., Wissocq, J.C. and Thellier, M. (1986). Lithium distribution in the brain of normal mice and of 'quaking' dysmyelinating mutants. *J. Neurochem.* **46**, 1318–1321.

Heydorn, H., Skanborg, P.Z., Gwozdz, R., Schmidt, J.D. and Wacks, M.E. (1977). Determination of lithium by instrumental neutron activation analysis. *J. Radioanal. Chem.* **37**, 155–168.

Hughes, M.S. (1988). Multinuclear NMR studies in lithium pharmacology. *In* "Lithium: Inorganic Pharmacology and Psychiatric Use" (ed. N.J. Birch), pp. 285–288. IRL Press, Oxford.

Hughes, M.S., Flavell, K.J. and Birch, N.J. (1988). Transport of lithium into human erythrocytes as studied by ^7Li nuclear magnetic resonance and atomic absorption spectroscopy. *Biochem. Soc. Trans.* **16**, 827–828.

Humbel, B.M. and Müller, M. (1984). Freeze-substitution and low-temperature embedding. *In* "Electron Microscopy 1984" (ed. A. Csanedy, P. Röhlich and D. Szabo), pp. 1789–1798. Programme Committee 8th European Congress on Electron Microscopy, Budapest, Hungary.

Hutton, W.C., Stephens, E.M. and Grisham, C.M. (1977). Lithium-7 nuclear magnetic resonance as a probe of structure and function of the monovalent cation site on pyruvate kinase. *Arch. Biochem. Biophys.* **184**, 166–171.

Janin, J. (1985). *In* "Méthodes Biophysiques pour l'Etude des Macromolécules", collection Méthodes. Hermann, Paris, France.

Kaeser, W. (1989). Freeze-substitution of plant tissues with a new medium containing dimethoxypropane. *J. Microsc.* **154**, 273–278.

Kaeser, W., Koyro, H.W. and Moor, H. (1989). Cryofixation of plant tissue without pretreatments. *J. Microsc.* **154**, 279–288.

Kaufmann, R., Hillenkamp, F., Nitsche, R., Schürmann, M. and Unsöld, E. (1975). Biomedical application of laser microprobe analysis. *J. Microsc. Biol. Cell.* **22**, 389–398.

Kaufmann, R., Hillenkamp, F. and Wechsung, R. (1979). The laser microprobe mass analyser (LAMMA): a new instrument for biomedical microprobe analysis. *Med. Prog. Technol.* **6**, 109–121.

Kimura, K., Oishi, H., Miura, T. and Shono, T. (1987). Lithium ion selective electrodes based on crown ethers for serum lithium assay. *Anal. Chem.* **59**, 2331–2334.

Lefèvre, R. (1975). La microanalyse par émission ionique secondaire et son application aux tissus calcifiés humains. *J. Microsc. Biol. Cell.* **22**, 335–348.

Lehmann, K. (1975). Die Kinetik von Lithium in Serum, Leber und Gehirn der Ratte. *Acta Biol. Med. Germ.* **34**, 1043–1047.

Lehn, J.M. (1978). Cryptates: the chemistry of macropolycyclic inclusion complexes. *Accounts. Chem. Res.* **11**, 49–57.

Lenk, R., Bonzon, M., Descouts, P. and Greppin, H. (1979). La résonance magnétique nucléaire: une approche nouvelle en biologie végétale. *Saussurea* **10**, 11–48.

Lloyd, J.R. and Field, F.H. (1981). Analysis of lithium using a commercial quadrupole mass spectrometer. *Biomed. Mass Spectrom.* **8**, 19–24.

Manning, D.C. and Slavin, W. (1962). Lithium isotope analysis by atomic absorption spectrophotometry. *At. Absorpt. Newslett.* **8**, 39–43.

Matusiewicz, H. (1982). Determination of natural levels of lithium and strontium in human blood serum by discrete injection and atomic emission spectrometry with a nitrous oxide-acetylene flame. *Anal. Chim. Acta* **136**, 215–223.

Menco, B.P.M. (1986). A survey of ultrarapid cryofixation methods with particular emphasis on applications to freeze-fracturing, freeze-etching, and freeze-substitution. *J. Electron Microsc. Techn.* **4**, 177–240.

Metzger, E., Ammann, D., Asper, R. and Simon, W. (1986). Ion selective liquid membrane electrode for the assay of lithium in blood serum. *Anal. Chem.* **58**, 132–135.

Morrison, J.M., Pritchard, H.D., Braude, M.C. and d'Aguanno, W. (1971). Plasma and brain lithium levels after lithium carbonate and lithium chloride administration by different routes in rats. *Proc. Soc. Exp. Biol. Med.* **137**, 889–892.

Müller, L.L. and Jacks, T.J. (1975). Rapid chemical dehydration of samples for electron microscopic examinations. *J. Histochem. Cytochem.* **23**, 107–110.

Nelson, S.C., Herman, M.M., Bensch, K.G., Barchas, J.D. and Sher, R. (1976). Localization and quantitation of lithium in rat tissue following intraperitoneal injections of lithium chloride. I. Thyroid, thymus, heart, kidney, adrenal and testis. *Exp. Mol. Pathol.* **25**, 38–48.

Nelson, S.C., Herman, M.M., Bensch, K.G. and Barchas, J.D. (1980). Localization and quantitation of lithium in rat tissue following intraperitoneal injections of lithium chloride. II. Brain. *J. Pharmacol. Exp. Ther.* **212**, 11–15.

Omura, I. and Morito, N. (1958). On the measurement of isotope abundance of lithium with mass spectrometer. *J. Phys. Soc. Japan* **13**, 659.

Pacey, G.E., Wu, Y.P. and Sasaki, K. (1987). Selective determination of lithium in biological fluids using flow injection analysis. *Anal. Biochem.* **160**, 243–250.

Pannetier, R. (1965). *In* "Vade Mecum du Technicien". Maisonneuve S.A. Pub., Moulin-les-Metz.

Partridge, S., Hughes, M.S., Thomas, G.M.H. and Birch, N.J. (1988). Lithium transport in erythrocytes. *Biochem. Soc. Trans.* **16**, 205–206.

Pedersen, C.J. (1967). Cyclic polyethers and their complexes with metal salts. *J. Amer. Chem. Soc.* **82**, 7017–7036.

Pierce, T.B. (1975). The nuclear microprobe. *J. Microsc. Biol. Cell.* **22**, 349–356.

Plattner, H. and Bachmann, L. (1982). Cryofixation: a tool in biological ultrastructural research. *Int. Rev. Cytol.* **79**, 237–304.

Plenge, P., Mellerup, E.T. and Norgaard, T. (1981). Functional and structural rat kidney changes caused by peroral or parenteral lithium treatment. *Acta Psychiat. Scand.* **63**, 303–313.

Pretorius, R. and Coetzee, P. (1972). Isotopic analysis of lithium by prompt alpha measurement during deuteron irradiation. *J. Radioanal. Chem.* **12**, 301–311.

Räisänen, J. and Lappalainen, R. (1986). Analysis of lithium using external proton beams. *Nucl. Inst. Meth. Phys. Res.* **15**, 546–549.

Renshaw, P.F. and Wicklund, S. (1988). *In vivo* measurement of lithium in humans by nuclear magnetic resonance spectroscopy. *Biol. Psychiat.* **23**, 465–475.

Renshaw, P.F., Haselgrove, J.C., Leigh, J.S. and Chance, B. (1985). *In vivo* nuclear magnetic resonance imaging of lithium. *Magn. Reson. Med.* **2**, 512–516.

Renshaw, P.F., Haselgrove, J.C., Bolinger, L., Chance, B. and Leigh, J.S. (1986). Relaxation and imaging of lithium *in vivo*. *Magn. Reson. Im.* **4**, 193–198.

Robards, A.W. and Sleytr, U.B. (1985). Low temperature methods in biological electron microscopy. *In* "Practical Methods in Electron Microscopy" (ed. A.M. Glauert), Vol. 10. Elsevier, Amsterdam.

Rocks, B.F., Sherwood, R.A. and Riley, C. (1982). Direct determination of therapeutic concentrations of lithium in serum by flow-injection analysis with atomic absorption spectroscopic detection. *Clin. Chem.* **28**, 440–443.

Schou, M. (1956). Lithiumterapi ved mani. *Nord. Med.* **7**, 790–794.

Schou, M. (1957). Biology and pharmacology of the lithium ion. *Pharmacol. Rev.* **9**, 17–58.

Schröder, W.C. (1981). Quantitative LAMMA analysis of biological specimens I Standards. II. Isotope labelling. *Fresenius Z. Anal. Chem.* **308**, 212–217.

Selinger, D., Hailer, A.W., Numberger, J.I., Simmons, S. and Gershon, E.S. (1982). A new method for the use of salivary lithium concentrations as an indicator of plasma lithium levels. *Biol. Psychiat.* **17**, 99–102.

Shanzer, A., Samuel, D. and Korenstein, R. (1983). Lipophilic lithium ion carriers. *J. Amer. Chem. Soc.* **105**, 3815–3818.

Sherman, W.R., Munsell, L.Y. and Wong, Y.H.H. (1984). Differential uptake of lithium isotopes by rat cerebral cortex and its effect on inositol phosphate metabolism. *J. Neurochem.* **42**, 880–882.

Siebers, R.W.L. and Maling, T.J.B. (1988). Flame photometric measurement of lithium concentrations in erythrocytes during determination of sodium-lithium countertransport rate. *Clin. Chem.* **34**, 1360–1361.

Sitte, H., Edelmann, L. and Neumann, K. (1987). Cryofixation without pretreatment at ambient pressure. *In* "Cryotechniques in Biological Electron Microscopy" (ed. A. Steinbrecht and K. Zierold), pp. 87–113. Springer-Verlag, Berlin.

Steinbrecht, R.A. and Müller, M. (1987). Freeze-substitution and freeze-drying. *In* "Cryotechniques in Biological Electron Microscopy" (ed. A. Steinbrecht and K. Zierold), pp. 149–174. Springer-Verlag, Berlin.

Stokes, P.E., Stoll, M., Okamoto, M. and Triana, E. (1988). Preliminary clinical and animal experience with lithium-7 isotope. *In* "Lithium: Inorganic Pharmacology and Psychiatric Use" (ed. N.J. Birch), pp. 303–304. IRL Press, Oxford.

Thellier, M., Stelz, T. and Wissocq, J.C. (1976a). Detection of stable isotopes of lithium or boron with the help of a (n, α) nuclear reaction. Application to the use of ^{6}Li as a tracer for unidirectional flux measurements and to the microlocalization of lithium in animal histologic preparations. *Biochim. Biophys. Acta* **437**, 604–627.

Thellier, M., Stelz, T. and Wissocq, J.C. (1976b). Radioautography using homogeneous detectors. *J. Microsc. Biol. Cell.* **27**, 157–168.

Thellier, M., Wissocq, J.C. and Heurteaux, C. (1980a). Quantitative microlocation of lithium in the brain by a (n, α) nuclear reaction. *Nature* **283**, 299–302.

Thellier, M., Heurteaux, C. and Wissocq, J.C. (1980b). Quantitative study of the distribution of lithium in the mouse brain for various doses of lithium given to the animal. *Brain Res.* **199**, 175–196.

Thellier, M., Heurteaux, C., Galle, P., Jouen, F., Colonna, L. and Wissocq, J.C. (1989). A method for the analysis of lithium in liquid droplets using isotopic dilution and SIMS measurements. *J. Trace Elem. Electrolytes Health Dis.* **3**, 35–37.

Thomas, G.M.H. and Olufunwa, R. (1988). Studies of lithium ion transport in isolated rat hepatocytes using ^{7}Li NMR. *In* "Lithium: Inorganic Pharmacology and Psychiatric Use" (ed. N.J. Birch), pp. 289–291. IRL Press, Oxford.

Thomas, G.M.H., Hughes, M.S., Partridge, S., Olufunwa, R.I., Marr, G. and Birch, N.J. (1988). NMR studies of lithium ion transport in isolated hepatocytes. *Biochem. Soc. Trans.* **16**, 208.

Thomas, R.C. (1978). Ion-sensitive intracellular microelectrodes. *In* "Biological Technique Series" (ed. J.E. Treherne and P.M. Rubery), pp. 61–63. Academic Press, London.

Thomas, R.C., Simon, W. and Oehme, M. (1975). Lithium accumulation by snail neurons measured by a new lithium ion-sensitive microelectrode. *Nature* **258**, 754–756.

Trautner, E.M., Morris, K., Noack, C.H. and Gershon, S. (1955). The excretion and retention of ingested lithium and its effect on the ionic balance of man. *Med. J. Aust.* **2**, 280–291.

Tremillon, B. (1965). *In* "Monographies de Chimie Minérale" (ed. A. Chrétien). Gauthier-Villars, Paris.

Trevisan, M., Ostrow, D., Cooper, R., Liu, K., Sparks, S. and Stamler, J. (1981). Methodological assessment of assays for red cell sodium concentration and sodium-dependent lithium efflux. *Clin. Chim. Acta* **116**, 319–329.

Ullrich, A., Baierl, P. and ten Bruggencate, G. (1980). Extracellular potassium in rat cerebellar cortex during acute and chronic lithium application. *Brain Res.* **192**, 287–290.

Vartsky, D., La Monte, A., Ellis, K.J., Yasumura, S. and Cohn, S.H. (1985). A proposed method for *in vivo* determination of lithium in human brain. *Phys. Med. Biol.* **30**, 1225–1236.

Vogt, H., Heinen, H.J., Meier, S. and Wechsung, R. (1981). LAMMA 500: principle and technical description of the instrument. *Fresenius Z. Anal. Chem.* **308**, 195–200.

Wheat, J.A. (1971). Isotopic analysis of lithium by atomic absorption spectrophotometry. *Appl. Spectrosc.* **25**, 328–330.

Wiernik, W. and Amiel, S. (1969). Lithium isotopic analysis by simultaneous measurements of ^{8}Li and ^{17}N. *Trans. Amer. Nucl. Soc.* **12**, 518.

Wissocq, J.C., Stelz, T., Heurteaux, C., Bisconte, J.C. and Thellier, M. (1979). Application of a (n,α) nuclear reaction to the microlocalization of lithium in the mouse brains. *J. Histochem. Cytochem.* **27**, 1462–1470.

Wissocq, J.C., Heurteaux, C., Hennequin, E. and Thellier, M. (1985). Microlocating lithium in the mouse embryo by use of a (n,α) nuclear reaction. *Roux's Arch. Dev. Biol.* **194**, 433–435.

Xie, R.Y. and Christian, G.D. (1987). Lithium ion-sensitive electrodes containing

TOPO: determination of serum lithium by flow injection analysis. *Analyst* **112**, 61–64.

Zaidel, A.N. and Korennoi, E.P. (1961). Spectral determination of the isotopic composition and concentration of lithium in solutions. *Opt. Spectrosk.* **10**, 299–302.

Zamenhof, R.G., Clement, S., Lin, K., Lui, C., Ziegelmiller, D. and Harling, O.K. (1989). Monte Carlo treatment planning and high-resolution alpha-track auto-radiography for neutron capture therapy. *Strahlenther. Onkol.* **165**, 188–192.

Zhuang, Y. and Qi, D. (1989). Response characteristics of lithium ion-selective microelectrode in biological body fluids. *Yingyong Huaxue*, **6**, 63–66.

Zhukov, A.F., Erne, D., Ammann, D., Güggi, M., Pretsch, E. and Simon, W. (1981). Improved lithium ion-selective electrode based on a lipophilic diamide as neutral carrier. *Anal. Chim. Acta* **131**, 117–122.

5 Studies on Li$^+$ Transport using ^{7}Li and ^{6}Li Nuclear Magnetic Resonance

FRANK G. RIDDELL

Department of Chemistry, The University, St Andrews, Fife KY16 9ST, UK

Several metabolic processes that are inhibited by lithium have been identified and suggested as the likely mechanism by which lithium exerts its pharmacological effects, including the widely canvassed inhibition of the hydrolysis of inositol phosphate. Despite this, none has definitely been ascribed as the actual mode for the therapeutic action of lithium. Other hypotheses as to the mode of action centre around observations that lithium transport behaviour in manic-depressive patients is different from that in normals. These observations suggest the involvement of cell membranes or of membrane transport systems in the mode of action of lithium. A study of membrane transport processes, for which nuclear magnetic resonance (NMR) is rapidly proving itself to be a most effective and remarkably informative tool, is therefore of great importance in understanding lithium therapy. This chapter reviews the use of Li$^+$ NMR in the study of transport processes in whole organisms, in cells and in phospholipid vesicle systems. The results are related, wherever possible, to their significance in elucidating the mechanism of action of lithium in the treatment of manic-depressive disorders.

NMR is nowadays the most powerful spectroscopic technique available for the study of chemical problems. It is increasingly being used in studies of lithium pharmacology (Birch, 1988). The advantages of the NMR technique are many, but the principal ones are that it is non-invasive and it is nuclear-specific. Spectroscopic techniques previously employed in the study of the inorganic pharmacology of lithium include flame photometry and atomic absorption. Concentrations have also been studied using ion-specific electrodes and neutron irradiation. These techniques are invasive, destructive or undesirable due to radioactivity and some are potentially subject to interference from other substrates. The non-invasive, non-destructive and nuclear-specific nature of NMR and lack of interference from other substrates

LITHIUM AND THE CELL
ISBN 0-12-099300-7

makes it an ideal technique for the study of lithium in biological systems. Modern instrumentation allows determinations of concentrations as low as 0.1 mM to be made with an accuracy of better than 10% in a matter of minutes. Higher concentrations are correspondingly easier to measure, alternatively longer accumulation times lead to greater accuracy.

Natural lithium contains two isotopes ^6Li (7.42%) and ^7Li (92.58%). Both isotopes of lithium are visible using NMR spectroscopy (Lambert and Riddell, 1982). The isotope most commonly observed is ^7Li because of its higher natural abundance and higher magnetogyric ratio which give rise to much greater receptivity. Relative to the receptivity of natural abundance $^{13}C = 1$ the receptivities are 3.58 (^6Li) and 1.54×10^3 (^7Li). Isotopically enriched ^6Li can be obtained relatively cheaply which allows more ready observation of this nucleus, although there are further disadvantages to the use of this isotope due to its relaxation behaviour. The operating frequencies for observation relative to $^1H = 100$ MHz are 14.716 MHz (^6Li) and 38.864 MHz (^7Li).

Both isotopes of lithium are quadrupolar. The spin quantum number I values are 1 (^6Li) and 3/2 (^7Li) and the quadrupole moments Q are -8×10^{-4} and -4.5×10^{-2} (10^{-28} m^2), respectively. The quadrupole moment of ^6Li is the smallest known and that of ^7Li is amongst the smallest.

Quadrupolar nuclei have available the quadrupolar relaxation mechanism which in most cases is their principal mode of nuclear relaxation (Lambert and Riddell, 1982). The rate of quadrupolar relaxation depends upon four main factors: the spin quantum number I, the magnitude of the quadrupole moment Q, the electric field gradient at the nucleus arising from the chemistry of the site occupied by the Li, and the rate at which the ion is tumbling. The very small values of Q for the lithium isotopes dominate their behaviour, making quadrupolar relaxation very inefficient (Lambert and Riddell, 1982). In aqueous solution dipolar relaxation with the ^1H in water, which is also relatively inefficient under these circumstances, is more important for both isotopes. Sharp lines are observed with slow relaxation rates and correspondingly long nuclear relaxation times (T_1 and T_2). In H_2O solution our measurements on lithium chloride give T_1 values of c. 20 s (^7Li) and c. 156 s (^6Li) (Bruker AM300 spectrometer at 303 K: F.G. Riddell and A. Patel, unpubl. obsv.). In D_2O solution, dipolar relaxation with the ^1H in water is no longer available and relaxation times are correspondingly longer, up to 1000 s for ^6Li (Wehrli, 1976). The relaxation time determines the rate at which data can be accumulated. Thus, data accumulation for ^6Li can be extremely slow.

However, when Li^+ is placed in an asymmetric chemical environment where it is subject to an appreciable electric field gradient, the quadrupolar mechanism becomes more important, relaxation rates increase and relaxation times decrease. Thus, although $^7Li^+$ has a T_1 value of c. 20 s in water, inside human erythrocytes T_1 is reduced to between 4 and 8 s due to interactions with proteins, membranes and other species inside the cell. The erythrocyte is a relatively simple biological system and for more complex biological systems the T_1 values for $^7Li^+$ will undoubtedly be lower. In the case of ^{23}Na and ^{39}K NMR, where the nuclei have larger quadrupole moments, the dominant quadrupolar relaxation can lead to loss of some or all of the signal intensity. This has never been observed for either isotope of lithium.

Living biological systems are invariably compartmentalized and to study them by NMR some means of contrasting between the intracompartmental and the extracompartmental signals is needed. In common with all the other alkali metals apart from caesium, the chemical shift of Li in aqueous solution is virtually independent of the accompanying anion(s). This means that in biological systems some other means must be found to distinguish the intracellular Li^+ from the extracellular Li^+. Two types of reagent are available to cause such contrast: shift reagents and relaxation agents (Riddell and Southon, 1987; Ramasamy *et al.*, 1989). Shift reagents rely upon bringing the Li^+ ions near a paramagnetic lanthanide such as dysprosium or terbium. Relaxation agents rely upon placing some of the Li^+ ions (which are exchanging rapidly with the bulk of the lithium) into a region where they are subject to a large quadrupolar interaction and are thereby relaxed. Figure 1 shows calculated spectra for a system containing 15% intracellular or intravesicular $^7Li^+$ containing either a shift or a relaxation agent.

Dysprosium tripolyphosphate forms a very effective shift reagent giving shifts of up to 40 ppm and is the contrast reagent of choice for Li^+. Other ligands have also been investigated (Ramasamy *et al.*, 1989). Lanthanum and

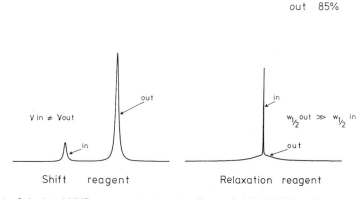

FIG. 1 Calculated NMR spectra showing the effects of shift and relaxation agents on the spectra of systems of cells or vesicles. Calculations are for 15% of the nuclei 'in' unaffected by the added contrast reagent and 85% of the nuclei 'out' in contact with the contrast reagent.

lutetium tripolyphosphates have been investigated as relaxation agents for $^7Li^+$ but are much less effective with this metal than with the other alkali metals (Riddell and Southon, 1987). Figure 2 shows the ^7Li spectrum of the erythrocytes from a patient undergoing Li^+ treatment gently packed down in a medium containing a small amount of the dysprosium tripolyphosphate shift reagent and 1 mM LiCl. The intracellular [Li^+] is measured to be 0.5 mM.

Relaxation time differences can be employed under two regimens: either the use of an introduced relaxation agent to differentiate between the sites or the use of natural relaxation time differences between intra- and extracellular Li^+, providing they differ by a substantial amount. In both cases it is possible to use a modified inversion recovery technique in which the slower relaxing signal is nulled to allow only the rapidly relaxing signal to be observed (Seo *et al.*, 1987; Espanol *et al.*, 1989). This technique employs a 180° pulse to invert both magnetization vectors. As both magnetizations relax back to their normal state each will pass a null point with no net magnetization along the direction of the magnetic field. When the more slowly relaxing magnetization has reached this point the more rapidly relaxing magnetization will be almost completely back at its equilibrium position. Thus if a monitoring pulse is applied at this point only the magnetization from the rapidly relaxing signal will be observed. The experiment can be applied in the opposite sense with only the slowly relaxing magnetization being detected. A correction factor

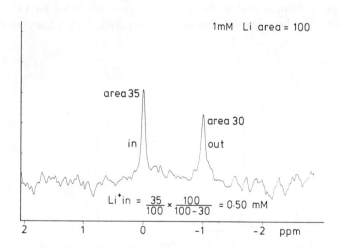

FIG. 2 ^7Li NMR spectrum at 116.34 MHz of the gently packed down erythrocytes from a patient undergoing Li treatment. The shift reagent employed is dysprosium tripolyphosphate, the 'out' signal occupies 30% of the volume and is 1 mM in Li^+. The intracellular concentration is 0.5 mM. Spectral accumulation time approximately 50 min.

must be applied in both cases to allow for incomplete relaxation of the observed signal.

The advantages of the non-invasive nature of NMR spectroscopy are ideally seen in the work of Renshaw *et al.* on measurements of tissue levels of Li^+. Following successful experiments in animals (Renshaw *et al.*, 1985, 1986), this group turned their attention to *in vivo* ^7Li NMR in man (Renshaw and Wicklund, 1988). They used a 1 m bore 1.8 T superconducting magnet capable of containing a human body, equipped with a double tuned 11.5 cm surface coil to perform the measurements. The standards used were phantoms made from agarose containing known concentrations of lithium. Spectra from the phantoms were compared with spectra obtained from skeletal muscle in the calf and from brain tissue from the back of the head of volunteers who had been taking doses of lithium carbonate. Corrections were applied to observed signal intensities to take account of relaxation times and interpulse intervals. Serum lithium levels were monitored by standard spectroscopic techniques.

Signals from the $^7Li^+$ could be clearly observed and quantified. Typical spectra are displayed in Fig. 3. For long-term treated patients brain and muscle lithium levels were considerably smaller (*c.* 30%) than serum levels. Uptake rates into muscle and brain were considerably less than uptake rates into the blood serum.

The importance of this work is three-fold. First, it demonstrates that lithium can be detected in selected parts of the human body using NMR techniques. Second, it shows that NMR will allow concentrations of lithium in various

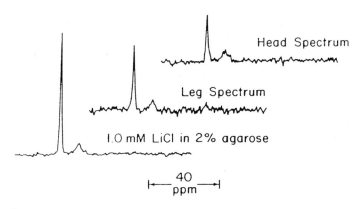

FIG. 3 ^7Li NMR spectra from the head and calf of a normal 75-kg volunteer after 4 days of 1200 mg day^{-1} Li_2CO_3. For comparison, the spectrum arising from an agarose phantom containing 1 mM LiCl is also shown. The smaller resonance in each spectrum is from a shifted reference standard. Reproduced with permission from Renshaw and Wicklund (1988), *J. Biol. Psychiat.* **23**, 465–475.

selected parts of the human body to be measured, and third, it provides a technique for measuring rates of uptake of lithium into specific parts of the body such as the brain. Since the work is of a highly technical nature it is unlikely that this technique will ever be applied routinely. However, further developments of this technique as a research tool are to be expected as the number of NMR imaging magnets increases.

Whilst there is still uncertainty over the mode of action of lithium in controlling manic-depressive illness, there is little doubt that its site of action is in the brain. In order to get into the brain, lithium must be absorbed through the gastrointestinal tract into the blood stream and from there it must penetrate the blood–brain barrier. Information about lithium traversing membranes and in particular the blood–brain barrier is therefore of critical importance.

Membrane transport of lithium can be studied by a variety of NMR techniques depending on the time-scale of the transport process. All the NMR techniques depend upon the use of vesicle or cell systems and an NMR shift reagent to separate the internal and external resonances. For very rapid processes (rates $> c.\ 5\,\mathrm{s}^{-1}$) dynamic NMR line-broadening can be employed. For intermediate cases where the time-scale is about the same as the NMR relaxation time, magnetization transfer can be employed. For ^{7}Li this means rates of $< c.\ 5\,\mathrm{s}^{-1}$. For ^{6}Li, where the relaxation time is about 100 times longer, correspondingly slower rates are in principle measurable, although such experiments have not yet been reported. For very slow exchange processes (half-life 20 min or longer) exchange can be followed by measuring signal intensities as a function of time. Such an experiment might involve ^{7}Li uptake into a previously Li$^+$-free system, or could involve ^{23}Na/^{7}Li or ^{7}Li/^{6}Li exchange. Adjustment of the experimental conditions can bring the rates of most exchange processes into one or more of these time-scale regimens.

The best model available for membrane transport in human cells is the erythrocyte, and it is believed that the erythrocyte is a good model for the blood–brain barrier. In particular the erythrocyte is readily available, is a well characterized cell system of simple morphology, and shares similar ion transport systems to neurons. A fair number of studies have taken place on lithium transport into and out of erythrocytes using the destructive and invasive techniques discussed earlier. The advantages of NMR are, however, now being recognized and increasing numbers of studies of lithium transport in erythrocytes are appearing. The area has been reviewed by Espanol *et al.* (1989).

Espanol and Mota de Freitas showed that ^{7}Li NMR could be used successfully to measure the intracellular concentration of Li$^+$ inside human erythrocytes (Espanol and Mota de Freitas, 1987). They showed that a low concentration of the shift reagent dysprosium tripolyphosphate ($c.\ 3\,\mathrm{mM}$) was sufficient to generate an appreciable chemical shift difference (*c.* 3 ppm), even in the presence of other physiologically relevant cations that might be expected to bind more strongly to the shift reagent such as Na$^+$, K$^+$, Mg^{2+} and Ca^{2+}. Addition of the ionophore monensin to lithium-loaded erythrocytes resuspended in a low Li$^+$/high K$^+$ medium resulted in a rapid efflux of the internal Li$^+$ (Fig. 4). They concluded that ^{7}Li NMR was an effective method for the measurement of intracellular [Li$^+$] and for measuring transport rates of Li$^+$.

Caution should be applied when following the experimental procedure for spectral acquisition described by these authors. The 'in' and 'out' signals had relaxation times of 4.9 and 0.15 s, respectively. The observations were accomplished by use of 45° pulses with a repetition rate of 7.5 s. This repetition rate is insufficient to allow the 'in' signal to relax completely, leading to a loss of about 15% of signal intensity. The rapidly relaxing 'out' signal suffers no intensity loss. The current author prefers a data acquisition regimen in which 90° pulses are used with a repetition rate of five times the longest relaxation time. Less than 1% of signal intensity is lost from the slowest relaxing signal if this regimen is adopted.

The same group have now extended their work into a comparison of lithium

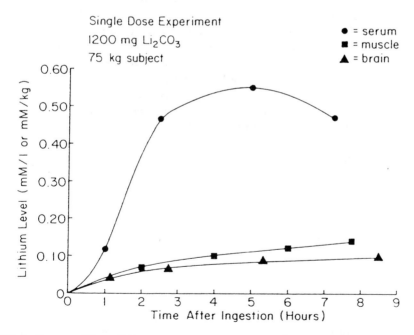

FIG. 4   Serum and tissue lithium concentrations following a single 1200 mg dose of Li$_2$CO$_3$ to a normal 75-kg volunteer followed by NMR imaging of ^{7}Li. Reproduced with permission from Renshaw and Wicklund (1988), *J. Biol. Psychiat.* **23**, 465–475.

transport rates out of the erythrocytes of manic-depressive and normal patients (Espanol *et al.*, 1989; Mota de Freitas *et al.*, 1990). Erythrocytes were loaded with lithium to about 1 mM and the rate of efflux of lithium as it exchanged with sodium was measured both by atomic absorption and by NMR techniques. Both methods gave very similar results. The efflux of lithium from the erythrocytes of the psychiatric patients was significantly lower than from the erythrocytes of normal patients. This result agrees with that of earlier workers who employed invasive analytical techniques and suggests abnormalities in the Na^+/Li^+ exchange mechanism. These differences may arise as a direct consequence of the disorder or may be acquired as a result of lithium therapy.

The work of Espanol and Mota de Freitas has been followed up in the author's own laboratory with studies of the concentrations of Li^+ inside the erythrocytes of patients undergoing lithium therapy. Studies have also been made of the rate of uptake of Li^+ into human erythrocytes and of Li^+/Li^+ exchange across the erythrocyte wall (Riddell *et al.*, 1990). Figure 2 shows a typical spectrum obtained from the erythrocytes of a patient undergoing lithium therapy.

When human erythrocytes from a 'normal' subject were incubated to maintain their viability in a medium that contained 2 mM Li^+ and a low concentration of shift reagent, slow uptake of lithium was observed by ^7Li NMR (Fig. 5). After 24 h the intracellular concentration was c. 0.36 mM and the external medium was changed so that the lithium was isotopically enriched (94%) in ^6Li. A rapid reduction in intracellular lithium was observed with the rate constant for lithium loss being 16 times greater than the rate constant for lithium uptake (Fig. 6). These results were explained on the basis of lithium uptake arising from a lithium/sodium exchange mechanism and the loss of the intracellular ^7Li signal arising from $^7Li^+(in)/^6Li^+(out)$ exchange. The Li^+/Li^+ exchange reaction is 16 times faster than the Li^+/Na^+ exchange. The potential of the NMR method for non-destructive monitoring of lithium transport in erythrocytes is clearly displayed by these results and those of Mota de Freitas and colleagues (Espanol and Mota de Freitas; 1987; Espanol *et al.*, 1989; Mota de Freitas *et al.*, 1990).

The transport of Li^+ mediated by ionophores has also been studied in vesicle systems using ^7Li NMR. Vesicles can be prepared in a variety of sizes and from a wide range of lipids by several techniques. Vesicles thus form excellent model systems in which the effects of changes in metal ion concentration, membrane composition and transporting agent can be studied. Ionophores that mediate the transport of metal ions through membranes are molecules that form a complex with the metal ion, allowing it to be taken into the interior of the membrane and thus transported through the membrane (Pressman, 1976). The work already reported has only scratched the surface of this very powerful technique for following transport processes.

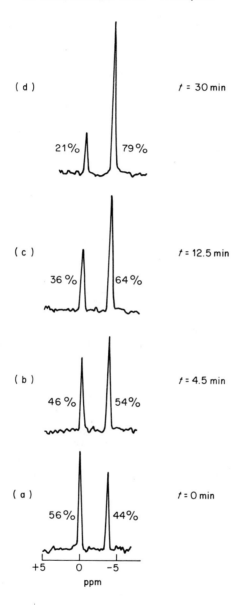

FIG. 5 Time dependence of the ^{7}Li NMR spectrum of a suspension of erythrocytes that had been incubated overnight with 150 mM LiCl and 5 mM HEPES at pH 7.5. The spectra are labelled with the time elapsed after the suspension was made 0.06 mM in monensin. The normalized percentage areas are shown in the figure and show the rapid monensin-mediated efflux of Li⁺. Reproduced with permission from Espanol and Mota de Freitas (1987), *Inorg. Chem.* **26**, 4356–4359. Copyright 1987 American Chemical Society.

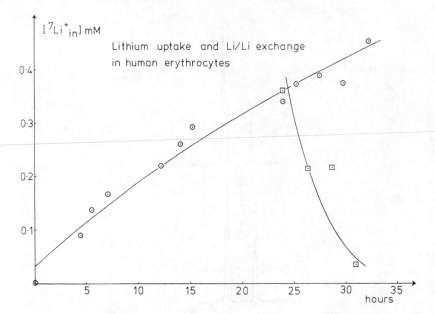

FIG. 6 The growth of the internal concentration of lithium in the erythrocytes of a 'normal' volunteer incubated in a medium with $[Li^+] = 2\,mM$, followed by ^7Li NMR. After 24 h half of the sample was incubated in a ^6Li-containing medium and the internal ^7Li signal decayed rapidly. From Riddell *et al.* (1990).

Shinar and co-workers reported the transport of Li^+ and Na^+ ions through phosphatidylcholine bilayers mediated by the ionophore shown in structure (**1**) (Shinar *et al.*, 1986). Vesicles containing NaCl, LiCl or a mixture of the two were prepared by dialytic detergent removal and the extravesicular medium was changed to aqueous KCl by dialysis. Shift reagent was added followed by the ionophore. The intensity of the intravesicular signal was followed against time using ^{23}Na or ^7Li NMR. The ionophore exhibited a transport rate selectivity for Li^+ over Na^+ of a factor of between 20 and 40 but was rather less effective at transporting lithium than the naturally occurring ionophore monensin.

The lithium transporting abilities of the commercial ionophores monensin and M139603 have been studied by the author's group (Riddell *et al.*, 1987; Riddell and Arumugam, 1988, 1989). Because the rates of transport are slightly too slow to study by the conventional NMR line-broadening technique used for Na$^+$ and K$^+$ transport, magnetization transfer was employed. The rates measured are, however, considerably faster than those measured for (**1**) by Shinar and colleagues.

Magnetization transfer involves measuring the rate of transfer of a magnetic label from one site to the other (Riddell *et al.*, 1987). First there is a specific inversion of the population of one of the lines in the ^{7}Li spectrum (invariably the larger out peak) to generate the magnetic label. Then there is a variable delay during which chemical exchange allows the inverted magnetization to be transferred from one site to the other, reducing the intensity of the non-inverted peak. The reduction in height of the non-inverted peak due to the introduction of inverted magnetization is monitored. If exchange is taking place at a rate comparable with the nuclear relaxation rates of the two sites, the rate of chemical exchange can be measured conveniently (Morris and Freeman, 1978).

This work has allowed a detailed examination of the mechanism of ionophore-mediated metal ion transport through biological membranes and has given insights into the role of the membrane in these processes. The postulated mechanism of ionophore-mediated transport of alkali metal ions is shown in Fig. 7 (Painter and Pressman, 1982; Riddell *et al.*, 1988). The ionophore starts in the membrane surface as its anion and forms a complex with the metal ion with a rate constant k_f. The neutral complex then dissolves

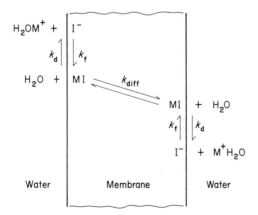

FIG. 7   Mechanism for the transport of a metal ion (M$^+$) through a membrane by an anionic ionophore I$^-$. The individual phases are association (k_f), diffusion (k_{diff}) and dissociation (k_d).

in the interior of the membrane and diffuses across to the other surface with a rate constant k_{diff}. The complex then dissociates at the opposite surface with a rate constant k_d.

The results obtained for transport through phosphatidylcholine membranes are entirely consistent with the model shown in Fig. 7. From an analysis of the concentration dependence of the transport rates it is possible to obtain values of the association and dissociation rates k_f and k_d in the membrane surface. In all cases for lithium transport diffusion is rapid with respect to the surface processes. M139603 (Riddell and Arumugam, 1989) transports lithium more rapidly than monensin (Riddell and Arumugam, 1988). The rate-determining step in both cases is dissociation of the ionophore/metal complex.

In the case of monensin-mediated transport a study was made of the rate variation upon the incorporation of positive and negative charges in the surface of the phosphatidylcholine bilayer. Positive charge was introduced by use of cetyl pyridinium ion, and negative charge was introduced by use of phosphatidylserine. Both were present at 5% of the total lipid. Placing positive charge in the surface reduces the formation rate by a factor of about three, but hardly alters the dissociation rate. Placing negative charge in the surface hardly alters the formation rate but speeds up the dissociation rate by a factor of about

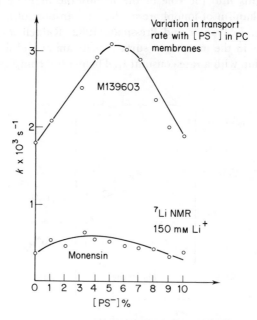

FIG. 8   Variation in $^7Li^+$ transport rate mediated by the ionophores monensin and M139603 as the concentration of phosphatidylserine in the membrane is changed.

three. Additional unpublished results (F.G. Riddell and S. Arumugam) show that the transport rate for both monensin and M139603 varies with the amount of phosphatidylserine in the membrane, reaching a maximum in both cases at about 5% incorporation (Fig. 8).

Recently there has been increasing evidence that phospholipid composition is highly conserved and that certain phospholipid compositions are vital for the form and function of the membrane. Minor changes in, for example, the fatty acyl content of membrane lipids can have profound consequences for the cell, organ or organism. The lithium transport results in model systems shed some light on this generality. They show how minor changes in the composition of biological membranes can have substantial effects on the rates of membrane-related chemical processes and thus affect the function of the membrane. The consequences for an understanding of the role of lithium in the control of bipolar affective disorder remain to be explored.

References

Birch, N.J. (ed.) (1988). "Lithium: Inorganic Pharmacology and Psychiatric Use". IRL Press, Oxford.

Espanol, M.C. and Mota De Freitas, D. (1987). ^{7}Li NMR studies of lithium transport in human erythrocytes. *Inorg. Chem.* **26**, 4356–4359.

Espanol, M.T., Ramasamy, R. and Mota de Freitas, D. (1989). *In* "Measurement of Lithium Transport Across Human Erythrocyte Membranes by ^{7}Li NMR Spectroscopy, Biological and Synthetic Membranes" (ed. A. Butterfield), pp. 33–43. Alan R. Liss Inc., New York.

Lambert, J.B. and Riddell, F.G. (eds) (1982). "The Multinuclear Approach to NMR Spectroscopy". Reidel, Dordrecht.

Morris, G.A. and Freeman, R. (1978). Selective excitation in Fourier transform nuclear magnetic resonance. *J. Magn. Reson.* **29**, 433–462.

Mota de Freitas, D., Silberberg, J., Espanol, M.T., Abraha, A., Dorus, W. Elenz, E. and Whang, W. (1990). Measurement of lithium transport in RBC from psychiatric patients on lithium therapy and normal controls by ^{7}Li NMR spectroscopy. *Biol. Psychiat.*, submitted.

Painter, G.R. and Pressman, B.C. (1982). Dynamic aspects of ionophore-mediated membrane transport. *Topics Curr. Chem.* **101**, 83–110.

Pressman, B.C. (1976). Biological application of ionophores. *Ann. Rev. Biochem.* **45**, 501–530.

Ramasamy, R., Espanol, M.C., Long, K.M., Mota de Freitas, D. and Geraldes, C.F.G.C. (1989). Aqueous shift reagents for ^{7}Li⁺ NMR transport studies in cells. *Inorg. Chim. Acta* **163**, 41–52.

Renshaw, P.F. and Wicklund, S. (1988). *In vivo* measurements of lithium in humans by nuclear magnetic resonance spectroscopy. *Biol. Psychiat.* **23**, 465–475.

Renshaw, P.F., Haselgrove, J.C., Leigh, J.S. and Chance, B. (1985). *In vivo* nuclear magnetic resonance imaging of lithium. *Magn. Reson. Med.* **2**, 512–516.

Renshaw, P.F., Haselgrove, J.C., Bolinger, L., Chance, B. and Leigh, J.S. (1986). Relaxation and imaging of lithium *in vivo*. *Magn. Reson. Im.* **4**, 193–198.

Riddell, F.G. and Arumugam, S. (1988). Surface charge effects upon membrane transport processes: the effects of surface charge on the monensin-mediated transport of lithium ions through phospholipid bilayers studied by ^{7}Li-NMR spectroscopy. *Biochim. Biophys. Acta* **945**, 65–72.

Riddell, F.G. and Arumugam, S. (1989). The transport of Li$^+$, Na$^+$ and K$^+$ ions through phospholipid bilayers mediated by the antibiotic M139603 studied by ^{7}Li, ^{23}Na- and ^{39}K-NMR. *Biochim. Biophys. Acta* **984**, 6–10.

Riddell, F.G. and Southon, T.E. (1987). Contrast reagents for the NMR spectra of the alkali metals. *Inorg. Chim. Acta* **136**, 133–137.

Riddell, F.G., Arumugam, S. and Cox, B.G. (1987). Ion transport through phospholipid bilayers studied by magnetisation transfer; membrane transport of lithium mediated by monensin. *J. Chem. Soc., Chem. Commun.* 1890–1891.

Riddell, F.G., Arumugam, S., Brophy, P.J., Cox, B.G., Payne, M.C.H. and Southon, T.E. (1988). The nigericin-mediated transport of sodium and potassium ions through phospholipid bilayers studied by ^{23}Na and ^{39}K NMR spectroscopy. *J. Amer. Chem. Soc.* **110**, 734–738.

Riddell, F.G., Patel, A. and Hughes, M.S. (1990). Lithium uptake rate and lithium: lithium exchange rate in human erythrocytes at a nearly pharmacologically normal level monitored by ^{7}Li NMR. *J. Inorg. Biochem.* **39**, 187–192.

Seo, Y., Murakami, M., Susuki, E. and Watari, H. (1987). A new method to discriminate intracellular and extracellular potassium without the use of chemical shift reagents. *J. Magn. Reson.* **75**, 529–533.

Shinar, H., Navon, G. and Klaui, W. (1986). Novel organometallic ionophore with a specificity toward Li$^+$. *J. Amer. Chem. Soc.* **108**, 5005–5006.

Wehrli, F. (1976). Temperature dependent spin lattice relaxation of lithium-6 in aqueous lithium chloride. *J. Magn. Reson.* **23**, 527–532.

6 Effect of Lithium on Viral Replication

SHARON RANDALL, CHRISTOPHER E. HARTLEY,
ALEXANDER BUCHAN, SARAH LANCASTER and
GORDON R. B. SKINNER

*Department of Medical Microbiology and Vaccine Research
Foundation (Department of Medical Microbiology), The Medical
School, University of Birmingham, Birmingham B15 2TJ, UK*

Introduction

The use of lithium as an antiviral agent is a relatively recent notion introduced about 10 years ago following several studies which demonstrated inhibition of replication of certain viruses under particular experimental conditions (Skinner *et al.*, 1980). Moreover, patients receiving lithium therapy for psychiatric disorders have altered patterns of recurrent oral or genital herpes simplex virus (HSV) infections (Lieb, 1979; Hakerem, 1983; Skinner, 1983).

Our own work has explored the effects of lithium chloride and dilithium succinate on the replication of several DNA and RNA viruses. Our spectrum of DNA viruses included vaccinia, adenovirus and herpes viruses; HSV type 1, HSV type 2, bovine herpes virus 2 (bovine mammillitis virus), equine herpes virus 1 (equine abortion virus), pseudorabies virus, canine herpes virus, cytomegalovirus and Epstein–Barr virus; and RNA viruses influenza A and encephalomyocarditis virus (EMC). Lithium inhibited the replication of all the DNA viruses tested but did not affect replication of these RNA viruses. Other researchers however have reported inhibition of measles virus replication by lithium (Cernescu *et al.*, 1988).

The effect of lithium on intracellular macromolecular events—in particular the synthesis of DNA, polypeptides and antigens—has been examined for HSV (Skinner *et al.*, 1980; Buchan *et al.*, 1989). Lithium had no viricidal effect on extracellular herpes simplex virus particles at concentrations which inhibited replication.

Viruses are obligate intracellular parasites and so consideration will be given to the effects of lithium on the cell and then on virus replication.

Effects of Lithium on the Cell

Under certain conditions, the growth of some cell types is stimulated by lithium. Lithium treatment of BALB/c 3T3 (mouse) cell cultures brought about an increase in both the proportion of cells initiating DNA synthesis and (within the limited period of culture) an increase in cell numbers. Similar effects were obtained by the addition of insulin or epidermal growth factor and when both lithium and either insulin or epidermal growth factor were added the effects were additive. There was, however, no effect of lithium on Madin–Darby canine kidney cells (Rybak and Stockdale, 1981).

Hori and Oka (1979) examined the effects of lithium and various lactogenic hormones on nucleic acid and protein synthesis in mouse mammary epithelial organ culture. DNA and RNA synthesis was stimulated by lithium both in the absence and in the presence of insulin. There is, however, some disagreement with the earlier work of Turkington (1968), who found that insulin-induced DNA synthesis, in the same species and type of tissue, was inhibited (95%) by a similar concentration of lithium (15 mM compared with 20 mM). Though it is not clear what effect lithium alone had on DNA synthesis at these concentrations, it was inhibited by a higher and possibly cytotoxic concentration (50 mM). Turkington determined that lithium acted in the G1 phase of the cell cycle to delay the onset of DNA synthesis, but that lithium had no effect on the rate of DNA synthesis during the S phase of the cell cycle, once it had been initiated. If the 'G1 phase' block were operative under Hori's experimental conditions, then the enhanced DNA synthesis he observed must have been due to an effect of lithium on those cells in which DNA synthesis was already initiated. Nevertheless, the net effects of lithium on DNA synthesis in mouse mammary epithelium do conflict and there is no obvious explanation for this.

Recently, lithium was found to inhibit the replication of Friend erythro-leukaemia cells (Gallicchio, 1985). Moreover, cell differentiation was inhibited but it is unclear at what point in the cell cycle differentiation was initiated and difficult to relate this directly to the work of Turkington.

The hormone-like effects of lithium on cell replication *in vitro* are therefore dependent at least upon the type of tissue under investigation and there must be other influencing factors. It is interesting to note, for example, that lithium-induced DNA synthesis in BALB/c 3T3 cells was dependent upon the presence of physiological concentrations of sodium and potassium.

Various effects of lithium on the activities of intracellular enzymes have been documented. For example, vasopressin-stimulated adenylate cyclase activity in renal epithelial cells is inhibited by lithium. Cyclic-AMP production is decreased in lithium-treated tissue due to competition between lithium ions and magnesium ions for the magnesium-dependent GTP binding protein (Goldberg *et al.*, 1988). In rat hepatocytes lithium stimulated both glycogen synthase and glycogen phosphorylase, the rate-limiting enzymes in glycogen metabolism (Bosch *et al.*, 1986). The precise mechanism by which lithium activated glycogen synthase has yet to be identified and in the absence of any evidence for lithium-stimulated glycogen synthase phosphatase or glycogen synthase kinase activities the authors postulated the 'existence of some intermediary step between the initial action of lithium and the final modification of phosphatase or kinase activity' (or both). There is evidence, therefore, that lithium can bring about both stimulatory and inhibitory effects on intracellular enzyme activities, though the mechanisms are not always understood.

The availability of purified reagents facilitates precise evaluation of enzymic properties and requirements although caution is of course necessary when extrapolating the data to the physiological situation where the conditions are less clearly defined. One of the enzymes involved in the somewhat complex metabolism of phosphatidyl inositols, for example, has recently been isolated and substrate specificities and cationic requirements have been precisely determined *in vitro*. This enzyme, designated inositol-polyphosphate-1-phosphatase, is dependent upon magnesium ions, is stimulated by the monovalent cations sodium and potassium, is inhibited by the divalent cations of calcium and manganese and uncompetitively inhibited by lithium ions; other enzymes involved in the metabolism of phosphatidyl inositols are in fact affected by calcium ions, some inhibited and others stimulated (Inhorn and Majerus, 1987). The influences of cations on the enzyme-catalysed reactions within this single metabolic pathway are evidently intricate and those reactions must be susceptible to any variation in the intracellular cationic environment.

While no biochemical or physiological system has yet been found that is lithium dependent, the lithium ion is clearly very active biologically. Lithium gains access into the cell via transport systems which are 'primarily' concerned with the transport of other ions and so any lithium import, unless immediately counterbalanced by its export, will likely perturb the intracellular and subcellular ionic environments, if only due to maintenance of electrochemical stasis. The effect of lithium on the fundamental physiology of the cell should be borne in mind when considering the mechanism of inhibition of virus replication by lithium.

Effects of Lithium on Virus Replication *In Vitro*

Herpes simplex virus growth cycle

Before considering the effect of lithium on virus growth, it is necessary to describe the mechanism of virus replication, and as most of our studies have concerned HSV we will focus on this virus. HSV replication occurs efficiently

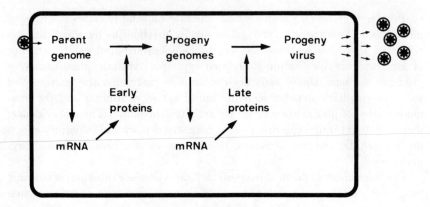

FIG. 1 Schematic diagram of the major events during HSV replication *in vitro*.

in vitro and a schematic representation outlining the major events is shown in Fig. 1. Initiation of infection occurs through interaction between glycoproteins in the virion envelope and cell surface receptors. Upon release of the capsid into the cell, the virus genome is uncoated and transported to the nucleus where transcription of genes occurs. During this period (the eclipse phase) and until the synthesis of new infectious virus, there is a decline in the level of infectious virus. Gene expression results in the sequential synthesis of groups of polypeptides loosely denoted 'early' proteins and 'late' proteins. After synthesis of 'early' proteins the virus genome is replicated. Transcription of progeny genomes increases 'late' mRNA expression and therefore the synthesis of 'late' proteins.

There are numerous enhanced or new enzymic activities in HSV-infected cells, several of which are involved in viral DNA synthesis. Some are known to be encoded by the virus genome, e.g. DNA polymerase and thymidine kinase. HSV DNA is itself essential for virus infectivity and so these enzymes and the reactions which they catalyse have become a focus of attention in the development of herpes virus chemotherapy. Many of the 'late' gene products are structural components of the virion and are transported to the nucleus where newly synthesised genomes and structural components accumulate. After capsid assembly and DNA packaging, the virion acquires a glycoprotein envelope and is transported to the cytoplasmic membrane and released from the cell. When lithium is introduced into the HSV-infected cell there is inhibition of synthesis of infectious virus. While viral DNA synthesis is abolished, viral and host cell polypeptide synthesis continues.

There was a dose-dependent inhibition of HSV replication at various concentrations of lithium succinate in HSV-infected baby hamster kidney

(BHK) cell monolayers; replication was reduced to approximately 30% at 5 mM dilithium succinate and was completely inhibited at 30 mM. Similar results were found with vaccinia and adenovirus, but the RNA viruses influenza and EMC continued to replicate at these concentrations. The virus yields from a typical experiment using dilithium succinate are shown in Fig. 2. This inhibitory action is specific to the lithium ion, as similar inhibition was found with molar equivalent concentrations of lithium chloride and lithium sulphate but not with molar equivalent sodium or potassium chloride or sodium succinate. Virus replication is inhibited only if lithium is added to the culture within the first 6 h of the growth cycle (Skinner *et al.*, 1980). HSV replication

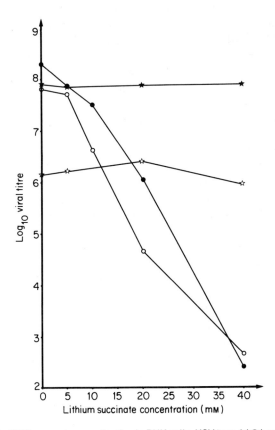

FIG. 2   Effect of lithium on virus replication in BHK cells. HSV type 1 (●) and HSV type 2 (○) show decreasing virus yields with increasing lithium concentration; influenza A (☆) and EMC virus (★) yields are not affected.

was inhibited by 60 mM lithium to a level that was equivalent to eclipse phase titre. Cernescu *et al.* (1988) reported inhibition of replication of herpes simplex virus, but only when host cells (human embryo lung cells, HEL) were treated with 1–10 mM lithium for 1 h before infection, then removed. Moreover, maximal inhibition was obtained when virus infection postceded lithium treatment by 18–24 h. This is surprising, as inhibition of HSV replication in BHK cells by lithium is a rapidly reversible phenomenon (Skinner *et al.*, 1980). It is necessary to postulate an irreversible, non-lethal event in the lithium-treated HEL cells; if those findings were confirmed this would certainly be an exciting basis for further research.

Polypeptide synthesis

Polypeptide synthesis in HSV type 1-infected BHK cells in the presence of lithium chloride has been investigated by polyacrylamide gel electrophoresis (PAGE). HSV-infected (and control uninfected) BHK cells were metabolically labelled with [^{35}S]methionine, which becomes incorporated into all methionine-containing polypeptides, or [^{14}C]glucosamine, which is incorporated specifically into glycoproteins. Polypeptides were separated by PAGE and examined by autoradiography (C. E. Hartley, PhD thesis, in prep.).

There was apparently complete synthesis of host polypeptides in uninfected cells cultured with lithium at a concentration as high as 60 mM. As a consequence of HSV infection, host cell polypeptide synthesis was suppressed. We have observed, however, that in the presence of lithium, suppression of host cell polypeptide synthesis was less marked.

Recently, a reduced rate of protein synthesis in the presence of lithium concentrations of 20–30 mM was reported in HSV type 1-infected human endothelial cells by Ziaie and Kefalides (1989); it is interesting that at 30 mM lithium virus-induced 'shut-off' of host cell proteins was reduced. Ziaie and Kefalides noticed increased synthesis of host cell thrombospondin and plasminogen activator inhibitor 1 in particular, and to a lesser extent of fibronectin and type IV collagen. They suggest this differential influence of lithium on host protein synthesis may assist in further studies of gene expression in virus-infected mammalian cells.

In HSV-infected cells in the presence of lithium, there was synthesis of all virus-specific polypeptides as adjudged by [^{35}S]methionine labelling. However, some polypeptides were synthesised in decreased amounts and others in increased amounts compared with lithium-free culture. All of the HSV glycoproteins were synthesised in the presence of lithium at virus inhibitory concentrations but in decreased amounts. The synthesis of HSV glycoprotein C (gC) was particularly affected, reaching only 10% of control levels (Fig. 3). These results are consistent with the fact that the amounts of certain

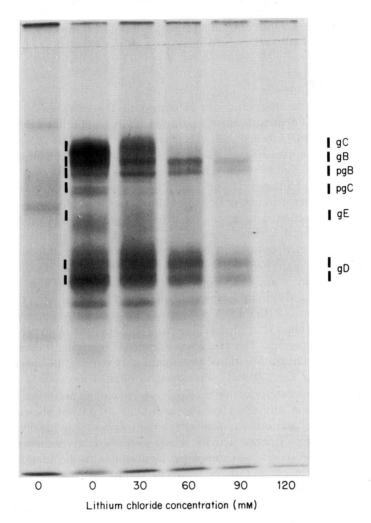

FIG. 3 The effect of lithium on HSV glycoprotein synthesis. The first track shows the polypeptide profile of uninfected BHK cells, the following five tracks show HSV-infected BHK cells with increasing concentrations of lithium. The positions of various virus specific glycoproteins, gB, gC, gD, gE and the precursors, pgB, pgC, are shown.

polypeptides synthesised in infected cells in the presence of inhibitors of viral DNA synthesis are reduced (Powell *et al.*, 1975). It should be noted that the efficient synthesis of glycoprotein C is dependent upon prior replication of viral DNA.

It was also clear that virus polypeptides retain their antigenic reactivity at lithium concentrations of 30 and 60 mM; by immunodiffusion tests the full complement of virus antigens was detected, albeit in reduced amounts.

DNA synthesis

We investigated the effect of lithium on herpes virus DNA synthesis by caesium chloride buoyant density gradient separation of virus and cellular DNA from infected BHK cells labelled with [^{3}H] thymidine. There was reduced viral DNA synthesis in the presence of 7.5 mM lithium with complete inhibition at 30 mM lithium (Fig. 4). Host cell DNA synthesis was unaffected at these concentrations of lithium. These results were confirmed by quantitative extraction and denaturation of DNA from lithium-treated HSV-infected cells followed by detection using a ^{32}P-labelled whole genome virus DNA probe. In 30 mM lithium, less than 'input levels' of DNA were detected while, in lithium-free infected cells, DNA levels were approximately 100-fold above input level. Two of the herpes virus coded enzymes involved in DNA synthesis are thymidine kinase and DNA polymerase. HSV thymidine kinase synthesis in tissue culture and activity *in vitro* were unaffected by lithium. On the other hand, the synthesis of HSV DNA polymerase was reduced by approximately 50% in 60 mM lithium. The effect of lithium on the activity of DNA polymerase *in vitro* is more complex. This effect was examined by adding lithium to the enzyme assay mixture. Additional potassium, a cofactor of the enzyme and normally present (at 200 mM), was added to parallel assays as a control. Increased salt concentrations inhibited polymerase activity and it was unclear whether there was any effect specific to lithium (data not shown). The effects of lithium and other biologically active cations, individually and in combination, on polymerase activity *in vitro* were therefore examined. HSV DNA polymerase activity was stimulated by all the salts, including lithium, to a maximum at concentrations between 100 mM and 200 mM (Fig. 5). Polymerase was most active with potassium and least active with lithium; the divalent cation calcium had no effect. The ionic conditions prevailing in lithium-treated cells is unknown and the effects of potassium, sodium and lithium, in combination, on polymerase activity *in vitro* was examined by adding various concentrations of each salt (two at a time) to the enzyme assay mixture (Fig. 6). All combinations of cations stimulated polymerase activity and polymerase activity was maximal at a total cation concentration of between 100 mM and 200 mM. However, combinations containing lithium showed least stimulation. We are currently looking at the effects of various cations on HSV replication and DNA synthesis.

In summary, while lithium reduced both synthesis of DNA polymerase in tissue culture and activity of DNA polymerase *in vitro*, each by approximately 50%, this scarcely explains the total inhibition of virus DNA synthesis at these concentrations of lithium.

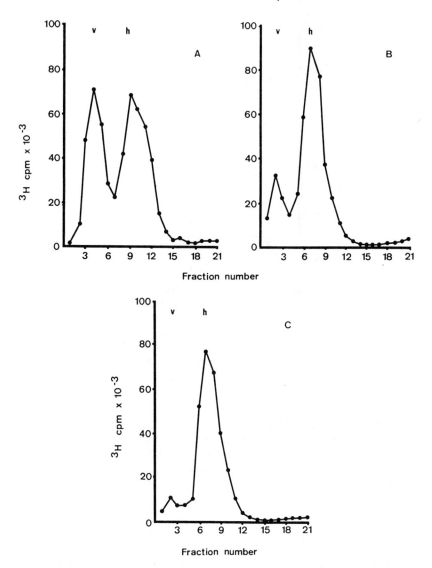

FIG. 4 The effect of lithium on the synthesis of HSV DNA examined by caesium chloride density gradient separation. Virus DNA (v) and host DNA (h) are indicated. (A) control, (B) 7.5 mM lithium and (C) 15 mM lithium. Values given as radioactive counts per minute ^{3}H cpm.

Possible modes of action of lithium

The exact mechanism of action of lithium is unknown. Preliminary findings in our laboratory indicate that the inhibitory effects of lithium on HSV replication in tissue culture can be moderated by increasing the potassium

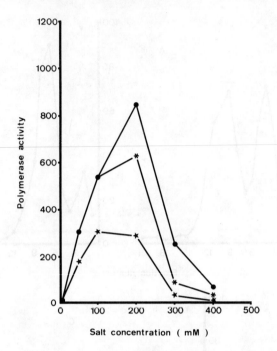

FIG. 5   The effect of potassium, sodium and lithium on HSV DNA polymerase activity *in vitro*.
Potassium chloride (●), sodium chloride (*) and lithium chloride (★).

concentration of the culture medium. Altering calcium, sodium or magnesium concentrations had no effect. Furthermore, virus replication can be inhibited by potassium starvation of host cells and 'recovered' by potassium replacement. We are currently investigating the hypothesis that inhibition of virus replication by lithium is due to an altered intracellular potassium content.

Other mechanisms of the antiviral action of lithium have been proposed. It has been hypothesised that lithium might have a modifying effect on high energy phosphate compounds, either by replacement of sodium or potassium cations in Na^+-K^+-ATPase (King *et al.*, 1969). Bach (1987) has suggested that replacement of magnesium ions by lithium may affect ATP formation and so influence DNA replication. This is an interesting suggestion but, as mentioned above, the inhibitory effects of lithium on HSV replication in tissue culture were not affected by altering magnesium concentration.

Another proposition suggests that lithium has a non-specific inhibitory role on phosphoinositide metabolism, reducing amounts of inositol-1-phosphatase and allowing accumulation of inositol in the cell. This in turn reduces the binding of α and β interferons to receptors which may result in enhanced

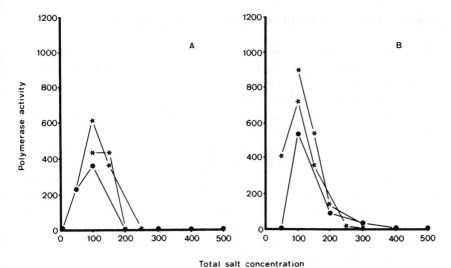

Total salt concentration

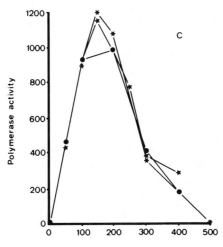

Total salt concentration

FIG. 6   The effect of combinations of potassium, sodium and lithium on DNA polymerase activity *in vitro*. (A) Lithium chloride and potassium chloride: 0 mм KCl (●), 50 mм KCl (★), and 100 mм KCl (*). (B) Lithium chloride and sodium chloride: 0 mм NaCl (●), 50 mм NaCl (★), and 100 mм NaCl (*). (C) Potassium chloride and sodium chloride: 0 mм KCl (●), 50 mм KCl (★) and 100 mм KCl (*). All total salt concentrations are in mм.

antiviral activity mediated through these interferons (Cernescu *et al.*, 1988). We think this is unlikely, at least in BHK cells, since we have found that lithium does not affect RNA viruses, which are susceptible to interferon.

Effects of Lithium on Herpes Simplex Virus Replication *In Vivo*

Herpes viruses, unlike most viruses, have the ability to lie dormant in the nerve ganglia of the host until reactivated by environmental, immunological or other unknown factors. The inhibition of HSV replication *in vitro* by lithium coupled with the apparent remission of HSV attacks in patients on lithium therapy poses the question: can lithium block the reactivation of HSV from cells that are latently infected? Trousdale *et al.* (1984) studied the effect of lithium on latency and reactivation of HSV infection in rabbits following intra-ocular inoculation. Nine infected rabbits received $1-5$ mg ml^{-1} lithium in drinking water from day 28–63. The dose was adjusted to achieve the equivalent therapeutic level in humans. There was no significant difference in reactivation rate between lithium-treated and control animals, nor was there any significant difference between the numbers of animals from which virus could be rescued. Trousdale concluded that lithium did not prevent reactivation of latent HSV-1 infection.

In patients receiving systemic lithium therapy, however, it has been noticed that recurrence of HSV infection can be inhibited. Lieb (1979) reported that on cessation of lithium treatment HSV infection reappeared in two patients. Return to lithium therapy for one patient prevented further herpetic attacks. There has also been limited success with a lithium ointment for topical treatment of herpetic lesions (Skinner *et al.*, 1980). Patients using this ointment have reported reduced time of healing of lesions, reduced pain associated with lesions, and they have been found to excrete less virus from lesions than patients not receiving lithium treatment.

While it is likely that intracellular mechanisms as described above may be operative *in vivo*, lithium might also enhance the immune response via prostaglandin synthesis. Horrobin and Lieb (1981) have suggested that lithium prevents rapid mobilization of prostaglandin E_1 precursor (dihomogammalinolenic acid), with a decrease in prostaglandin synthesis and T-lymphocyte activity. The same authors indicated a precedent for such a mechanism; namely, acetylsalicylic acid, which inhibits prostaglandin synthesis, also inhibits replication of herpes simplex virus *in vitro*.

It may be, therefore, that *in vivo*, lithium operates by intracellular inhibition of virus DNA synthesis and enhanced T-lymphocyte activity.

Summary

Lithium inhibits the replication of DNA viruses vaccinia, adenovirus, HSV type 1, HSV type 2, bovine herpes virus 2, equine herpes virus 1, pseudorabies virus, canine herpes virus, cytomegalovirus and Epstein–Barr virus but not

RNA viruses influenza A or encephalomyocarditis. The inhibition of HSV replication is mediated via an effect on viral DNA synthesis. The molecular mechanism is unclear but probably occurs at a 'site' that is potassium-dependent since inhibition is moderated by potassium (but not by sodium, calcium or magnesium). Further, virus replication is similarly inhibited by potassium-starvation and recovered by potassium replacement.

There is some evidence that lithium is effective in the treatment of HSV infections of the whole animal where other mechanisms might also be operative.

Acknowledgement

We are grateful to Mrs Jane Gardiner for her patient and expert secretarial assistance.

References

Bach, R.O. (1987). Lithium and viruses. *Med. Hypoth.* **23**, 157–170.

Bosch, F., Gomez-Foix, A.M., Arino, J. and Guinovart, J.J. (1986). Effects of lithium ions on glycogen synthase and phosphorylase in rat hepatocytes. *J. Biol. Chem.* **261**, 16 927–16 931.

Buchan, A., Randall, S., Hartley, C.E., Skinner, G.R.B. and Fuller, A. (1989). Effect of lithium salts on the replication of viruses and non-viral micro-organisms. *In* "Lithium: Inorganic Pharmacology and Psychiatric Use" (ed. N.J. Birch), pp. 83–90. IRL Press, Oxford.

Cernescu, C., Popescu, L., Constantinesui, S.T. and Cernescu, S. (1988). Antiviral effect of lithium chloride. *Rev. Roum. Med. Virol.* **39**, 93–101.

Gallicchio, V.S. (1985). Inhibition of dimethyl sulphoxide-induced Friend erythro-leukaemia cell differentiation *in vitro* by lithium chloride. *Exp. Cell Biol.* **53**, 287–293.

Goldberg, H., Clayman, P. and Skorecki, K. (1988). Mechanism of lithium inhibition of vasopressin-sensitive adenylate cyclase in cultured renal epithelial cells. *Amer. J. Physiol.* **225**, F995–1002.

Hakerem, B. (1983). Lithium in herpes simplex. *Lancet* **ii**, 516.

Hori, C. and Oka, T. (1979). Induction by lithium ion of multiplication of mouse mammary epithelium in culture. *Proc. Natl. Acad. Sci. USA* **76**, 2823–2827.

Horrobin, D. and Lieb, J. (1981). A biochemical basis for the actions of lithium on behaviour and immunity: relapsing and remitting disorders of inflammation and immunity such as multiple sclerosis or recurrent herpes as manic depression of the immune system. *Med. Hypoth.* **7**, 891–905.

Inhorn, R.C. and Majerus, P.W. (1987). Inositol polyphosphate 1-phosphatase from calf brain. *J. Biol. Chem.* **262**, 15 946–15 952.

King, L.J., Carl, J.L., Archer, E.G. and Castellanet, M. (1969). Effects of lithium on brain energy reserves and cations *in vivo. J. Pharmacol. Exp. Ther.* **168**, 163–170.

Lieb, J. (1979). Remission of recurrent herpes infection during therapy with lithium. *New Engl. J. Med.* **301**, 942.

Powell, K.L., Purifoy, D.J.M. and Courtney, R.J. (1975). The synthesis of herpes simplex virus proteins in the absence of virus DNA synthesis. *Biochem. Biophys. Res. Commun.* **66**, 262–271.

Rybak, S.M. and Stockdale, F.E. (1981). Growth effects of lithium chloride in BALB/c 3T3 fibroblasts and Madin–Darby canine kidney epithelial cells. *Exp. Cell Res.* **136**, 263–270.

Skinner, G.R.B. (1983). Lithium ointment for genital herpes. *Lancet* **ii**, 288.

Skinner, G.R.B., Hartley, C.E., Buchan, A., Harper, L. and Gallimore, P. (1980). The effect of lithium chloride on the replication of herpes simplex virus. *Med. Microbiol. Immunol.* **168**, 139–148.

Trousdale, M.D., Gordon, Y.J., Peters, A.C.B., Gropen, T.I., Nelson, E. and Nesburn, A.B. (1984). Evaluation of lithium as an inhibitory agent of herpes simplex virus in cell cultures and during reactivation of latent infection in rabbits. *Antimicrob. Agents Chemother.* **25**, 522–523.

Turkington, R.W. (1968). Cation inhibition of DNA synthesis in mammary epithelial cells *in vitro*. *Experimentia* **24**, 226–228.

Ziaie, Z. and Kefalides, N.A. (1989). Lithium chloride restores host protein synthesis in herpes simplex virus-infected endothelial cells. *Biochem. Biophys. Res. Commun.* **160**, 1073–1078.

7 Effect of Lithium on Human Neurotransmitter Receptor Systems and G Proteins

ROBERT H. BELMAKER, SOPHIA AVISSAR and GABI SCHREIBER

Beer Sheva Mental Health Centre, Ida and Solomon Stern Psychiatry Research Unit, Ben Gurion University, P.O. Box 4600, Beer Sheva, Israel

The therapeutic effects of lithium in manic-depressive illness remain a central mystery of psychiatry. Theories of lithium action are more varied, more contradictory and more short-lived than in almost any other area of psychopharmacology. Antipsychotic neuroleptic drugs clearly block dopamine receptors in proportion to their clinically effective doses. Antidepressant compounds inhibit monoamine uptake; while this theory has numerous internal contradictions, it has been successful in predicting clinical efficacy from biochemical data for numerous compounds. Anxiolytic compounds have clear effects on benzodiazepine receptors and these receptor effects parallel the clinically effective doses. Pain-killing medications mimic endogenous substances at specific opiate receptors. Lithium, however, achieves stabilization of mood, both in mania and in depression, without clear blocking effects on any known receptor. Specific binding of lithium has not been found for any specific biological molecule.

Studies of the effects of lithium on known neurotransmitter receptors must be reviewed with caution. No dose-related, Scatchard-interpretable binding effects have been found for lithium at any receptor: β-adrenergic, α-adrenergic, dopamine, GABA receptor, opiate receptor, 5HT receptor subtypes, muscarinic cholinergic or nicotinic cholinergic, or imipramine binding site (serotonin uptake receptor) (Bunney and Garland-Bunney, 1987). This does not mean that an 'endogenous lithium receptor' will never be found, but this possibility is unlikely given the millimolar concentrations of lithium in tissue required for clinical effectiveness.

Studies of lithium *in vitro* have occasionally reported effects of lithium on receptor binding, but only when lithium concentrations were 10–50 times therapeutic concentrations and the effects could be due to replacement of sodium (Battaglia *et al.*, 1983). Chronic *in vivo* studies of lithium have occasionally been reported to cause changes in receptor B_{max}. Rosenblatt *et al.* (1979) reported that chronic lithium caused a small increase in α-noradrenergic binding and a small decrease in β-adrenergic binding, but Maggi and Enna (1980) could not replicate these findings. Schultz *et al.* (1981) reported electrophysiological data suggesting increased responsiveness to noradrenaline in rat cerebellum cells after chronic lithium treatment. Serotonin receptors have been reported to be mildly reduced by chronic lithium treatment in hippocampus but not in cortex (Maggi and Enna, 1980; Tanimoto *et al.*, 1983), but these studies have not yet been performed using more specific ligands for serotonin receptor subtypes.

In contrast to the paucity of lithium effects on neurotransmitter receptors, marked lithium effects have been found on post-receptor second-messenger systems (Forn and Valdecasas, 1971). Cyclic-AMP accumulation induced by several hormones and neurotransmitters is inhibited by lithium at therapeutic or near therapeutic concentrations, both *in vivo* and *in vitro* (Ebstein *et al.*, 1980). Noradrenaline-sensitive and adenosine-sensitive cyclic-AMP accumulation in rat cortical slices are inhibited significantly by $1-2$ mM Li^+, whereas human brain noradrenaline-sensitive cyclic-AMP accumulation is clearly inhibited by 1 mM lithium (in cortical slices prepared from healthy edges of surgically removed tumours) (Newman *et al.*, 1983). The exact mechanism by which lithium inhibition of cyclic-AMP accumulation could lead to mood stabilization is of course not known. It is possible that intervention by lithium distal to various receptors is essential to the stabilization of affect, since simple receptor-blocking of various cyclic-AMP-linked receptors (i.e. propranolol for β-adrenergic receptors) does not have lithium-like effects on mood. Simultaneous inhibition of cyclic-AMP accumulation by two or more neurotransmitters or inhibition of both pre- and post-synaptic cyclic-AMP generating systems are speculative hypotheses to explain the uniqueness of lithium effects compared with drugs that directly block cyclic-AMP-linked receptors.

The fact that rat brain noradrenaline-sensitive cyclic-AMP accumulation is less sensitive to lithium inhibition than human brain noradrenaline-sensitive cyclic-AMP accumulation raises the question of tissue differences and species differences. Clearly, not all neurotransmitter and hormone-stimulated cyclic-AMP accumulation is inhibited *in vivo* by lithium at therapeutic concentrations, or lithium would be intolerably toxic. The importance of human evidence for lithium effects on second messenger systems is clear.

We were able to evaluate the plasma cyclic-AMP response to epinephrine injection *in vivo* in patients on therapeutic doses of lithium (Belmaker *et al.*, 1980). Eight manic-depressive patients consented to participate. All were euthymic and approaching discharge at the time of study and were free of physical illness, including heart disease, hypertension, or diabetes. All other drugs had been discontinued at least 5 days prior to study and the patients had lithium levels ranging from 0.54 to 1.38 mM (mean = 1.04) on the morning of the examination. Controls were nine hospital staff members free of physical illness and one euthymic drug-free schizo-affective patient. The mean age of the drug-free control group was 29 (range 22 to 40) with seven males and three females.

Epinephrine (0.5 mg) was injected subcutaneously as described previously (Ebstein *et al.*, 1976). Blood samples were taken from an indwelling catheter in the antecubital vein before epinephrine injection and every 10 min thereafter for 60 min. The blood samples were anticoagulated with EDTA, plasma separated at 5°C, and stored at −70°C until assay for cyclic-GMP and cyclic-AMP using assay kits supplied by Radiochemical Centre, Amersham, England.

Figure 1 illustrates the results. These results demonstrate that Li *in vivo* at therapeutic concentrations inhibits β-adrenergic sensitive cyclic-AMP accumulation. No data were available on the sensitivity of this receptor–cyclic-AMP generating system in the bipolar manic-depressive patients before lithium therapy. A later study, however, was able to examine β-adrenergic cyclic-AMP accumulation in a small number of drug-free patients of various diagnoses and controls, using i.v. salbutamol as a receptor agonist (Belmaker *et al.*, 1982). Figure 2 illustrates these results. None of the differences are statistically significant but there is a trend for total cyclic-AMP accumulation to be greater in affective patients and especially bipolar patients.

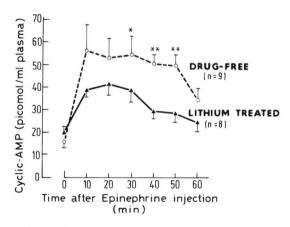

FIG. 1 The plasma cyclic-AMP responses to epinephrine in lithium-treated patients and controls. Reprinted by permission of Elsevier Science Publishing Company, Inc. from Belmaker *et al.* (1980). Copyright 1980 by the Society of Biological Psychiatry.

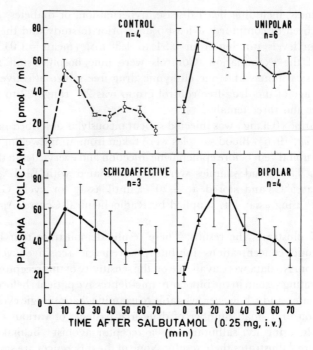

FIG. 2 The plasma cyclic-AMP response to i.v. salbutamol in controls and various patient groups. Reproduced from Belmaker *et al.* (1982), with permission.

The receptor–adenylate cyclase complex is known to consist of several components. A key step between the receptor and the catalytic unit is a membrane-bound protein that binds GTP upon agonist occupation of the receptor. The receptor–G protein–GTP complex then enhances the ability of the catalytic unit to convert ATP to cyclic-AMP. Recently we were able to measure lithium effects on agonist-induced rises in $[^3H]GTP$ binding in rat cortical membranes (Avissar *et al.*, 1988). Lithium at therapeutic concentrations clearly inhibits the agonist-induced increase in GTP binding. This suggests that lithium inhibition of noradrenaline-sensitive cyclic-AMP generation may be localized at the G protein step. Recent data suggest that lithium inhibition of agonist-induced rises in GTP binding is highly sensitive to Mg^{2+} concentrations (S. Avissar, D. Murphy and G. Schreiber, unpubl. obsv.), and it is thus difficult to extrapolate from *in vitro* results directly to *in vivo* microenvironments of varying Mg^{2+} concentrations. It is therefore important that we have recently been able to measure agonist-induced increases in binding for a GTP analogue *in vivo* in leukocytes of bipolar patients drug-free, or lithium-treated, and in normal controls (Schreiber *et al.*, 1990).

The group of drug-free manic patients consisted of seven males and three females, average age 35 (20–57) years, diagnosed as suffering from manic episode (DSM III-R). They had been drug-free for at least one month before the study because they had dropped out of outpatient follow-up or were newly diagnosed. Nine of the ten patients were newly hospitalized when blood was taken for the present study and the tenth was recommended hospitalization but refused.

The lithium-treated group consisted of seven males and three females, average age 36 (22–51) years, and were euthymic bipolar outpatients who had not received drug treatment other than lithium for at least one month. Lithium levels (0.62 ± 0.17 mM) were determined from the same sample of blood that was taken for the binding experiment.

All patients were diagnosed according to DSM-IIIR, and consented to a blood donation of 60 ml for the experiment. The healthy volunteer group consisted of six males and four females, average age 34 (28–42) years, from the medical staff of the Beersheva Mental Health Centre.

Leukocytes were isolated from 60 ml heparinized fresh blood, using Ficoll–Paque gradient according to Boyum (1968). Cells were homogenized in 25 mM Tris-HCl, pH 7.4, and 1 mM dithiothreitol (DTT). The homogenate was passed through two layers of cheesecloth to remove debris, and the membranes were collected by further centrifugation at $8\,000 \times g$ for 10 min. Binding reactions were carried out for 10 min at room temperature in a final volume of 200 μl. The reaction buffer consisted of 25 mM Tris-HCl, pH 7.4, 1 mM ATP, 1 mM Mg^{2+}, 1 mM EGTA, and 1 mM DTT, with varying concentrations of $[^3H]Gpp(NH)p$ (0.05–5 μM). Reactions were started by adding 50 μg of membrane protein and terminated with 5 ml of ice-cold buffer containing 10 mM Tris-HCl, pH 7.4, and 100 mM NaCl and filtering through GF/C Whatman filters. Filters were subsequently washed twice with 3 ml of cold buffer and taken for scintillation counting.

Table 1 represents the results using one agonist, isoproterenol. Drug-free manic patients have a significantly increased number of GTP binding sites (B_{max}) compared with lithium-treated bipolar patients, who are no different from controls. The K_d (affinity constant) is also increased in these patients, although an increase of this magnitude is less significant than the case for B_{max}. K_d, however, is increased in both lithium-treated and in drug-free bipolar patients. Similar data were obtained using carbamylcholine as agonist, suggesting that the increases in GTP cannot be explained by an increased number of receptors. These data are consistent with the suggestion of Fig. 2 that isoproterenol-induced cyclic-AMP accumulation may be increased in bipolar patients, and provide a possible explanation of this increased accumulation by postulating a change at the G protein step. The data also strongly support the concept that Li *in vivo* in humans under therapeutic

TABLE 1 [^{3}H]Gpp(NH)p binding in manic patients with and without Li-treatment and in controls in the presence of isoproterenol.

	Normals		Manics drug-free		Li-treated bipolars	
	K_d*	B_{max}†	K_d	B_{max}	K_d	B_{max}
	1.0	0.3	1.6	0.86	0.98	0.61
	0.75	0.45	1.39	0.56	1.44	0.52
	0.93	0.4	2.0	0.78	1.24	0.28
	0.74	0.42	1.8	0.78	0.98	0.38
	0.85	0.34	1.42	0.66	2.1	0.48
	0.79	0.3	2.49	0.85	1.31	0.42
	0.91	0.43	1.26	0.69	1.8	0.36
	0.85	0.36	1.43	0.64	1.9	0.5
	1.0	0.4	1.33	0.77	0.75	0.42
$\bar{x} \times$ s.e.	(0.86 ± 0.9)	(0.39 ± 0.06)	(1.59 ± 0.4)	(0.74 ± 0.09)	(1.35 ± 0.4)	(0.46 ± 0.11)

* micromoles/litre.
† pmoles/mg protein.

conditions inhibits agonist-induced increases in GTP binding. Such effects on second messenger systems may be essential to lithium's effects on neurotransmitter function and thus in manic-depressive illness.

References

Avissar, S., Schreiber, G., Danon, A. and Belmaker, R.H. (1988). Lithium inhibits adrenergic and cholinergic increases in GTP binding in rat cortex. *Nature* **331**, 440–442.

Battaglia, G., Shannon, M. and Teitler, M. (1983). Modulation of brain S2 serotonin receptors by lithium, sodium and potassium chloride. *Life Sci.* **32**, 2597–2601.

Belmaker, R.H., Kon, M., Ebstein, R.P. and Dasberg, H. (1980). Partial inhibition by lithium of the epinephrine-stimulated rise in plasma cyclic GMP in humans. *Biol. Psychiat.* **15**, 1, 3–8.

Belmaker, R.H., Lerer, B. and Zohar, J. (1982). Salbutamol treatment of depression. *In* "Typical and Atypical Antidepressants" (ed. E. Costa and G. Racagni), pp. 181–193. Raven Press, New York.

Boyum, A. (1968). Separation of leukocytes from blood and bone marrow. *Scand. J. Clin. Lab. Invest.* **21** (Suppl. 97), 7.

Bunney, W.E. Jr and Garland-Bunney, B.L. (1987). Mechanisms of action of lithium in affective illness: basic and clinical implications. *In* "Psychopharmacology: The Third Generation of Progress" (ed. H.Y. Meltzer), pp. 553–565. Raven Press, New York.

Ebstein, R.P., Belmaker, R.H., Grunhaus, L. and Rimon, R. (1976). Lithium inhibition of adrenaline-stimulated adenylate cyclase in humans. *Nature* **259**, 411.

Ebstein, R.P., Hermoni, M. and Belmaker, R.H. (1980). The effect of lithium on noradrenaline-induced cyclic AMP accumulation in rat brain: inhibition after chronic treatment and absence of super-sensitivity. *J. Pharmacol. Exp. Ther.* **213**, 161–167.

Forn, J. and Valdecasas, F.G. (1971). Effects of lithium on brain adenyl cyclase activity. *Biochem. Pharmacol.* **20**, 2773.

Maggi, A. and Enna, S.J. (1980). Regional alterations in rat brain neurotransmitter systems following chronic lithium treatment. *J. Neurochem.* **34**, 888–892.

Newman, M., Klein, E., Birmaher, B., Feinsod, M. and Belmaker, R.H. (1983). Lithium at therapeutic concentrations inhibits human brain noradrenaline-sensitive cyclic AMP accumulation. *Brain Res.* **278**, 380–381.

Rosenblatt, J.E., Pert, C.B., Tallman, J.F., Pert, A. and Bunney, W.E. Jr (1979). The effect of imipramine and lithium on alpha- and beta-receptor binding in rat brain. *Brain Res.* **160**, 186–191.

Schreiber, G., Avissar, S., Danon, A. and Belmaker, R.H. (1990). G Protein function is increased in leukocytes of patients with mania. *Biol. Psychiat.* (In press).

Schultz, J.E., Siggins, G.R., Schocker, F.W., Turck, M. and Bloom, F.E. (1981). Effects of prolonged treatment with lithium and tricyclic antidepressants on discharge frequency, norepinephrine responses on beta receptor binding in rat cerebellum: electrophysiological and biochemical comparison. *J. Pharmacol. Exp. Ther.* **216**, 28–38.

Tanimoto, K., Maeda, K. and Terada, T. (1983). Inhibitory effect of lithium on neuroleptic and serotonin receptors in rat brain. *Brain Res.* **265**, 148–151.

8 Lithium and the Phosphoinositide Signalling System

WILLIAM R. SHERMAN

Washington University School of Medicine, Department of Psychiatry, St Louis, MO 63110, USA

The Findings of James H. Allison, 1971–1979

Inositol levels decrease in cerebral cortex of rats treated with lithium

In 1969 James H. Allison was a resident in the Department of Psychiatry of Washington University School of Medicine and a Research Fellow with Mark A. Stewart, a psychiatrist who had received postdoctoral training in Oliver H. Lowry's laboratory in this University. Stewart and I had been working together using gas chromatographic methods for analysing polyol levels in tissues and had begun to focus our attention on *myo*-inositol. *myo*-Inositol had been known for some years to be a component of lipids that are termed collectively as phosphoinositides. Those lipids had first been shown by Hokin and Hokin (1953) to be involved in agonist-stimulated processes, including neural events involved in cell–cell signalling.

It was known at the time that Allison began his research that lithium affected the transport of several carbohydrates, reducing the uptake of *myo*-inositol in lens and kidney, for example (see references in Allison and Stewart, 1971). Nothing was yet known about brain in this regard and thus, in a series of experiments of his own design, Allison injected rats subcutaneously with either NaCl or LiCl at a level of 10 mmol kg^{-1} of body weight. The result was a decrease in cerebral cortex levels of *myo*-inositol in the LiCl-treated rats. The maximum decrease of 30% persisted from 6 to 24 h (Fig. 1; Allison and Stewart, 1971) and did not completely return to control levels even after 48 h. The serum levels of *myo*-inositol were increased during this period with a time course that paralleled the *myo*-inositol decrease in brain. At its maximum, the increase in the serum concentration of *myo*-inositol was 50% above that of NaCl-injected rats.

LITHIUM AND THE CELL
ISBN 0-12-099300-7

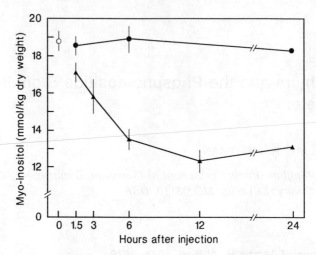

FIG. 1 The effect of lithium on *myo*-inositol levels in rat cerebral cortex. ○, non-injected ($n = 6$); ●, NaCl injected ($n = 4$); ▲, LiCl injected (10 mmol kg^{-1} body weight, $n = 4$). Each value is the mean ± SE. This is the first evidence for an effect of lithium on brain inositol levels. Reproduced by permission from *Nature* Vol. **233**, pp. 267–268, Copyright © 1971 Macmillan Magazines.

   The effect on *myo*-inositol levels was not confined to cerebral cortex, for Allison found similar decreases in other brain regions and in the adrenal gland, where the decrease was even larger (Allison, 1978; Sherman *et al.*, 1986).

   The effect of lithium on cerebral cortex *myo*-inositol levels was subsequently found to be blocked by centrally-acting anticholinergic agents such as atropine (Allison and Blisner, 1976). Anticholinergic agents that do not penetrate brain, e.g. methylatropine, were ineffective. Neither type of anticholinergic agent blocked the lithium-induced increase of inositol in serum. Thus, while the nervous system effects of lithium on inositol metabolism involved cholinergic processes, the peripheral effects appeared not to. The basis of the peripheral effects have not been further studied and are perhaps related to inositol transport.

   At that time, the blockage of the lithium-induced decrease in brain *myo*-inositol levels by anticholinergic agents was very puzzling, suggesting that the effect of lithium in brain was mediated by a cholinergic process. The only suggestion in support of that idea was that iontophoretically administered lithium somehow stimulated cholinoceptive neurons (Haas and Ryall, 1977). As will be discussed later, this may be a partial explanation for the psychotropic actions of lithium.

myo-Inositol 1-phosphate is elevated in cerebral cortex of rats treated with lithium

During the analysis of samples of cerebral cortex from rats treated with lithium, Allison observed that the decrease in *myo*-inositol was accompanied by an increase in a substance that eluted from the chromatographic column at a later time. In a collaboration with Allison that substance was identified as *myo*-inositol 1-phosphate (Allison *et al.*, 1976).

The elevation of *myo*-inositol 1-phosphate in cerebral cortex following lithium treatment is a much more sensitive tool for assessment of the effects of lithium than is inositol because it is the result of an increase over a small control level. Normal cortical levels of *myo*-inositol in these experiments are ≈ 25 mmol kg^{-1} dry weight of tissue, while the *myo*-inositol 1-phosphate level is ≈ 0.3 mmol kg^{-1}. Thus a small increase in *myo*-inositol 1-phosphate is much more easily seen than the same molar decrease in *myo*-inositol. Thus Allison was able to show that increases in *myo*-inositol 1-phosphate were measurable with LiCl doses as small as 1 mmol kg^{-1}.

The increase in *myo*-inositol 1-phosphate levels in rat cerebral cortex was also found to be blocked by atropine, and atropine by itself lowered control levels of the inositol phosphate by 40%. It thus seemed that the increase in the inositol phosphate and the decrease in *myo*-inositol were metabolically related.

In further exploration of the cholinergic aspect of lithium's actions, Allison found that the subcutaneous administration of both pilocarpine, a direct-acting cholinomimetric, or the cholinesterase inhibitor physostigmine, brought about 1.2- to 1.8-fold increases in *myo*-inositol 1-phosphate levels in rat cerebral cortex. These findings strengthened the idea that a cholinergic process was involved in the lithium effect (Allison, 1978).

In 1978 Jim Allison was convinced that the *myo*-inositol 1-phosphate that was being observed in these experiments was derived from phosphatidylinositol, thus involving the increasingly important research area of phosphoinositide metabolism. He did not live to see that proven, for he died in the following year.

Following Allison's death my laboratory began to devote increasing amounts of time to the problem. In a study that was suggested by him, and begun while he was living, we attempted to correlate the anatomical distribution of cholinergic neurons in cerebral cortex with the lithium effect on *myo*-inositol levels. In this experiment we treated rats with a single injection of lithium and measured *myo*-inositol in cortical and cerebellar layers (Allison *et al.*, 1980). We found a similar decrease in *myo*-inositol levels in each of layers I through VIA of cortex, whereas no decrease was seen in subcortical white matter or in any of three layers of cerebellum. The distribution of the lithium effect on *myo*-inositol levels was found to be similar to that of muscarinic receptors in cerebral cortex as revealed by radiography (Kuhar and

Yamamura, 1976). These receptors are of low density in corpus callosum, and in cerebellum there is only a small amount of cholinergic innervation (see references in Allison *et al.*, 1980). These studies were not corroborated by measuring *myo*-inositol 1-phosphate because the sensitivity of its measurement was inadequate at that time.

Two main objectives of the laboratory at the time of Allison's death were to test the sensitivity of inositol monophosphatase* to lithium and to establish the biochemical precursor of the *myo*-inositol 1-phosphate that lithium caused to be elevated.

The Inhibition of Inositol Monophosphatase by Lithium

Inositol monophosphatase is inhibited by clinically relevant levels of lithium

Inositol monophosphatase was first identified as an enzyme in the pathway of the *de novo* synthesis of inositol in which D-glucose 6-phosphate is converted first to L-*myo*-inositol 1-phosphate by *myo*-inositol 1-phosphate synthase and then, via inositol monophosphatase, to free *myo*-inositol (Chen and Charalampous, 1966; Eisenberg, 1967).

Following the lead of Naccarato *et al.* (1974) that 250 mM LiCl completely blocked the enzyme, we found that a partially purified preparation of inositol monophosphatase from bovine brain was 50% inhibited by a lithium concentration of 1 mM, comparable to the plasma level sought in the treatment of manic-depressive illness ($K_i \approx 1$ mM) (Hallcher and Sherman, 1980). The inhibition by lithium was unique among the Group IA metals and took place in 250 mM K^+ (Fig. 2). The mechanism of the inhibition was unusual, being uncompetitive. As will be discussed later, uncompetitive inhibition is distinctively and importantly different from competitive inhibition in the context of the possible mechanism of lithium's psychopharmacology.

We found the bovine brain inositol monophosphatase to be completely dependent on Mg^{2+}, to have a molecular weight (M_r) of 58 800, and to be competitively inhibited by Ca^{2+} (K_i 18 μM), Mn^{2+} (K_i 2 μM) and *myo*-inositol 2-phosphate (K_i 0.3 mM). Fluoride was also found to be an inhibitor with a concentration profile identical to that of lithium, a fact that should be considered in studies of the effects of G proteins on phosphoinositide

---

* Because the enzyme originally known as *myo*-inositol 1-phosphate phosphatase hydrolyses all of the *myo*-inositol monophosphates, except *myo*-inositol 2-phosphate, it has been suggested that a more appropriate name for the enzyme is inositol monophosphatase. That name will be used in this review.

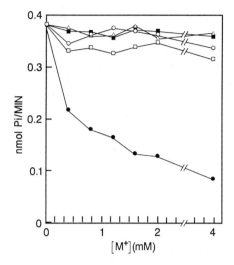

FIG. 2 Effects of lithium and other Periodic Table Group IA metals as well as ammonium ions on a partially purified preparation of bovine brain inositol monophosphatase. The ordinate axis is the amount of inorganic phosphate released per minute from *myo*-inositol 1-phosphate by the enzyme. The abscissa is the concentration of the cation used in the experiment: ●, LiCl; □, NaCl. The other symbols are for NH$_4$Cl, RbCl and CsCl; there are no significant differences between these. The incubations were carried out in buffer containing 250 mM KCl. Lithium is uniquely inhibitory with a half-maximal effect at 0.8 mM, i.e. within the plasma concentration range sought in the treatment of manic-depressive illness. From Hallcher and Sherman (1980), with permission.

metabolism. As Eisenberg (1967) had previously shown, both D- and L-*myo*-inositol 1-phosphate were substrates* for the enzyme. We also found the hydrolysis of both enantiomers to be inhibited to the same degree by lithium.

Recent studies on inositol monophosphatase

Takimoto *et al.* (1985) purified the monophosphatase to apparent homogeneity from rat brain and found it to have a molecular weight (M_r) of 55 000 and apparently to be a homodimer. The inhibitory and kinetic properties of the purified rat enzyme were very much like those of the bovine enzyme as we determined them.

---

* The inositol phosphates produced by phospholipase C hydrolysis of the phosphoinositide lipids are optically active and have absolute conformations in the D-series. Since all but one of the inositol phosphates to be discussed are derived from the lipids the use of the D- prefix will be omitted in this review. The L- notation will only be used when discussing the product of the *de novo* synthesis of L-*myo*-inositol 1-phosphate. This avoids the confusion that can arise in the numbering of these substances, i.e. L-*myo*-inositol 1-phosphate is identical to D-*myo*-inositol 3-phosphate.

A group at Merck, Sharp & Dohme, England, first reported the purification of bovine brain inositol monophosphatase to apparent homogeneity (Gee *et al.*, 1988). They found the enzyme to be a homodimer with a molecular weight of 54 000 and to have a maximal specific activity (with enzyme purified almost 1000-fold from brain) of 5.7 μmol of *myo*-inositol 1-phosphate hydrolysed per mg protein per min. The Merck group has successfully sequenced and cloned bovine inositol monophosphatase with good expression by *Escherichia coli* and they find it to be unlike any protein yet sequenced (Diehl *et al.*, 1990). The bovine enzyme has also been purified to apparent homogeneity by Attwood *et al.* (1988).

Two studies on inositol monophosphatase suggest that the mechanism of *myo*-inositol 1-phosphate hydrolysis involves transfer of the phosphate moiety of the substrate to form a phosphorylated enzyme, and that it is the hydrolysis of this intermediate that is inhibited by lithium (Shute *et al.*, 1988; Jackson *et al.*, 1989).

Inositol monophosphatase has been identified in human erythrocytes (Agam and Livne, 1988) and in transformed lymphocytes from humans (Jarvis, M.R. *et al.*, unpublished observations). In both instances the enzyme is inhibited by lithium with a K_i of ≈ 1 mM. In the latter study the human-derived inositol monophosphatase was also found to be inhibited by Ca^{2+} and Mn^{2+} and the K_m of *myo*-inositol 1-phosphate was ≈ 0.1 mM, similar to the bovine and rat enzymes.

Inositol monophosphatase from rats treated chronically with lithium

It is important in the context of the therapeutic actions of lithium to know if the inhibitory effect on inositol monophosphatase is compensated by an increase in enzyme activity following prolonged treatment. If this were to happen it would suggest that a therapeutic effect of lithium does not reside in this enzyme.

We have found that the activity of brain inositol monophosphatase of rat is unaltered by chronic lithium treatment. Four separate experiments were performed where lithium was given either orally or by injection for between 22 and 52 days. The animals were then sacrificed, the brains removed, and either a simple homogenate supernatant or an ammonium sulphate precipitate was assayed for lithium-sensitive hydrolysis of *myo*-inositol 1-phosphate. In none of these experiments was there any change in enzyme activity when compared with controls (Honchar *et al.*, 1989). These results are consistent with what we observed in two earlier studies in which *myo*-inositol levels were decreased and *myo*-inositol 1-phosphate levels were increased, even after prolonged exposure of the rats to lithium (Sherman *et al.*, 1981, 1985a). Casebolt and Jope (1989) have also found no decrease in the lithium sensitivity of inositol monophosphatase activity in brain slices from rats given lithium for 30 days.

The above findings differ from those of a study by Renshaw *et al.* (1986), who reported that a rat brain homogenate activity which hydrolyses *myo*-inositol 1-phosphate increases over time as rats are fed a lithium-containing diet. When the lithium-sensitive component of the hydrolytic activity was compared with the lithium-insensitive phosphatase activity that was present in the preparation, it was the lithium-sensitive activity that increased with time. In spite of many similarities in the protocols used in our experiments and those of the Renshaw group, we are unable to explain the discrepancy.

Evidence that Most of the *myo*-Inositol 1-phosphate in Brain of Lithium-treated Rats Originates in the Phosphoinositides

At the time that we found lithium to inhibit inositol monophosphatase, the metabolic source of the lithium-responsive brain *myo*-inositol 1-phosphate was unknown. The answer lay in its enantiomeric composition. D-*myo*-Inositol 1-phosphate is produced by the action of a phospholipase C on phosphatidyl-inositol, while L-*myo*-inositol 1-phosphate is the product of *de novo* synthesis as described on p. 124. We developed a method to separate the enantiomers of *myo*-inositol 1-phosphate using gas chromatography and found that, in control rat cerebral cortex, both D- and L-*myo*-inositol 1-phosphate were present, with the D-form accounting for 79% of the total (Sherman *et al.*, 1981). Treatment of rats with $5-7 \, \mathrm{mmol \, kg^{-1}}$ (s.c.) of lithium daily for 4–6 days resulted in an increase in both the D- and the L-enantiomers with the D-form making up 93% of the total.

The lithium-inhibited inositol monophosphatase does not discriminate between the two enantiomers of *myo*-inositol 1-phosphate and the V_{max} ratio for the hydrolysis of the L- and D-forms is 1.3 (Hallcher and Sherman, 1980). Thus differences in the rates of hydrolysis of D- and L-*myo*-inositol 1-phosphate by the enzyme do not account for the differences in the amounts of the two enantiomers in brain; rather it is a reflection of the rates at which they are formed and of the effect of uncompetitive inhibition on the hydrolysis of *myo*-inositol 1-phosphate.

Uncompetitive Inhibition of Inositol Monophosphatase by Lithium: A Unique Property

Uncompetitive inhibition of an enzyme is uncommon in Nature. Cornish-Bowden (1986) has pointed out that this form of inhibition will have a much larger effect on substrate concentration than does competitive inhibition. A comparison of the two forms of inhibition is given in Fig. 3, taken from Cornish-Bowden (1986). Figure 3 illustrates the concentration of a theoretical substrate that is necessary to overcome the inhibition of a given concentration of inhibitor so as to maintain a constant rate of metabolism, i.e. the situation that must be maintained in any metabolic system for biological processes to continue.

In competitive inhibition the substrate and inhibitor levels have a linear relationship over a wide concentration range: an incremental increase in inhibitor will be overcome by a similar increment of substrate. In the case of uncompetitive inhibition, at a level of inhibitor close to the K_i, the concentration of substrate necessary to maintain flux increases steeply and at higher inhibitor levels a further increase cannot overcome the inhibition at all. In the case of inositol monophosphates, produced by phosphoinositide

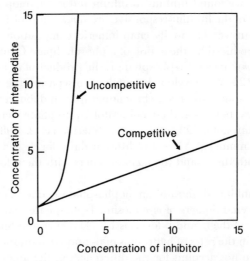

FIG. 3 Computer simulation of competitive and uncompetitive inhibition. Both forms of inhibition have the same initial slope. In competitive inhibition the relationship between the concentration of the inhibitor (e.g. lithium) and that of the metabolic intermediate (*myo*-inositol 1-phosphate) is linear over a wide range of concentrations. In uncompetitive inhibition, at inhibitor concentrations close to the K_i, small changes in the concentration of either the intermediate or the inhibitor result in a disproportionate increase in the concentration of the intermediate to restore the metabolism to the original rate. Further increases in lithium require further disproportionate increases in substrate to achieve the same metabolic level. At some concentration of lithium, not far from the K_i, the inhibition may not be overcome. In the case where inositol monophosphates are being produced, even at low lithium concentrations the substrate must rise to very high levels (and inositol levels fall in proportion) before the inhibition can be overcome. Reprinted by permission of the publisher from Cornish-Bowden (1986), Copyright © 1986 by Elsevier Science Publishing Co., Inc.

metabolism following stimulation, as they are produced at increasing rates, their concentration increases dramatically in order to restore flux through the enzyme. The result is that *myo*-inositol levels fall markedly at the same time.

Cornish-Bowden (1986) has pointed out that uncompetitive inhibition is very rare in Nature, probably because of the large effects it has on the concentrations of metabolites. Such effects would be expected to be extremely disruptive metabolically, even toxic in some cases. It might be expected that enzymes with structures that are susceptible to this kind of inhibition would have been selected against during evolution. Those enzymes that did survive evolution with the property of being uncompetitively inhibited by a natural agent would be expected to be extremely sensitive to those inhibitors. This closely describes the pharmacological properties of lithium; a drug with a narrow therapeutic range and with extreme toxicity at the upper end of that range.

An example of the effect of lithium on the inositol monophosphates *in vivo* is given by Gee *et al.* (1988). Using data from our work they calculated that in untreated rats the *myo*-inositol 1-phosphate level is ≈ 0.044 mM, a level that would support inositol monophosphatase activity at 22% of its maximum rate. In rats given 10 mmol kg^{-1} of LiCl the brain lithium level is 4.2 mM and the *myo*-inositol 1-phosphate level must rise to 1.5 mM, a 34-fold increase, in order to restore steady-state metabolism. Yet the inositol monophosphatase is operating at nearly the same functional level, 19%, of the maximal rate. Thus, in order for the cell to maintain the flux through this enzyme it must increase *myo*-inositol 1-phosphate levels dramatically, causing *myo*-inositol levels to decrease in proportion.

The Inositol Depletion Theory of the Mechanism of Lithium's Action in the Treatment of Manic-depressive Illness

In 1982 Berridge, Downes and Hanley published a study in which they examined the effect of stimulating brain slices and parotid gland fragments in the presence of lithium. Muscarinic, H$_1$-histaminergic, α-adrenergic and V$_1$-vasopressin agonists and antagonists were examined, agents that act on receptors known to result in the hydrolysis of phosphatidylinositol with the production of *myo*-inositol 1-phosphate. These receptors were known to mobilize calcium, thus they are distinguished from receptors that operate through the agency of cyclic AMP (for a comprehensive early review of this subject see Michell, 1975).

In the Berridge *et al.* (1982) experiments the phosphoinositides in the tissues were labelled with [^{3}H]*myo*-inositol, stimulated, and the tritium-labelled inositol monophosphates that were produced were counted following separation

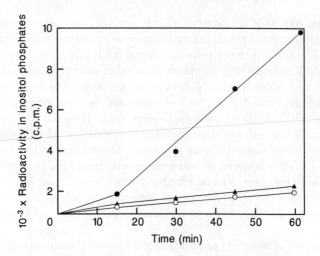

FIG. 4 The effects of carbachol and lithium on the labelling of inositol phosphates in cerebral cortex slices. ○, no additions; ▲, with 0.1 mM carbachol, the same result is obtained with 10 mM LiCl; ●, with 0.1 mM carbachol and 10 mM LiCl. Carbachol stimulation in the absence of lithium produces a slow accumulation of inositol phosphates that is no different from that achieved by lithium alone or no addition, i.e. the inositol phosphates are being hydrolysed as fast as they are formed by the stimulus. In the presence of lithium the inositol phosphates accumulate to a much larger extent, and they serve as a measure of stimulation of phosphoinositide metabolism. From Berridge et al. (1982), with permission.

by ion-exchange chromatography. The results, shown in Fig. 4, show that, in the absence of lithium, the amounts of inositol phosphates obtained following stimulation are very small, having been hydrolysed by inositol mono-phosphatase. The innovation that made these experiments useful as detectors of phosphoinositide hydrolysis was to include 10 mM LiCl in the incubations, thus blocking the hydrolysis of the inositol monophosphates (Fig. 4). This method has become widely used to detect receptor response to this class of agonists and antagonists *in vitro* and was an important step in revealing the second-messenger role of inositol trisphosphate.

While we had recognized that lithium might interfere with phosphoinositide metabolism, it was the Berridge group that realized that the way that interference might play a role in the psychopharmacology of lithium could be by interfering with the resynthesis of the phosphoinositides. In their words:

Perhaps the most important point to emerge from this study is that the ability of Li$^+$ to distort inositol metabolism is related to the receptor-mediated turnover of inositol phospholipid head groups. In a quiescent cell, or when a cell is stimulated at normal agonist concentrations where turnover is low, there is very little accumulation of *myo*-inositol 1-phosphate, but this accumulates rapidly with increasing receptor occupancy. This implies that the action of Li$^+$ *in vivo* may be rather selective in that it will be maximally effective against those cells whose inositol phospholipid-linked receptors are being abnormally stimulated. Thus, Li$^+$ could preferentially affect those receptor pathways that are abnormally active and this may account for its equal effectiveness in controlling both mania and depression.

The Discovery of the Second-messenger Role of *myo*-Inositol 1,4,5-trisphosphate and its Relationship to the Pharmacology of Lithium

By the time the Berridge *et al.* (1982) study was published it was clear that receptor occupancy by the appropriate agonist caused a rapid hydrolysis of the phosphoinositides by a phospholipase C, producing intracellular inositol phosphates and an increase in intracellular calcium levels. The likelihood that a second messenger was the internal signal for calcium mobilization had been speculated on, but it was not yet identified. That second messenger was shown to be *myo*-inositol 1,4,5-trisphosphate by Streb *et al.* (1983) in experiments in which permeabilized pancreatic acinar cells were incubated with micromolar concentrations of the inositol trisphosphate while calcium levels were monitored. While *myo*-inositol 1,4-bisphosphate and *myo*-inositol 1-phosphate were shown to be without effect in these experiments, *myo*-inositol 1,4,5-trisphosphate caused a rapid increase in cytosolic calcium. Thus a strong case was made that the link between receptor occupation and calcium release was phospholipase C cleavage of phosphatidylinositol 4,5-bisphosphate to produce *myo*-inositol 1,4,5-trisphosphate. Evidence was also given that the calcium was released from the same pool as that which carbachol acted upon, the endoplasmic reticulum.

The phosphoinositide signalling pathway is outlined in Fig. 5, where it is shown that there exists a cycle of metabolism starting with phosphatidylinositol 4,5-bisphosphate which, upon receptor occupation, produces *myo*-inositol 1,4,5-trisphosphate that is ultimately converted to inositol monophosphates. At the latter step lithium inhibition could interfere with the conversion of the monophosphates to free *myo*-inositol which is necessary for the resynthesis of the lipid source of the second messengers. This, in turn, might lead to a diminished response. If the effect of the decrease in cellular *myo*-inositol levels is to diminish agonist-dependent responses, then it has been suggested that the action of uncompetitive inhibition by lithium on neural pathways is analogous to an inertia reel automobile seat belt mechanism where the degree of braking increases in proportion to the forward momentum (Berridge *et al.*, 1989). Thus lithium would be an ideal drug in that it would be innocuous during periods of low metabolite flux, only being engaged when the stimulus-induced metabolism exceeded a certain level.

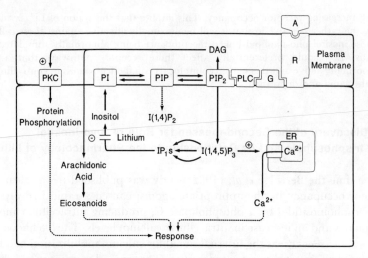

FIG. 5   A simplified representation of the phosphoinositide signalling pathway. Abbreviations: PKC, protein kinase C; PI, phosphatidylinositol; PIP, phosphatidylinositol 4-phosphate; PIP$_2$, phosphatidylinositol 4,5-bisphosphate; DAG, diacylglycerol; PLC, phospholipase C; G, a GTP-binding protein; R, receptor; A, a receptor-specific agonist; I(1,4)P$_2$, *myo*-inositol 1,4-bisphosphate; IP$_1$s, inositol monophosphates; I(1,4,5)P$_3$, *myo*-inositol 1,4,5-trisphosphate; ER, endoplasmic reticulum.

In this scheme occupation of the receptor by the agonist initiates hydrolysis of PIP$_2$ by PLC with the intermediacy of G. Some PIP is converted to I(1,4)P$_2$ (dashed arrow) and some PI is converted to inositol monophosphate (arrow not shown). I(1,4,5)P$_3$, acting through its receptor on the ER, causes calcium release to the cytosol, i.e. it is a second messenger for calcium mobilization. DAG activates PKC bringing about protein phosphorylation, another second-messenger function. DAG is also a source of eicosanoids. Together these events give rise to the specific response the agonist has called for.

The importance of these findings extends well beyond the subject of this chapter and many recent reviews are excellent sources for further reading (e.g. Berridge *et al.*, 1989; Downes, 1989; Chuang, 1989). As an indication of the extent of interest in this topic, in 1989 there were over 150 research reports published per month that were derived in one way or another from the discovery of the phosphoinositide-derived second messengers and related subjects.

A Proliferation of Inositol Phosphates Generated by the Phosphoinositide Signalling Pathway: Another Phosphatase is Inhibited by Lithium

The intense interest that grew from the inositol trisphosphate finding has resulted in the discovery of a number of different inositol mono- and polyphosphate metabolites that are produced by the signalling pathway.

There are 64 different *myo*-inositol phosphates theoretically possible without considering that many of them can exist as cyclic phosphate esters. Nineteen inositol phosphates have thus far been identified in mammalian cells, including three cyclic phosphates of inositol. Figure 6 shows nine of these products of phosphoinositide metabolism and indicates the enzymatic steps that are presently known to be inhibited by lithium.

At the time of writing, only *myo*-inositol 1,4,5-trisphosphate has been unequivocally shown to have a second-messenger role, although there is increasing evidence that *myo*-inositol 1,3,4,5-tetrakisphosphate may control the influx of external calcium into cells, perhaps to refill the stores of calcium depleted by the action of inositol trisphosphate (for a discussion and other references, see Changya *et al.*, 1989). What role any of the other inositol phosphates may play is presently unknown.

Lithium inhibits a second enzyme in the phosphoinositide signalling pathway, Enzyme 2 in Fig. 6, which has been named inositol polyphosphate 1-phosphatase (Inhorn and Majerus, 1987). Enzyme 2 acts on two substrates, *myo*-inositol 1,4-bisphosphate to produce *myo*-inositol 4-phosphate, and *myo*-inositol 1,3,4-trisphosphate to form *myo*-inositol 3,4-bisphosphate. The effect of lithium on the hydrolysis of the two substrates, as well as the affinities of these inositol phosphates for the enzyme, are different. Enzyme 2 hydrolyses *myo*-inositol 1,4-bisphosphate with an apparent K_m of 5 μM and *myo*-inositol

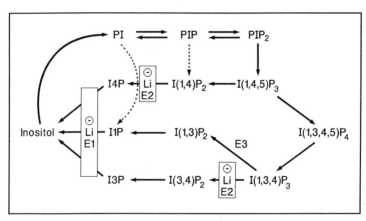

FIG. 6 Scheme showing some of the metabolic products of the phosphoinositide signalling system and those enzymes presently known to be inhibited by lithium. Abbreviations: I1P, *myo*-inositol 1-phosphate, (4-phosphate, 3-phosphate); I(1,4)P_2 *myo*-inositol 1,4-bisphosphate (1,3-bisphosphate, 3,4-bisphosphate); I(1,4,5)P_3, *myo*-inositol 1,4,5-trisphosphate (1,3,4-trisphosphate); I(1,3,4,5)P_4, *myo*-inositol 1,3,4,5-tetrakisphosphate; E1, inositol monophosphatase; E2, inositol polyphosphate 1-phosphatase; E3, inositol polyphosphate 4-phosphatase. The dashed arrows represent pathways the contributions of which appear to be variable with different tissues and with respect to time following stimulation.

1,3,4-trisphosphate with an apparent K_m of 20 μM (Inhorn and Majerus, 1987). The K_is of lithium for the two substrates also differ: for the bisphosphate it is 4.5 mM; for the trisphosphate it is 0.521 mM.

myo-Inositol 1,3,4-trisphosphate appears also to be acted on by Enzyme 3. Enzyme 3, named inositol polyphosphate 4-phosphatase (Bansal *et al.*, 1987), is lithium-insensitive and produces *myo*-inositol 1,3-bisphosphate from *myo*-inositol 1,3,4-trisphosphate. As can be seen in Fig. 6, it is possible that lithium could shunt the metabolism of the 1,3,4-trisphosphate toward the formation of *myo*-inositol 1,3-bisphosphate. This could then change the product ratios of *myo*-inositol 1- and 3-phosphate, enhancing the formation of the latter. The consequences of such shunting are unknown.

The inhibition of Enzyme 2 by lithium is also uncompetitive (Inhorn and Majerus, 1987), suggesting a structural relationship to inositol monophosphatase, which is most interesting from the standpoint of the evolution of these proteins. Because the mechanism is uncompetitive, the same situation exists with respect to a reinforcing effect on the inhibition of hydrolysis of *myo*-inositol 1,4-bisphosphate. A possible site for an indirect action of lithium on cellular function has been found in that *myo*-inositol 1,4-bisphosphate activates DNA polymerase α, perhaps by converting a low-affinity form to its high-affinity isozyme (Sylvia *et al.*, 1989).

The Biological Basis for the Increased Levels of Inositol Phosphates in Brain of Lithium-treated Rats

It is now reasonably certain that the effect of lithium on *myo*-inositol 1-phosphate levels *in vivo* is the result of ongoing neural activity in brain, a significant portion of which is cholinergic in nature. Consistent with this is evidence that the anaesthetic halothane, the hypnotic phenobarbital and the anticonvulsant diazepam all cause a lowering of *myo*-inositol 1-phosphate levels in cerebral cortex (Sherman *et al.*, 1985b, 1986). Alcohol has a similar effect (Allison and Cicero, 1980).

In our original studies only *myo*-inositol 1-phosphate was measured. Subsequently each of the *myo*-inositol monophosphates has been found in brain. Inositol 1-, 4- and 5-phosphate are present in a concentration ratio of 10:1:0.2 and all three increase in concentration and in parallel following lithium treatment (Sherman *et al.*, 1985a; Ackermann *et al.*, 1987). Figure 6 shows that *myo*-inositol 4-phosphate is the product of the action of Enzyme 2 on *myo*-inositol 1,4-bisphosphate, which, as far as is known at this time, is the sole source of the 4-isomer and thus should be uniquely a product of polyphosphoinositide metabolism.

myo-Inositol 2-phosphate is also present in brain. However, its levels do not change following lithium administration (Sherman, 1989). The origins of *myo*-inositol 5-phosphate and 2-phosphate are unknown.

When LiCl is administered subcutaneously to rats in increasing doses, and the animals are sacrificed 20 h later, the brain lithium levels are found to have increased linearly with doses of from 2 to 17 mmol kg^{-1} body weight (achieving cerebral cortex levels of 1.5 to 43.5 mmol kg^{-1} dry weight; Sherman *et al.*, 1985a). At the highest dose some deaths occur. Over this range of lithium concentrations the *myo*-inositol 1-phosphate brain levels increase, in an approximately linear manner, from a control level of ≈ 0.2 mmol kg^{-1} dry weight, to a plateau of 8 mmol kg^{-1} dry weight at a LiCl dose of 9 mmol kg^{-1} body weight (corresponding to a brain lithium level of 16 mmol kg^{-1} dry weight). *myo*-Inositol levels fall in parallel from ≈ 24 mmol kg^{-1} dry weight of cortex to ≈ 14 mmol kg^{-1}, where they also reach a plateau at a lithium dose of ≈ 9 mmol kg^{-1}. Further increases in the LiCl dose and the lithium brain level bring about no further change in inositol or in *myo*-inositol 1-phosphate. It is unknown whether this plateau results from toxicity or from a compromise in phosphoinositide signalling, perhaps because of decreased inositol availability for resynthesis of second messenger. Further stimulation with pilocarpine (see the next section) causes almost a two-fold additional increase in *myo*-inositol 1-phosphate, again reaching a plateau at a dose of ≈ 9 mmol kg^{-1} LiCl (Sherman *et al.*, 1986). This suggests that additional cells are stimulated to produce *myo*-inositol 1-phosphate, and these cells are subjected to the same limitations as those exposed to lithium alone.

The Effects of Drugs on the Phosphoinositide Signalling Pathway in Lithium-treated Rats

Measures of stimulated phosphoinositide metabolism *in vivo*

If the direct-acting cholinomimetic pilocarpine (30 mg (kg body weight)$^{-1}$) is given to rats subcutaneously, by 1 h *myo*-inositol 1-phosphate levels in cerebral cortex are 0.5 mmol kg^{-1}, about twice the control level (Fig. 7). When rats are given a single dose of LiCl (3 mmol kg^{-1}), acutely, by 24 h the cerebral cortex level of *myo*-inositol 1-phosphate is 0.75 mmol kg^{-1} dry weight. At this time the cortical lithium level is 2.5 mmol kg^{-1} dry weight (≈ 0.5 mmol kg^{-1} wet weight). This brain level corresponds to a plasma concentration of about 0.2 mmol l^{-1} of serum, i.e. a very low dose in the therapeutic context. When 30 mg kg^{-1} of pilocarpine is given to rats that have received 3 mmol kg^{-1} of LiCl 24 h earlier, the level of *myo*-inositol 1-phosphate rises to 7 mmol kg^{-1} within 30 min and the effect is sustained

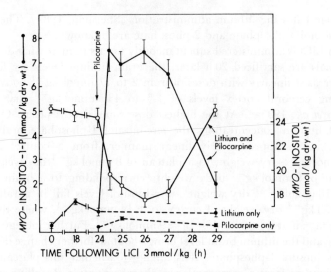

FIG. 7 The effect of administering 30 mg kg^{-1} of pilocarpine (a direct-acting muscarinic cholinergic agonist) to untreated rats and to rats that had, 24 h earlier, received a 3 mmol kg^{-1} dose of LiCl. Closed circles (●) represent the *myo*-inositol 1-phosphate levels and open circles (○) the levels of *myo*-inositol (both in mmol kg^{-1} of cerebral cortex, dry weight ± SE, $n = 6$ for each data point). Note that there are two ordinate axes and they are scaled differently. When the same dose of pilocarpine is given to the lithium-treated rats the *myo*-inositol 1-phosphate level has already reached a maximum by 30 min, the first data point taken. *myo*-Inositol levels fall with a slight delay, perhaps due to some of the inositol being used for the resynthesis of phosphatidylinositol while the *myo*-inositol 1-phosphate builds up. In addition, the other inositol phosphates produced on stimulation may not have been hydrolysed to the monophosphate stage. From Sherman *et al.* (1985b), with permission.

for at least 2.5 h (Fig. 7; Sherman *et al.*, 1985b). The *myo*-inositol level in this experiment fell from 24 mmol kg^{-1} to 19.5 mmol kg^{-1} in 30 min and to 17.5 mmol kg^{-1} by 2 h after the pilocarpine was administered. The 2 mmol difference between the 30 min and the 2 h inositol level may be due to a lag in the hydrolysis of more highly phosphorylated inositols that are generated by the stimulus, as well as to further loss of inositol by uptake into phosphatidylinositol to maintain phosphoinositide levels. Thus an appreciable *myo*-inositol deficit is rapidly created by the stimulus and it is sustained for the duration of the *myo*-inositol 1-phosphate elevation.

The above experiment may be analogous to the effects of lithium on normal and on manic-depressive patients. In the case of normal levels of neural stimulation (i.e. a normal level of cerebral cortex activity in the rat), a small dose of lithium has a negligible effect on *myo*-inositol 1-phosphate levels. However, in the situation where there is additional stimulation brought about by pilocarpine, the increase in *myo*-inositol 1-phosphate and the decrease in inositol are both rapid and large. As in the case of lithium administered by itself, it is not known whether the resulting inositol depletion has any effect on function. Perhaps the comparison of the response to lithium alone with that to lithium and pilocarpine is a model for normally-functioning and overstimulated neural metabolism. If so, it is clear that large inositol deficits can be produced quickly, perhaps with the effect of dampening responses that depend on the phosphoinositide pathway.

If, in Fig. 7, atropine is administered 30 min before the pilocarpine, it blunts the cholinergic stimulation and, by 1 h, there is less than a two-fold increase in *myo*-inositol 1-phosphate above control levels. If, instead, the atropine is administered 1 h *after* the pilocarpine, the *myo*-inositol 1-phosphate level falls from its maximum to 50% of the lithium/pilocarpine level, i.e. the response is reduced but not eliminated (Sherman *et al.*, 1985b).

The events in Fig. 7 are an example of the use of lithium as a tool to reveal activation of the phosphoinositide signalling pathway *in vivo*, something that is difficult to achieve with radioisotope labelling. Other drugs which bring about lithium-enhanced increases in cerebral cortex levels of *myo*-inositol 1-phosphate include mecamylamine and pempidine (nicotinic cholinergic blockers that do not elicit seizures in the presence of lithium). Each of these effects is reversed in some degree by atropine, indicating a muscarinic cholinergic component, perhaps distal to the first site of action of the agent.

Seizures result from the coadministration of lithium and cholinomimetics

The administration of either pilocarpine (30 mg kg^{-1} of body weight) or the anticholinesterase physostigmine (0.4 mg kg^{-1}) to rats that have been treated with 3 mmol kg^{-1} of LiCl 24 h earlier results in sustained limbic seizures and widespread brain damage (Honchar *et al.*, 1983). Increases in *myo*-inositol 1-phosphate accompany treatments with both cholinomimetics and, with pilocarpine (physostigmine not tested), the seizures are blocked by prior administration of atropine.

Once seizures have begun the *myo*-inositol 1-phosphate elevation that remains in the presence of atropine may be the result of the continuing seizures (seizures produced in rats by metrazole and kainic acid also cause *myo*-inositol 1-phosphate increases that are only partially reduced by atropine; Sherman *et al.*, 1985b). Diazepam (20 mg kg^{-1} of body weight), given 30 min before the pilocarpine, prevents the seizures, but not the increase in *myo*-inositol 1-phosphate, which is about half the magnitude seen with lithium–pilocarpine alone. If the anticonvulsant diazepam is given after seizures have begun, the behavioural pattern of the seizures stops and the brain concentration of *myo*-inositol 1-phosphate falls to half the lithium–pilocarpine level. Thus the increase in *myo*-inositol 1-phosphate that is due to cholinergic stimulation remains and the balance of the response is attributable to stimulation resulting from the seizures.

The induction of seizures by the coadministration of lithium and cholino-mimetics is cause for some concern in the clinical context. Cholinomimetics have been advocated as therapy for mania, senile dimentia and other disorders. Some insecticides are also cholinomimetics. Thus in some cases patients taking lithium may be exposed to these agents, which may call for caution (for a discussion, see Honchar *et al.*, 1983 and Evans *et al.*, 1990). It is not known whether these concerns are appropriate for humans and it may be that species differences are protective. For example, mice do not appear to undergo seizures with these drugs, even with larger doses (W.R. Sherman and M.P. Honchar, unpubl. obs.).

An Explanation of the Enhancement of Seizure Susceptibility by Lithium and Another Possible Mechanism of Lithium's Psychotropic Action

The mechanism of lithium's seizure potentiation may be related to its psychotropic actions in a manner involving *myo*-inositol depletion. Using rat hippocampal slices, Evans *et al.* (1990) have found that perfused pilocarpine ($0.5-10\ \mu M$) increases postsynaptic neuronal excitability while simultaneously decreasing synaptic transmission (seen as diminished E-EPSPs). This occurs by means of presynaptic inhibition by the muscarinic agonist. At pilocarpine concentrations of $1\ \mu M$ or less, presynaptic inhibition predominates over postsynaptic excitatory effects. When lithium is administered alone ($1-5\ mM$) synaptic transmission is enhanced, as evidenced by increased E-EPSPs and population spikes. This may occur by enhancing transmitter release. While lithium does not alter the postsynaptic excitatory action of pilocarpine, it does block the presynaptic inhibitory action of the cholinomimetic, leading to markedly increased transmission and thus to seizure activity. This enhance-ment of transmission is abolished by phorbol-12,13-dibutyrate ($0.5-2\ \mu M$), and thus there is the apparent involvement of protein kinase C. The phorbol ester does not affect muscarinic presynaptic inhibition. Pretreatment with the protein kinase C inhibitors sphingosine and H-7 blocked the ability of lithium to enhance synaptic transmission. It thus seems that lithium enhances synaptic transmission through a mechanism in common with or by acting on protein kinase C. One possibility is by means of the activation of protein kinase C by diacylglycerol, which may accumulate in cells that have *myo*-inositol levels too low to sustain resynthesis of phosphatidylinositol. This aspect of a potential inositol depletion effect is illustrated in Fig. 8 and discussed on p. 144.

These findings not only offer an explanation for the seizure-producing effects of lithium–pilocarpine, but also suggest that modulation of neural function at presynaptic terminals may allow lithium to alter signalling via many transmitters of the central nervous system.

The presynaptic disinhibition by lithium raises questions about compli-cations that might result from the administration of electroconvulsive therapy (ECT) to patients undergoing lithium treatment. Penney *et al.* (1990) present evidence that the period of confusion that follows ECT is prolonged in patients receiving ECT concurrent with lithium. Nevertheless, the length of stay in hospital of these patients was not extended beyond that of patients that were treated with ECT alone. The authors suggest that some caution is appropriate in the combined therapy. The literature on this subject is reviewed in Penney *et al.* (1990).

Studies on the Inositol Depletion Hypothesis *In Vivo*

Lithium by itself has no measurable effect on the phosphoinositides *in vivo*

Experiments investigating the effects of acutely administered lithium on phosphoinositide metabolism *in vivo* have revealed no change in the levels of any of the phosphoinositides in cerebral cortex or in whole brain (Sherman *et al.*, 1985a). Neither are the rates of ^{32}P uptake into the phosphoinositides affected by lithium.

Because of the well-known post-mortem lability of phosphatidylinositol 4,5-bisphosphate, these experiments were carried out using decapitation into liquid nitrogen, with dissection of the cerebral cortex at $-17°$C, or microwave fixation (3.5 s exposure). With a 10 mmol kg^{-1} dose of LiCl, which lowered cerebral cortex inositol levels to 58% of control, the uptake of ^{32}P into each of the phosphoinositides was no different than with the controls. Neither were the levels of the phosphoinositides changed. A study was also carried out with mice, using an acute dose of 15 mmol kg^{-1} LiCl and sacrifice by total immersion in liquid nitrogen, a procedure which is superior in maintaining ATP levels in brain (Sherman *et al.*, 1986). Again, no decreases in the levels of phosphatidylinositol or phosphatidylinositol 4,5-bisphosphate were measurable.

We have also been unable to find an effect of chronically administered lithium on the phosphoinositide lipids. In experiments with four groups of rats given lithium for from 22 to 52 days, there were no differences between the brain levels of the three phosphoinositides of the chronic versus the control groups (Honchar *et al.*, 1989). This contrasts with a study by Joseph *et al.* (1987), who found a 24% decrease in phosphatidylinositol relative to several other phospholipids following chronic administration of LiCl. Our disagreement with the Joseph *et al.* (1987) study may arise because their data are based on the amount of phosphatidylinositol measured as a percentage of the total of several lipids. In our study, we measured the amount per unit brain weight. Thus a redistribution of phospholipid levels could decrease the relative amount of phosphatidylinositol in the Joseph *et al.* study. Confirmation that the phosphoinositides do not change following lithium treatment is found in a study by Kendall and Nahorski (1987), who reported no difference in [^{3}H]*myo*-inositol incorporation into a phospholipid extract from rats treated acutely or chronically with lithium.

Lithium administered with seizure-producing cholinomimetics causes phosphoinositide levels to decrease

We have also measured the levels of the phosphoinositides in experiments where pilocarpine was administered in the presence of lithium to create a stimulus-induced inositol depletion to about 40% of that of controls without lithium (Sherman *et al.*, 1986). In these experiments seizuregenic doses of the drugs caused a 25–50% decrease in the three phosphoinositides in one of two experiments and a 20% decrease in phosphatidylinositol 4,5-bisphosphate alone in the second experiment. At doses of the two drugs that did not cause seizures there was an average 20% decrease in the lipids; however, the results did not reach statistical significance. While these experiments may have demonstrated that the lipid precursors of the phosphoinositide signalling system can be depleted if inositol levels are drastically reduced, they also involved massive stimulation and toxic effects. Under such pathological conditions the interpretation of the results is clearly not unambiguous.

There is no difference in the *in vivo* effect of drugs on *myo*-inositol 1-phosphate levels in rats treated with lithium chronically versus acutely

When rats fed a lithium-containing diet over 22 days were sacrificed at 7, 11 and 22 days the cerebral cortex *myo*-inositol 1-phosphate and inositol levels were proportional to the lithium concentration in each case. To the extent that *myo*-inositol 1-phosphate levels in the presence of lithium are a measure of neural activity in the phosphoinositide signalling pathway, no changes resulting from the chronic administration of lithium were found (Sherman *et al.*, 1985a). We have also measured the response to stimulation by pilocarpine, physostigmine and pargyline (a monoamine oxidase inhibitor) in rats that had received lithium for 27 to 39 days. Cerebral cortex *myo*-inositol 1-phosphate was found to be unchanged in the chronic versus the control groups following treatment with each of the drugs (Sherman *et al.*, 1986;

Honchar *et al.*, 1990). In contrast, similar experiments will be described later (p. 147) in which diminished levels of inositol monophosphates are formed on stimulation of cerebral cortex slices from chronically treated rats.

An *in vivo* effect with a fuctional correlate

A criterion of an effect of inositol depletion is that any functional deficit created by the depletion should be restored to normal by supplementing the system with *myo*-inositol. The only *in vivo* example of this is the restoration of the motor nerve conduction velocity deficit of diabetic rats with dietary *myo*-inositol, which is thought to be related to phosphoinositide metabolism (Greene *et al.*, 1975). For a brief discussion of the controversy over this subject, see Sherman (1989). There is greater evidence for the effects of depletion and the subsequent restoration of function by *myo*-inositol supplementation in *in vitro* experiments.

Unanswered questions about the inositol depletion theory

Why is there no measurable effect of lithium on the uptake of $^{32}PO_4^{3-}$ or on the levels of the phosphoinositides *in vivo*, in non-seizuregenic experiments, even when large *myo*-inositol deficits are achieved (Sherman *et al.*, 1986)? Similarly, why is there no measurable change in response as measured by changes in *myo*-inositol 1-phosphate levels when comparing acute and chronic lithium administration (Honchar *et al.*, 1990)? The reasons for this lack of an effect need to be understood if the *myo*-inositol depletion theory (or any of the other effects of lithium on signalling to be discussed in subsequent sections) are to be accepted as mechanisms for lithium's therapeutic effects. If the degree of depletion is insufficient to affect lipid levels or responses when measured in the acute experiments, where large doses of LiCl are used, as well as in experiments with chronic administration of lithium, where the *myo*-inositol decrease is much less, then what is the evidence for an effect on signalling? It is possible that there is a small and analytically inaccessible pool of lipid that is the active species in the signalling process. This requires that most of the lipid that is measured in these experiments be unresponsive in the signalling context. Based on our measurements, that active pool would have to be less than 15% of the total lipid, an amount that could easily remain unmeasured considering the analytical difficulties involved.

If true, the above considerations produce a surprising conclusion: If the remaining 85% of lipid is stimulus-inactive, then the source of *myo*-inositol 1-phosphate we see in acute lithium experiments must be a small amount of phospholipid. In the original Allison and Stewart (1971) study the level of *myo*-inositol was reduced by about $1.5 \, \text{mmol kg}^{-1}$ wet weight of cerebral cortex. If what is withheld from the free inositol pool as inositol phosphates all comes from phosphatidylinositol 4,5-bisphosphate, then 15% of the cortex level of that lipid may be responsible for the inositol decrement. Our values for the tissue concentration of phosphatidylinositol 4,5-bisphosphate in cerebral cortex of rats are $0.22 \, \text{mmol kg}^{-1}$ wet weight (Sherman *et al.*, 1985a). If the inositol decrement comes entirely from an active pool of lipid that is 15% of the total, the pool size may be only 15% of 0.22 mmol, or only 33 μmol of phosphatidylinositol 4,5-bisphosphate per kg of cortex. Since these are steady-state conditions, for this small an amount of lipid to maintain such a large deficit of inositol it must be undergoing hydrolysis and resynthesis of phosphatidylinositol at a furious rate.

The idea of an inositol deficit being sustained in brain in itself should be questioned. It might be expected that any deficit would be compensated for by transport or by synthesis. This is not evident in either Allison's experiments, or in our own where lithium was given acutely and where a rebound effect might be expected to be evident as a transient supernormal tissue level of inositol. It is also not evident in our experiment where lithium was given chronically and inositol was measured in brain at several times during the treatment (Sherman *et al.*, 1985a). In those studies the *myo*-inositol levels remained significantly depressed throughout the study, with inositol and *myo*-inositol 1-phosphate levels proportional to the tissue lithium concentration. Furthermore, no rebound was measurable when the animals were removed from the diet. The possibility exists that some compensatory *myo*-inositol uptake or synthesis does occur in the acute experiments, during the period when lithium levels are returning to normal following treatment, and that this increase is not seen because it is merged with the changes in inositol brought about by the diminishing lithium levels.

The subject of inositol homeostasis has been addressed in recent reviews (Holub, 1986; Sherman, 1989; Downes, 1989). The known sources of inositol in brain are: synthesis from glucose 6-phosphate, which appears to be localized in the vasculature, and not in parenchymal cells (Wong *et al.*, 1987); synthesis from other inositol isomers which are present only in very low concentration (Hipps *et al.*, 1982; Sherman, 1989); transport from blood (Spector, 1988 and references therein); and inositol released from phosphoinositide metabolism. Brain has no mechanism for the metabolic degradation of inositol; the only known site of inositol catabolism is in kidney, where it is converted into glucuronic acid. Thus, in one respect, inositol is a conserved substrate in brain, making local decrements of inositol dependent on the sources mentioned. It is nevertheless surprising that, given the decreases in inositol levels that result from lithium treatment, no compensating uptake of inositol has been observed following acute or chronic lithium treatment.

In Vitro Studies Showing an Effect Apparently Due to Inositol Depletion

An experiment with inositol-dependent restoration of function

The earliest report of a functional deficit in a phosphoinositide-linked receptor that could be restored by *myo*-inositol replacement was performed using a serotonin response of the blowfly salivary gland (Berridge and Fain, 1979; Fain and Berridge, 1979). If the gland is stimulated with 1 μM serotonin there is an increase in the uptake of $^{45}Ca^{2+}$ from the incubation medium to the gland saliva. With time that uptake declines, but it can be restored by incubation with 2 mM *myo*-inositol. If the glands are labelled with [^{3}H]*myo*-inositol and then stimulated there is a loss of 80% of the labelled phosphatidyl-inositol. In these experiments measurement of the total mass of gland phosphatidylinositol showed no change; however, it was found that, of the 135 pmol of phosphatidylinositol in each gland, only 9 pmol had to be restored by incubation with *myo*-inositol for the serotonin-responsive calcium flux to be restored. Thus these experiments are an example of a functional pool of one of the phosphoinositides that is a small fraction of the total mass of this lipid.

Brain slices from lithium-naive rats show attenuation of carbachol-stimulated inositol trisphosphate and tetrakisphosphate accumulation in the presence of lithium

When [^{3}H]inositol-labelled rat brain slices are stimulated with carbachol in the presence of 5 mM lithium there is an initial increase in inositol tetra-kisphosphate that decreases after a 10 min delay (Batty and Nahorski, 1987). Whitworth and Kendall (1988) confirmed that finding and found that the delayed decrease in inositol tetrakisphosphate occurred with carbachol stimulation, but not with noradrenaline, histamine or with 31 mM KCl. The IC_{50} of LiCl in these experiments was 0.1 mM. Using either [^{3}H]inositol-labelled rat brain slices, or measurement of the mass of inositol trisphosphate with a radioreceptor method (Challiss *et al.*, 1988), Kennedy *et al.* (1989) found that *myo*-inositol 1,4,5-trisphosphate underwent a similar increase and, after a delay, a decrease when stimulated with carbachol in the presence of lithium. The IC_{50} concentrations for lithium were found to be 0.3 mM for the effect on the trisphosphate and confirmed to be 0.1 mM for the tetra-kisphosphate. Interestingly, the decrease in inositol 1,4,5-trisphosphate was prevented by a 60 min incubation with 10 mM *myo*-inositol prior to carbachol stimulation. If the inositol is added 20 min after stimulation, once the trisphosphate level has decreased, there is no restoration of the levels. The requirement for preincubation with inositol may be due to depletion of its levels during slice preparation. For example, we found that incubation with

10 mM, but not 0.1 mM, inositol was required to restore brain slices to *in vivo* inositol concentrations (Sherman *et al.*, 1986). Possibly stimulation of slices that have subnormal inositol levels, in the presence of lithium, is the reason for this effect. Other possibilities are discussed in Kennedy *et al.* (1989).

Diacylglycerol elevation may be one of the results of inositol depletion: potential effects on protein kinase C

The details of how *myo*-inositol depletion could interfere with phosphatidyl-inositol synthesis and thus with the production of phosphatidylinositol 4,5-bisphosphate and its two second messenger products are illustrated in Figs 5 and 6. In Fig. 8 it can be seen that both of the second messengers, inositol trisphosphate and diacylglycerol, are also involved in the resynthesis of phosphatidylinositol, i.e. both diacylglycerol and inositol are recycled into the phosphoinositides.

The synthesis of phosphatidylinositol involves the conversion of diacyl-glycerol to phosphatidic acid by phosphorylation and the transfer of the phosphatidate to *myo*-inositol with the intermediacy of cytidine monophosphate-phosphatidate (CMP-PA). The CMP-PA is synthesised from cytidine triphosphate and phosphatidic acid. An important point is that diacylglycerol might remain unmetabolized in a situation with inositol depletion and thus rise in concentration and stimulate protein kinase C.

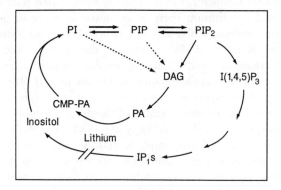

FIG. 8 Inositol depletion can perturb the diacylglycerol leg of the phosphoinositide signalling pathway. Abbreviations: as in Fig. 5 and; PA, phosphatidic acid (diacylglycerol phosphate); CMP-PA, cytidine monophosphoryl-phosphatidic acid. The synthesis of phosphatidylinositol involves the transfer of the phosphatidic acid moiety of CMP-PA to *myo*-inositol. When lithium causes a decrease in the amount of inositol the levels of CMP-PA increase (see text). Diacylglycerol levels have also been found to increase in this situation (see text). Since diacylglycerol is the second messenger for protein kinase C (Fig. 5) the response to an agonist could be increased stimulation of protein kinase C, another modality for altering responses.

GH$_3$ cells are a line of rat pituitary tumour cells that have receptors to thyrotropin-releasing hormone (TRH) that are coupled to the phosphoinositide signalling system. If these cells are stimulated in the presence of 10 mM LiCl, the *myo*-inositol level is lowered to 65% of normal, and the phosphatidylinositol level to 50% of the control value (Drummond and Raeburn, 1984). This is accompanied by about a 1.5-fold increase in the level of diacylglycerol. In spite of the 50% decrease in phosphatidylinositol there is no measurable change in the level of phosphatidylinositol 4,5-bisphosphate. Thus this precursor of second messengers may be spared while diacylglycerol increases.

In a similar set of experiments, Downes and Stone (1986) followed the labelling of the phosphoinositides and CMP-PA in rat parotid gland acinar cells. This was done by measuring the uptake of ^{32}P$_i$ when parotid slices were stimulated with carbachol in the presence and absence of lithium. The label in phosphatidylinositol was reduced by 48% when lithium was present in these experiments, while the rate of ^{32}P$_i$ uptake into CMP-PA was increased 16-fold by stimulation with lithium present. Both were restored to normal by incubation of the cells with *myo*-inositol. Half-maximal effects were obtained using 0.8–1.0 mM inositol. In these experiments phosphatidylinositol 4,5-bisphosphate was, again, unaffected by the phosphatidylinositol deficit. That the phosphoinositide signalling system was not functionally depleted may have been indicated by the observation that, when ^{86}Rb$^+$ efflux, a measure of parotid gland function, was measured, it was found to be unaltered when stimulated with carbachol in the presence of lithium.

A third study of this kind was reported by Godfrey (1989), who used rat cerebral cortex slices stimulated with carbachol with or without added lithium (10 mM). After 1 h of stimulation in the presence of lithium he found a greater than four-fold increase in CMP-PA over basal. Five minutes of stimulation was sufficient for a significant increase in CMP-PA. The increase was reversed by *myo*-inositol (half-maximal restoration with 0.8 mM inositol). The increase in CMP-PA accumulation was also observed with 25 mM K$^+$, with serotonin and with noradrenaline, all of which are stimuli of phosphoinositide turnover.

Lithium blocks an electrophysiological response

Worley *et al.* (1988) have reported an inhibition of a muscarinic response by lithium using a system where Schaffer collaterals are stimulated and CA1 population spikes are the functional measure of response. Muscarinic stimulation in this system blocks the inhibitory action of adenosine on synaptic transmission. This effect appears to involve protein kinase C activation since phorbol esters simulate the effect of cholinergic agonists. In these experiments adenosine blocks the population spike generated by electrical stimulation. That inhibitory effect of adenosine is prevented by incubating the slices for

1 h with carbachol. If 1 mM LiCl is added to the carbachol incubation the effect of adenosine to block the population spike is restored. If, at this point, a phorbol ester is added to the lithium/carbachol incubation, the adenosine block is overcome and the population spike is restored. This might occur in the following way. Incubation with carbachol in the absence of lithium generates second messengers that override the adenosine blockade. In the presence of lithium, second messengers, including diacylglycerol, are decreased and the adenosine blockade is restored. If a phorbol ester is added at this point the effect of lithium is bypassed and the adenosine-blocked population spike is again blunted.

The Worley *et al.* (1988) experiments appear to be giving a result that is the opposite of the lithium-induced increase in diacylglycerol described earlier (see p. 144). In the Worley *et al.* work, protein kinase C appears not to be activated in the presence of lithium, suggesting that the production of diacylglycerol is diminished, perhaps by a decrease in phosphoinositide levels as originally suggested by Berridge *et al.* (1982). The lithium-decreased formation of inositol phosphates described above (p. 143) could play a role as well.

Inhibition of (Na^+, K^+)-ATPase by inositol depletion

Working with an aortic preparation *in vitro*, Winegrad and his collaborators have described a component of resting (Na^+, K^+)-ATPase that is dependent on phosphatidylinositol turnover for normal function (Simmons and Winegrad, 1989 and references therein). Hyperglycaemia (10 mM for 60 minutes), in the presence of normal extracellular fluid levels of *myo*-inositol (70 μM), decreases the (Na^+, K^+)-ATPase activity. The inhibition was prevented by raising the *myo*-inositol level to 0.5 mM or by incubating in the presence of an aldose reductase inhibitor. If the aortic preparation is incubated under normoglycaemic (5 mM) conditions and with 70 μM *myo*-inositol the (Na^+, K^+)-ATPase activity was decreased by adding *scyllo*-inositol (0.5 mM), a competitive inhibitor of *myo*-inositol transport. An interesting aspect of this effect is that the phosphatidylinositol that is involved is not in a pool used for phosphatidylinositol 4,5-bisphosphate synthesis, but is involved in rapid basal phosphatidylinositol turnover. Furthermore, the hyperglycaemia that causes the decrease in (Na^+, K^+)-ATPase activity brings about no detectable decrease in *myo*-inositol levels (is it a very small pool?).

Inositol, lithium and teratogenesis

An interesting example of what appears to be a lithium-induced inositol depletion effect has been reported by Busa and Gimlich (1989). While the mechanism of the effect is unknown, lithium has a readily-demonstrable teratogenic effect on *Xenopus* embryos that is blocked by the coinjection of *myo*-inositol (for a review of this and other aspects of the biochemistry of lithium, see Berridge *et al.*, 1989).

In Vitro Effects on Tissues from Rats that have Received Chronic Lithium Treatment

Any effect of lithium that does not persist over an extended period of administration probably cannot be related to lithium's effects in the treatment of manic-depressive illness. It is also possible that acute effects of lithium are not related to the treatment of manic-depressive illness because there is usually a delay in the response of patients following the initiation of therapy. The latter is a much weaker criterion, however, since nothing is known of the reason for the delay.

Responses in brain slices from rats treated chronically with lithium

Kendall and Nahorski (1987) examined cerebral cortex slices from rats that had received lithium either acutely (a single 6.75 mmol kg^{-1} dose with sacrifice after 18 h) or for 14 days in their diet. In these experiments labelling of the slices with [^{3}H] *myo*-inositol was followed by stimulation and the response was measured as the counts in total inositol phosphates produced relative to controls. Responses to carbachol, K$^+$, histamine and serotonin were reduced in both the chronic and acute groups. The response to noradrenaline was reduced only in experiments with slices from chronically-treated rats.

Casebolt and Jope (1989) also report that there is decreased responsiveness of noradrenaline-stimulated cortical, hippocampal and striatal slices from rats given lithium for 30 days. These workers used [^{3}H]inositol labelling and inositol monophosphate analysis as a measure of response. The response to carbachol was diminished only in striatal slices. Kendall and Nahorski (1987) reported that, with slices prepared from rats fed lithium for two weeks and then placed on a lithium-free diet for 18 h, both the carbachol and the noradrenaline response became supernormal relative to the controls.

Elphick *et al.* (1988), using radiolabelling and changes in unseparated inositol phosphates as the measure, found that cerebral cortex slices from rats given lithium for 14 days and then stimulated with carbachol, noradrenaline and serotonin were less responsive than controls. A comparison with cerebral cortex slices from rats given carbamazepine for 2 weeks showed that carbamazepine did not produce decreased responses in phosphoinositide signalling when stimulated with the same agonists.

In a study by Godfrey *et al.* (1989a), brain slices were used from four groups of rats: those exposed to a single injection of LiCl (10 mmol kg^{-1}) for 4 or 24 h ('acute'), a group that received lithium (3 mmol kg^{-1} s.c.) twice daily for 3 days ('subacute'), and a group that received LiCl by injection for 16 days ('chronic'). Slices were again labelled with [^{3}H]*myo*-inositol. In neither of the acute groups was the release of total inositol phosphates reduced following stimulation with carbachol or serotonin (thus different from the results of Kendall and Nahorski, 1987) or noradrenaline. Both subacute and chronic lithium regimens, however, produced slices that had diminished responses to the three agonists as well as to 20 mM K$^+$ and to histamine. The EC$_{50}$ values for carbachol and serotonin were unchanged in the subacute group (i.e. the only one tested). The reduction in the responses to serotonin and carbachol cannot be explained by a reduction in the number of receptors since the K_d and B_{max} values for the binding of ketanserin (serotonergic) and quinuclidinyl benzilate (muscarinic) were unaffected by three days of lithium treatment (Godfrey *et al.*, 1989a).

Carbachol treatment of slices from the subacute lithium group of the Godfrey *et al.* (1989a) study showed that inositol mono-, bis-, tris- and tetrakisphosphates were all reduced by 50% or more when compared with controls. The decreases in inositol tris- and tetrakisphosphates found in brain slices from lithium-naive rats incubated with lithium have been discussed earlier (p. 143). Whether the effects reported by Godfrey *et al.* (1989a) are related, and why no change was seen in the (total) inositol phosphates from rats exposed to lithium for 24 h, is not yet known. Comparable lithium-dependent effects are being reported by groups using brain slices that are first exposed to lithium *in vitro* and with slices from rats that have had 3 days exposure to lithium. Thus the relationships between these two modes of lithium treatment are not yet clear.

While these effects could be the result of decreased phosphoinositide synthesis caused by lithium-induced inositol depletion, when the labelling of the lipids was examined, labelling in phosphatidylinositol 4,5-bisphosphate was *increased* by 30–40% in the chronic animals (Godfrey *et al.*, 1989a). In the same study, phosphatidylinositol and phosphatidylinositol 4-phosphate were also found to have increased labelling; however, the levels did not reach statistical significance. Taken together, the above-described results raise questions about the metabolic basis of these effects. Could the observed results be due to *myo*-inositol depletion? The increase in the labelling of the lipids might be the result of an increase in the specific activity of intracellular [^{3}H]inositol resulting from a reduction in intracellular inositol levels. However, if this were the case, then it would be expected that labelling in the inositol phosphates formed from the phosphoinositides would be increased, rather than decreased. If there are separate stimulus-responsive and unresponsive pools of the lipids, these arguments could be invalid, but evidence for such pools in brain has not yet been produced. There is even a suggestion that there is no depletion of intracellular *myo*-inositol in responsive cells—Kendall and Nahorski (1987) found that the lithium-diminished responses of brain slices were not restored by incubation in 2.5 mM *myo*-inositol. Paradoxically, the responses in that study were *further diminished* by incubation with inositol.

Effects of lithium on inositol trisphosphate and its receptors

The recent development of receptor binding assays for measuring the levels of *myo*-inositol 1,4,5-trisphosphate as well as the inositol trisphosphate receptor itself (e.g. Challiss *et al.*, 1988) have provided new opportunities to examine the effects of lithium on these elements of the phosphoinositide signalling system. Few earlier studies have measured the actual levels of inositol trisphosphate, but have used radionuclide labelling which leaves uncertainties in interpretation because of the difficulties in determining specific activities of the analysed substances. Using a binding assay, Godfrey *et al.* (1989b) found that when rats were exposed to lithium for 22–28 days, there was a 40–60% decrease in the basal level of *myo*-inositol 1,4,5-trisphosphate in cerebral cortex slices when compared with controls. The release of inositol trisphosphate following carbachol stimulation was even further attenuated. This confirms other observations regarding the effects of chronic lithium administration on inositol phosphates. However, in another study the B_{max} for *myo*-inositol 1,4,5-trisphosphate receptor binding in cerebral cortex was found to be $0.99 \pm 0.13 \, \text{pmol mg}^{-1}$ protein in the control group (NaCl-treated) and $1.8 \pm 0.09 \, \text{pmol mg}^{-1}$ protein in the chronic lithium group (P.P. Godfrey, pers. commun.). Thus the decreased production of *myo*-inositol 1,4,5-trisphosphate may be compensated for by an increase in the number of trisphosphate receptors during lithium treatment. This raises the question as to whether the increase in receptor number is sufficient to restore a full response or whether the net functional response is diminished because of a deficiency in inositol trisphosphate.

Effects of Lithium on Phosphoinositide Signalling that involve G Proteins

Litosch (1987), Casebolt and Jope (1989), Godfrey *et al.* (1989a) and Li *et al.* (1989) have all found that fluoride, a direct activator of guanine nucleotide binding proteins, stimulates the phosphoinositide signalling system in brain preparations. Non-metabolizable GTP analogues also stimulate phosphoinositide metabolism (Litosch, 1987; Li *et al.*, 1989).

Litosch (1987) found that 1 mM F^- maximally stimulates the phospholipase C hydrolysis of phosphatidylinositol 4,5-bisphosphate by rat cerebral cortex membranes. The same effect was seen with 0.1 mM Gpp(NH)p (a non-metabolized GTP analogue). When fluoride and Gpp(NH)p were used together the effect was only slightly greater than either alone, suggesting that they both act on the phospholipase C using the same mechanism, i.e. probably involving the same GTP-binding protein.

Godfrey *et al.* (1989a) have reported that when cerebral cortex membranes are incubated with 10 mM NaF a 66% increase in total [^{3}H]inositol phosphates results. Membranes prepared from rats treated with lithium for three days underwent only a 45% increase in inositol phosphates, a 32% decrease ($p < 0.01$) in the fluoride response in tissues from lithium-treated animals. Casebolt and Jope (1989) did not find that the fluoride effect was diminished in cortical slices from rats given lithium for 30 days. The reason for the differences between the studies is unknown.

Avissar *et al.* (1988) report an effect of lithium on G protein(s) of a membrane preparation from rat brain. Cerebral cortex membranes from lithium-naive rats were incubated in buffers with and without 0.6 mM lithium. When stimulated with isoproterenol (a β-adrenergic agonist) or carbachol the binding of [^{3}H]GTP was significantly increased. In both cases that increase in binding was abolished in the presence of lithium. This inhibition of [^{3}H]GTP binding was also found in washed cerebral cortex membranes from animals that had received lithium carbonate for 12–21 days. Withdrawal of lithium from the chronic group for 48 h restored the agonist-induced stimulation of GTP binding to normal. It is noteworthy that this effect of lithium is: (1) present in both *in vitro* incubations with lithium of tissues from lithium naive rats and; (2) in membranes isolated from chronically lithium-treated rats (i.e. in the latter case, not incubated with lithium and presumably freed of lithium during the preparation of the membranes). The effect disappears in preparations from rats following two days of lithium abstinence. This persistence of a lithium effect in the washed membranes may denote a lithium-induced change with a rapid onset but one that continues with a half-life of several hours.

The Avissar *et al.* (1988) study supports the evidence of G protein involvement by showing that GTP binding, following isoproterenol stimulation, is completely blocked by cholera toxin but not by pertussis toxin, characteristic of a G_s protein. Carbachol-stimulated binding of GTP was unaffected by cholera toxin and reduced by 40% by the action of pertussis toxin, characteristics of G_i and G_o proteins. These results confirm an earlier report by Ebstein *et al.* (1980), who showed that noradrenaline-stimulated cyclic AMP formation was reduced in rat brain following chronic lithium administration. Godfrey *et al.* (1989a) found that three days of lithium administration did not have any effect on cyclic AMP formation stimulated by isoproterenol, a finding that contrasts with the Ebstein *et al.* (1980) results and, indirectly, with the Avissar *et al.* (1988) study.

In spite of the conflicting evidence, there appears to be emerging the likelihood that lithium has an effect, via GTP-binding proteins, on both the cyclic AMP and the phosphoinositide second-messenger systems. Thus lithium may be able to modify signalling by both of these important pathways. The idea that bipolar illness involves a disturbance in the balance of both cholinergic and adrenergic signalling (Janowsky *et al.*, 1972) fits well with an effect of lithium on G proteins of both second-messenger systems.

Isotopes of Lithium, Their Uptake by Tissues and Their Effect on Inositol Monophosphatase

Lithium occurs naturally as two stable isotopes: ^{7}Li, with a typical abundance of 92.6%, and ^{6}Li comprising the balance. With the highly purified separate isotopes now available it is possible to address the question of whether one of them is more toxic than the other and whether there is any difference in their therapeutic properties. Lieberman *et al.* (1979a) have reported a greater decrease in the motility of rats given ^{6}Li, relative to rats given ^{7}Li. Alexander *et al.* (1980) reported a greater lethality of ^{6}Li; however, Parthasarathy and Eisenberg (1983) could not confirm that finding. Stokes *et al.* (1982) showed that, in the cat, equal doses of the two isotopes resulted in higher levels of ^{6}Li in cerebrospinal fluid and in plasma. It was proposed that, in therapy where the natural mixture of lithium is used, the greater uptake of ^{6}Li might contribute to a greater extent than ^{7}Li to the toxicity or efficacy of lithium therapy in manic-depressive illness.

We were curious to know if the two isotopes differentially inhibited inositol monophosphatase and if they had different effects on the formation of *myo*-inositol 1-phosphate in brain *in vivo*. When several concentrations of the individual purified isotopes were incubated with inositol monophosphatase *in vitro* their inhibition curves were found to be indistinguishable from one another, and from the naturally occurring mixture of lithium isotopes, over a range from 0.25 to 4 mM (Sherman *et al.*, 1984). This was also found by Parthasarathy and Eisenberg (1983). When rats were given the purified isotopes separately in approximately equivalent amounts (the ^{6}Li used contained 5.3% ^{7}Li and thus the mixture had to be given in that experiment) the effect of ^{6}Li was greater than that of the 7-isotope on both the *myo*-inositol 1-phosphate increase in cerebral cortex and on the *myo*-inositol decrease. When the tissue lithium levels were determined, it was found that ^{6}Li is taken up by cortex to the extent of about 1.5 times that of ^{7}Li, much greater than the

^{6}Li/^{7}Li ratio of 1.07 found in human erythrocytes by Lieberman *et al.* (1979b). When we gave the two isotopes separately to rats, with ^{6}Li two-thirds the dose of ^{7}Li, the resulting tissue levels of the two isotopes were the same. The resulting *myo*-inositol 1-phosphate levels were also the same in both experimental groups.

These results suggest that the different effects of the lithium isotopes seen by Lieberman *et al.* (1979b), Alexander *et al.* (1980), and Stokes *et al.* (1982) are due solely to differential uptake of the isotopes, i.e. if pure ^{6}Li is given at a dose two-thirds that of ^{7}Li, the effects are equivalent. The non-reproducibility of the differential toxicity (in peripheral tissues?) may be because the uptake ratio of the isotopes is smaller in the periphery than it is in cerebral cortex and the isotope effect is a slight one. The possibility that one of the isotopes of lithium might be more effective in the treatment of manic-depressive illness has not been evaluated.

Concluding Remarks

In vivo versus in vitro studies

Almost all of the published work on the effects of lithium on phosphoinositide metabolism *in vivo* has come from either Allison's studies or those that my collaborators and I have produced. The reason for this is largely due to the methodology we use. Allison's and my own early work employed gas chromatography to measure levels of *myo*-inositol and *myo*-inositol 1-phosphate. In more recent studies I have employed gas chromatography/mass spectrometry. Both methods give the same quantitative results; the latter is more sensitive and more specific with respect to molecular structure. The development of a more accessible radiometric assay for the inositol phosphates that began with Berridge *et al.* (1982) is now used in nearly every laboratory doing research in this area; it is very sensitive and the liquid chromatographic methods that have been developed give excellent separations. Radiometric studies are, however, very difficult to carry out *in vivo*.

There are some clear differences between our *in vivo* studies and some of the *in vitro* studies that I have discussed in this chapter. For example, we have not been able to find differences in the brain responses *in vivo* to agonists, as measured by changes in the levels of *myo*-inositol 1-phosphate, in rats given lithium chronically versus acutely. This contrasts with the results of Godfrey *et al.* (1989a), for example, who found the response to carbachol simulation, as measured by the formation of [^{3}H]inositol-labelled inositol mono-phosphates, to be significantly reduced in cerebral cortex slices from rats that had been treated with lithium for three days.

We see differences in the amounts of *myo*-inositol 1-phosphate formed on *in vitro* stimulation of brain slices versus results obtained *in vivo*. In comparisons we have made between the levels of *myo*-inositol 1-phosphate that result from cholinergic stimulation of brain slices and cholinergic stimulation of brain *in vivo*, we have found the *in vivo* response to be 10 times greater (Sherman *et al.*, 1986).

Some of the differences between the *in vivo* and *in vitro* studies may be due to incomplete knowledge about the specific activity of the inositol phosphates in the radiolabelling studies and, in our studies, ignorance about the rates of turnover of these molecules. Another possibility is that these differences are due to homeostasis *in vivo* that is absent *in vitro*. Such a restorative effect *in vivo* might obscure effects that are readily seen *in vitro*. Perhaps differences are due to differences in active pool sizes in the two conditions. There are many possible explanations but no definitive ones as yet.

The appearance of new sites of action of lithium in neural pathways in the past 10 years suggests that it would be naive to believe that the psycho-pharmacological effects of the ion are the result of a single point of action. Instead we are beginning to see a richness of activity that might provide a better explanation for the adjustment of the central nervous system brought about by lithium in moderating the mood swings of manic-depressive illness. At the beginning of 1990 the prospect for an understanding of both the cause and the treatment of that illness seems better than at any other time. In the next decade we will most certainly see new surprises and, hopefully, some good answers.

References

Ackermann, K.E., Gish, B.G., Honchar, M.P. and Sherman, W.R. (1987). Evidence that inositol 1-phosphate in brain of lithium-treated rats results mainly from phosphatidylinositol metabolism. *Biochem. J.* **242**, 517–524.

Agam, G. and Livne, A. (1988). Inositol-1-phosphatase of human erythrocytes is inhibited by therapeutic lithium concentrations. *Psychiat. Res.* **27**, 217–224.

Alexander, G.J., Lieberman, K.W. and Stokes, P. (1980). Differential lethality of lithium isotopes in mice. *Biol. Psychiat.* **15**, 469–471.

Allison, J.H. (1978). Lithium and brain *myo*-inositol metabolism. *In* "Cyclitols and Phosphoinositides" (ed. W.W. Wells and F. Eisenberg, Jr), pp. 507–519. Academic Press, New York.

Allison, J.H. and Blisner, M.E. (1976). Inhibition of the effect of lithium on brain inositol by atropine and scopolamine. *Biochem. Biophys. Res. Commun.* **68**, 1332–1338.

Allison, J.H. and Cicero, T.J. (1980). Alcohol acutely depresses *myo*-inositol 1-phosphate levels in the male rat cerebral cortex. *J. Pharmacol. Exp. Ther.* **213**, 24–27.

Allison, J.H. and Stewart, M.A. (1971). Reduced brain inositol in lithium treated rats. *Nature New Biol.* **233**, 267–268.

Allison, J.H., Blisner, M.E., Holland, W.H., Hipps, P.P. and Sherman, W.R. (1976). Increased brain myo-inositol 1-phosphate in lithium-treated rats. Biochem. Biophys. Res. Commun. **71**, 664–670.

Allison, J.H., Boshans, R.L., Hallcher, L.M., Packman, P.M. and Sherman, W.R. (1980). The effects of lithium on myo-inositol levels in layers of frontal cerebral cortex, in cerebellum, and in corpus callosum of the rat. J. Neurochem. **34**, 456–458.

Attwood, P.V., Ducep, J.B. and Chanal, M.C. (1988). Purification and properties of myo-inositol-1-phosphatase from bovine brain. Biochem. J. **253**, 387–394.

Avissar, S., Schreiber, G., Danon, A. and Belmaker, R.H. (1988). Lithium inhibits adrenergic and cholinergic increases in GTP binding in rat cortex. Nature **331**, 440–442.

Bansal, V.S., Inhorn, V.S. and Majerus, P.W. (1987). The metabolism of inositol 1,3,4-trisphosphate to inositol 1,3-bisphosphate. J. Biol. Chem. **262**, 9444–9447.

Batty, I. and Nahorski, S.R. (1987). Lithium inhibits muscarinic-receptor-stimulated inositol tetrakisphosphate accumulation in rat cerebral cortex. Biochem. J. **247**, 797–800.

Berridge, M.J. and Fain, J.N. (1979). Inhibition of phosphatidylinositol synthesis and the inactivation of calcium entry after prolonged exposure of the blowfly salivary gland to 5-hydroxytryptamine. Biochem. J. **178**, 59–69.

Berridge, M.J., Downes, C.P. and Hanley, M.R. (1982). Lithium amplifies agonist-dependent phosphatidylinositol responses in brain and salivary glands. Biochem. J. **206**, 587–595.

Berridge, M.J., Downes, C.P. and Hanley, M.R. (1989). Neural and developmental actions of lithium: a unifying hypothesis. Cell **59**, 411–419.

Busa, W.B. and Gimlich, R.L. (1989). Lithium-induced teratogenesis in frog embryos prevented by a polyphosphoinositide cycle intermediate or a diacylglycerol analog. Dev. Biol. **132**, 315–324.

Casebolt, T.L. and Jope, R.S. (1989). Long-term lithium treatment selectively reduces receptor-coupled inositol phospholipid hydrolysis in rat brain. Biol. Psychiat. **25**, 329–340.

Challiss, R.A.J., Batty, I.H. and Nahorski, S.R. (1988). Mass measurements of inositol (1,4,5) trisphosphate in rat cerebral cortex slices using a radioreceptor assay: effects of neurotransmitters and depolarization. Biochem. Biophys. Res. Commun. **157**, 684–691.

Changya, L., Gallacher, D.V., Irvine, R.F. and Petersen, O.H. (1989). Inositol 1,3,4,5-tetrakisphosphate and inositol 1,4,5-trisphosphate act by different mechanisms when controlling Ca^{2+} in mouse lacrimal acinar cells. FEBS Lett. **251**, 43–48.

Chen, I.W. and Charalampous, F.C. (1966). Biochemical studies on inositol IX. D-inositol 1-phosphate as intermediate in the biosynthesis of inositol from glucose 6-phosphate, and characteristics of two reactions in this biosynthesis. J. Biol. Chem. **241**, 2194–2199.

Chaung, D.M. (1989). Neurotransmitter receptors and phosphoinositide turnover. Ann. Rev. Pharmacol. Toxicol. **29**, 71–110.

Cornish-Bowden, A. (1986). Why is uncompetitive inhibition so rare? A possible explanation, with implications for the design of drugs and pesticides. FEBS Lett. **203**, 3–6.

Diehl, R.E., Gee, N.S., Whiting, P., Potter, J., Ragan, C.I., Linemeyer, D., Schoefer, R., Bennett, C. and Dixon, R.A.F. (1990). Cloning and expression of bovine brain inositol monophosphatase. J. Biol. Chem. **265**, 5946–5949.

Downes, C.P. (1989). The cellular functions of myo-inositol. Biochem. Soc. Trans. **17**, 259–268.

Downes, C.P. and Stone, M.A. (1986). Lithium induced reduction in intracellular inositol supply in cholinergically stimulated parotid gland. *Biochem. J.* **234**, 199–204.

Drummond, A.H. and Raeburn, C.A. (1984). The interaction of lithium with thyrotropin-releasing hormone-stimulated lipid metabolism in GH3 pituitary tumour cells. *Biochem. J.* **224**, 129–136.

Ebstein, R.P., Hermoni, M. and Belmaker, R.H. (1980). The effect of lithium on noradrenaline-induced cyclic AMP accumulation in rat brain: inhibition after chronic treatment and absence of supersensitivity. *J. Pharmacol. Exp. Ther.* **213**, 161.

Eisenberg, F., Jr (1967). D-*myo*-inositol 1-phosphate as product of cyclization of glucose 6-phosphate and substrate for a specific phosphatase in rat testis. *J. Biol. Chem.* **242**, 1375–1382.

Elphick, M., Taghavi, Z., Powell, T. and Godfrey, P.P. (1988). Alteration of inositol phospholipid metabolism in rat cortex by lithium but not carbamazepine. *Eur. J. Pharmacol.* **156**, 411–414.

Evans, M.S., Zorumski, C.F. and Clifford, D.B. (1990). Lithium enhances neuronal muscarinic excitation by presynaptic facilitation. *Neuroscience* **38**, 457–468.

Fain, J.N. and Berridge, M.J. (1979). Relationship between phosphatidylinositol synthesis and recovery of 5-hydroxytryptamine-responsive Ca^{2+} flux in blowfly salivary glands. *Biochem. J.* **180**, 655–661.

Gee, N.S., Ragan, C.I., Watling, K.J., Aspley, S., Jackson, R.G., Reid, G.G., Gani, D. and Shute, J.K. (1988). The purification and properties of *myo*-inositol monophosphatase from bovine brain. *Biochem. J.* **249**, 883–889.

Godfrey, P.P. (1989). Potentiation by lithium of CMP-phosphatidate formation in carbachol-stimulated rat cerebral-cortical slices and its reversal by *myo*-inositol. *Biochem. J.* **258**, 621–624.

Godfrey, P.P., McClue, S.J., White, A.M., Wood, A.J. and Grahame-Smith, D.G. (1989a). Subacute and chronic *in vivo* lithium treatment inhibits agonist- and sodium fluoride-stimulated inositol phosphate production in rat cortex. *J. Neurochem.* **52**, 498–506.

Godfrey, P.P., Varney, M.A., O'Callaghan, K., Taghavi, Z. and Watson, S.P. (1989b). Chronic lithium treatment decreases inositol 1,4,5-trisphosphate levels in rat cerebral cortex slices. *Br. J. Pharmacol.* **98**, 909P.

Greene, D.A., DeJesus, P.V., Jr and Winegrad, A.I. (1975). Effects of insulin and dietary myoinositol on impaired peripheral motor nerve conduction velocity in acute streptozotocin diabetes. *J. Clin. Invest.* **55**, 1326–1336.

Haas, H.L. and Ryall, R.W. (1977). An excitatory action of iontophoretically administered lithium on mammallian neurones. *Br. J. Pharmacol.* **60**, 185–195.

Hallcher, L.M. and Sherman, W.R. (1980). The effects of lithium ion and other agents on the activity of *myo*-inositol-1-phosphatase from bovine brain. *J. Biol. Chem.* **255**, 10896–10901.

Hipps, P.P., Ackermann, K.E. and Sherman, W.R. (1982). Inositol epimerase-inosose reductase from bovine brain. *Methods Enzymol.* (ed. W.A. Wood), **89**, 593–598.

Hokin, M.R. and Hokin, L.E. (1953). Enzyme secretion and the incorporation of P32 into phospholipides of pancreas slices. *J. Biol. Chem.* **203**, 967–977.

Holub, B.J. (1986). Metabolism and function of *myo*-inositol and inositol phospholipids. *Ann. Rev. Nutr.* **6**, 563–597.

Honchar, M.P., Olney, J.W. and Sherman, W.R. (1983). Systemic cholinergic agents induce seizures and brain damage in lithium-treated rats. *Science* **220**, 323–325.

Honchar, M.P., Ackermann, K.E. and Sherman, W.R. (1989). Chronically administered lithium alters neither *myo*-inositol monophosphatase activity nor phosphoinositide levels in rat brain. *J. Neurochem.* **53**, 590–594.

Honchar, M.P., Vogler, G.P., Gish, B.G. and Sherman, W.R. (1990). Evidence that phosphoinositide metabolism in rat cerebral cortex stimulated by pilocarpine, physostigmine and pargyline is not changed by chronic lithium treatment. *J. Neurochem.* **55**, 1521–1525.

Inhorn, R.C. and Majerus, P.W. (1987). Inositol polyphosphate 1-phosphatase from calf brain. Purification and inhibition by Li^+, Ca^{2+} and Mn^{2+}. *J. Biol. Chem.* **262**, 15 946–15 952.

Jackson, R.G., Gee, N.S. and Ragan, C.I. (1989). Modification of *myo*-inositol monophosphatase by the arginine-specific reagent, phenylglyoxal. *Biochem. J.* **264**, 419–422.

Janowsky, D.S., el-Yousef, M.K., Davis, J.M., and Sekerke, H.J. (1972). A cholinergic-adrenergic hypothesis of mania and depression. *Lancet* **ii**, 632–635.

Joseph, N.E., Renshaw, P.F. and Leigh, J.S., Jr (1987). Systemic lithium administration alters rat cerebral cortex phospholipids. *Biol. Psychiatr.* **22**, 540–544.

Kendall, D.A. and Nahorski, S.R. (1987). Acute and chronic lithium treatments influence agonist and depolarization-stimulated inositol phospholipid hydrolysis in rat cerebral cortex. *J. Pharmacol. Exp. Ther.* **241**, 1023–1027.

Kennedy, E.D., Challiss, R.A.J. and Nahorski, S.R. (1989). Lithium reduces the accumulation of inositol polyphosphate second messengers following cholinergic stimulation of cerebral cortex slices. *J. Neurochem.* **53**, 1652–1655.

Kuhar, M.J. and Yamamura, H.I. (1976). Localization of cholinergic muscarinic receptors in rat brain by light microscopic radioautography. *Brain Res.* **110**, 229–243.

Li, P.P., Chiu, A.S. and Warsh, J.J. (1989). Activation of phosphoinositide hydrolysis in rat cortical slices by guanine nucleotides and sodium fluoride. *Neurochem. Int.* **14**, 43–48.

Lieberman, K.W., Alexander, G.J. and Stokes, P. (1979a). Dissimilar effects of lithium isotopes on motility in rats. *Pharmacol. Biochem. Behav.* **10**, 933–935.

Lieberman, K.W., Stokes, P.E. and Kocsis, J. (1979b). Characteristics of the uptake of lithium isotopes into erythrocytes. *Biol. Psychiat.* **14**, 845–849.

Litosch, I. (1987). Guanine nucleotide and NaF stimulation of phospholipase C activity in rat cerebral-cortical membranes. *Biochem. J.* **244**, 35–40.

Michell, R.H. (1975). Inositol phospholipids and cell surface receptor function. *Biochim. Biophys. Acta* **415**, 81–147.

Naccarato, W.F., Ray, R.E. and Wells, W.W. (1974). Biosynthesis of *myo*-inositol in rat mammary gland. Isolation and properties of the enzymes. *Arch. Biochem. Biophys.* **64**, 194–201.

Parthasarathy, R. and Eisenberg, F., Jr (1983). Lack of differential lethality of lithium isotopes in mice. *Ann. N.Y. Acad. Sci.* **435**, 463–465.

Penney, J.F., Dinwiddie, S.H., Zorumski, C.F. and Wetzel, R.D. (1990). Concurrent and close temporal administration of lithium and ECT. *Convulsive Therapy* **6**, 139–145.

Renshaw, P.F., Joseph, N.E. and Leigh, J.S., Jr (1986). Chronic dietary lithium induces increased levels of *myo*-inositol 1-phosphatase activity in rat cerebral cortex homogenates. *Brain Res.* **380**, 401–404.

Sherman, W.R. (1989). Inositol homeostasis, lithium and diabetes. *In* "Inositol Lipids in Cell Signalling," (R.H. Michell, A.H. Drummond and C.P. Downes), pp. 39–79. Academic Press, London.

Sherman, W.R., Leavitt, A.L., Honchar, M.P., Hallcher, L.M. and Phillips, B.E. (1981). Evidence that lithium alters phosphoinositide metabolism: chronic administration elevates primarily D-*myo*-inositol 1-phosphate in cerebral cortex of the rat. *J. Neurochem.* **36**, 1947–1951.

Sherman, W.R., Munsell, L.Y. and Wong, Y.-H.H. (1984). Differential uptake of lithium isotopes by rat cerebral cortex and its effect on inositol phosphate metabolism. *J. Neurochem.* **42**, 880–882.

Sherman, W.R., Munsell, L.Y., Gish, B.G. and Honchar, M.P. (1985a). Effects of systemically administered lithium on phosphoinositide metabolism in rat brain, kidney, and testis. *J. Neurochem.* **44**, 798–807.

Sherman, W.R., Honchar, M.P. and Munsell, L.Y. (1985b). Detection of receptor-linked phosphoinositide metabolism in brain of lithium-treated rats. *In* "Inositol and Phosphoinositides: Metabolism and Regulation" (ed. J.E. Bleasdale, J. Eichberg and G. Hauser), pp. 49–65. Humana Press, Clifton, NJ.

Sherman, W.R., Gish, B.G., Honchar, M.P. and Munsell, L.Y. (1986). Effects of lithium on phosphoinositide metabolism *in vivo*. *Fed. Proc.* **46**, 2639–2646.

Shute, J.K., Baker, R., Billington, D.C. and Gani, D. (1988). Mechanism of the *myo*-inositol phosphatase reaction. *J. Chem. Soc. Chem. Commun.*, 626–628.

Simmons, D.A. and Winegrad, A.I. (1989). Mechanism of glucose-induced (Na^+, K^+)-ATPase inhibition in aortic wall of rabbits. *Diabetologia* **32**, 402–408.

Spector, R. (1988). *myo*-inositol transport through the blood–brain barrier. *Neurochem. Res.* **13**, 785–787.

Stokes, P.E., Okamoto, M., Lieberman, K.W., Alexander, G. and Triana, E. (1982). Stable isotopes of lithium: *in vivo* differential distribution between plasma and cerebrospinal fluid. *Biol. Psychiat.* **17**, 413–421.

Streb, H., Irvine, R.F., Berridge, M.J. and Schultz, I. (1983). Release of Ca^{2+} from a nonmitochondrial intracellular store in pancreatic acinar cells by inositol-1,4,5-trisphosphate. *Nature* **306**, 67–69.

Sylvia, V.L., Joe, C.O., Norman, J.O., Curtin, G.M., Tilley, R.D. and Busbee, D.L. (1989). Interaction of phosphatidylinositol-4-monophosphate with a low activity form of DNA polymerase alpha: a potential mechanism for enzyme activation. *Int. J. Biochem.* **21**, 347–353.

Takimoto, K., Okada, M., Matsuda, Y. and Nakagawa, H. (1985). Purification and properties of *myo*-inositol-1-phosphatase from rat brain. *J. Biochem.* **98**, 363–370.

Whitworth, P. and Kendall, D.A. (1988). Lithium selectively inhibits muscarinic receptor-stimulated inositol tetrakisphosphate accumulation in mouse cerebral cortex slices. *J. Neurochem.* **51**, 258–265.

Wong, Y.-H.H., Kalmbach, S.J., Hartman, B.K., and Sherman, W.R. (1987). Immunohistochemical staining and enzyme activity measurements show *myo*-inositol-1-phosphate synthase to be localized in the vasculature of brain. *J. Neurochem.* **48**, 1434–1442.

Worley, P.F., Heller, W.A., Synder, S.H. and Baraban, J.M. (1988). Lithium blocks a phosphoinositide-mediated cholinergic response in hippocampal slices. *Science* **239**, 1428–1429.

9 Lithium in the Cellular Environment

NICHOLAS J. BIRCH
Biomedical Research Laboratory, School of Health Sciences,
Wolverhampton Polytechnic, Wolverhampton WV1 1DJ, UK

One of the problems of a study of lithium action is the lack of precision in localization of the ion and in the measurement of its movements between cells and between tissues. This lack of precision is partly because lithium is a very mobile ion, partly because of its widespread distribution in the body and partly because of the difficulties of lithium analysis. Analytical problems stem not generally from the lack of sensitivity for its analysis, but from the interference of related metals and common anions present in large quantities in animal tissues. Thellier and Wissocq (Chapter 4, this volume) have described in detail a number of the useful analytical techniques for the biological study of lithium. In our laboratory we have tried to provide some answers of immediate clinical relevance, and for this we have exploited three relatively novel techniques, ^{6}Li atomic absorption spectroscopy (AAS), ^{7}Li nuclear magnetic resonance (NMR) and lithium-ion selective electrodes: each has been used in other spheres for many years, but has been available only recently for the study of lithium.

Table 1 shows the physical properties of lithium in relation to the closely related cations sodium, potassium, magnesium and calcium. It is clear that there are similarities between lithium and each of the other cations, but that

TABLE 1   Physical properties of Group I and Group II elements.

	Li	Na	K	Mg	Ca
Charge on ion	1+	1+	1+	2+	2+
Atomic radius (Å)	1.33	1.57	2.03	1.36	1.74
Ionic radius (Å)	0.60	0.95	1.33	0.65	0.99
Hydrated radius (Å)	3.40	2.76	2.32	4.67	3.21
Polarizing power z/r^2	2.80	1.12	0.56	4.70	2.05
Electronegativity	1.0	0.9	0.8	1.2	1.0

LITHIUM AND THE CELL
ISBN 0-12-099300-7

none of them is an exact match. The chemistry of lithium classically is described in relation to that of magnesium by the so-called 'diagonal relationship' (Birch, 1970, 1973a,b, 1976, 1978a,b). However, extensive biochemical studies have failed to prove a conclusive identity of lithium and magnesium effects (Phillips and Birch, 1990). Nevertheless, it is still the working hypothesis of this laboratory that magnesium is in some way involved in lithium pharmacology because of its chemical and biochemical similarity and because of its key role in biochemical processes. The explosion of interest in the relationship between lithium and phosphoinositide metabolism described so fully by Sherman (Chapter 8, this volume) highlights another area in which magnesium is intimately involved and in which lithium has major effects. Other chapters reveal other magnesium-dependent processes in which lithium interferes. Recent clinical evidence of changes in calcium and magnesium metabolism following lithium administration (Linder *et al.*, 1989) and the acute effects of lithium on the parathyroid gland causing release of PTH (Seely *et al.*, 1989) suggests that much is still to be learned of the lithium/ magnesium relationship in humans.

Administration and Pharmacokinetics of Lithium

Lithium carbonate is used predominantly in the preventive treatment of manic-depressive psychoses where the drug prevents both manic and depressive episodes. Early studies suggested that a target plasma concentration was required of between 0.8 and 1.6 mmol l^{-1}* for the effective prophylaxis of recurrent affective disorder (Schou, 1968). This figure was derived from the original studies on the use of lithium in the acute treatment of mania and represents a daily dose of about one or two grams of lithium carbonate.

More recently it has been recognized that a number of the more troublesome side effects could be ameliorated using a lower dose and our early studies, reported by Jerram and McDonald (1978), showed that at a 12-h lithium plasma concentration of 0.4 mmol l^{-1} there was no significant increase in relapse rate. These studies were extended (Hullin and Birch, 1979; Hullin, 1979) and have now received independent confirmation (reviewed by Abou Saleh and Coppen, 1989). It is now generally accepted that a target plasma lithium concentration of 0.4–0.8 mmol l^{-1} should be obtained 12 h after the last dose. The most usual daily dose is now 800 mg Li_2CO_3, though this depends on the patient's size and renal function and is individually assessed during the early stages of treatment. There has been no real determination of an optimum lithium dose: most of the studies have attempted to reduce

---

* For comparative purposes: the older literature uses the unit mEquiv l^{-1}, which is identical in value to the SI unit mmol l^{-1}.

lithium dose until some relapses occur. This does not mean that an even lower dose might not be optimal in some patients, reducing still further the incidence of side effects.

The role of so-called 'slow release' and 'sustained release' preparations is still unclear. Early studies (Birch *et al.*, 1974; Tyrer *et al.*, 1976) suggested that there was no clinical advantage in the use of 'slow release' preparations, though naturally this was disputed by the manufacturers. Subsequent history has shown that there is little, if any, advantage, though there are some slight pharmacokinetic differences between formulations. These results are discussed in detail by Phillips (Chapter 17, this volume).

A recent review has compared, using Bayesian techniques, the clinical prediction of plasma lithium resulting from test doses using 12 methods commonly employed in psychiatric practice. There was no significant overall advantage of any one test (Williams *et al.*, 1989).

Ion-Selective Electrode for the Determination of Lithium

It has long been a personal goal to develop a technique which would allow the rapid determination of lithium in the clinic in the presence of the patient. Much of the benefit of lithium is lost if compliance is not maintained and this requires the motivation of both the patient and of the psychiatrist. Present practices have meant that results of blood estimations frequently are not available to the patient and psychiatrist until some time after the interview because of the need to send samples to a remote laboratory for estimation. The consequent delay of despatch and receipt of the report may result in poor compliance.

Some preliminary experiments were carried out in the 1970s with a view to developing a rapid method for testing blood lithium, but these were not successful because of the lack of sensitivity of portable analytical techniques at that time. The potential for an ion-selective electrode was not then realizable because lithium glasses, which would have been required to produce a lithium glass electrode, are extremely brittle and not practical to manipulate. Modern ion-selective electrode technology relies on specific resins rather than the glass of the electrode to act as conductive agent.

These recent advances in electrode technology have brought to the market commercial ion-selective electrode (ISE) systems which allow the analysis of lithium. These have been used recently in the clinical situation with some success. The technology of the lithium ISE has been discussed in detail by Xie and Christian (1986). Four manufacturers of ISE for lithium now produce instruments which are suitable for routine laboratory use. We have evaluated three of these and have found them analytically to be suitable.

There is up to 6% difference between the atomic absorption measurement of standard lithium and that obtained by ion-selective electrode, but this is consistent within each instrument and may be obviated by the original calibration. The major difference between the instruments is in the running costs and in portability and our final selection has been made on these grounds. In our early report we evaluated one particular manufacturer's instrument and pointed out the advantages for the clinician of such an instrument (Phillips *et al.*, 1989).

Fifty-one samples were obtained randomly from batches of routine plasma lithium samples received by a clinical laboratory. We used these to check the reproducibility between the atomic absorption reference method and the ion-selective electrode (ISE). There is a very high correlation between the serum lithium determined by the two methods (Table 2) and a Student's '*t*' test indicates that there is no significant difference between methods. The analytical ISE technique in the laboratory is therefore acceptable. One advantage of the technique is that determination may be carried out on whole blood, thus avoiding the use of centrifuges in the clinical area with their consequent noise and potential for infective aerosols. Our study was actually carried out on separated plasma in order to be absolutely sure of the comparability of results. Further clinical trials are in progress to determine the value of whole blood versus plasma results.

We then tested the ion-selective electrode in a clinical setting. At two sessions of a lithium clinic, during which 21 patients were seen, we measured blood samples taken from each patient. The results were available within 10 minutes of phlebotomy and the psychiatrist was able to discuss them with the patient before a decision was made on any change of medication. A blood sample was retained for subsequent atomic absorption analysis on return to the laboratory. The results are also shown in Table 2 where it can be seen that they were broadly comparable to the laboratory based study. There was

TABLE 2   Comparison of plasma lithium determinations (in mmol l^{-1}) by AAS and ISE.

	Laboratory study $n = 51$	Clinic $n = 21$
Mean ($\pm$ SE) AAS	(a) 0.627 (0.044)	(a) 0.585 (0.032)
Mean ($\pm$ SE) ISE	(b) 0.639 (0.040)	(b) 0.610 (0.027)
Correlation coefficient (r)	+ 0.982	+ 0.967
$p <$	0.001	0.001
Paired 't' test (a) versus (b)	Not significant	Not significant

no significant difference between the atomic absorption and ISE results when tested with the Student's '*t*' test.

There are many advantages to the patient in the immediate assessment of blood lithium while the patient is still in the clinic. There is no possibility of confusion of samples and the almost instant return of a result means that the patient is aware that the monitoring has been successful. Changes in medication are made as a result of consideration of up to the moment information. Since each visit is potentially expensive (because of loss of earnings or travel costs), embarrassing or traumatic for the patient, it is of particular value that only one visit is required.

For the psychiatrist the advantages are that time is saved because only one interview is required. Dubious results can be repeated immediately and patients who have defaulted on treatment can be faced with this with confidence. Changes in medication may be more rapidly monitored and, in the initial stages of lithium treatment, the time taken to attain optimum plasma lithium can be reduced. With the added confidence of rapid results we can now consider a 'negotiated treatment plan' where the patient and psychiatrist agree to a plan of action for therapy which can be continuously monitored. The patient is then involved in the decision making. This can improve the perception of lithium as a drug since there is no longer the 'threat' that this is a life-long therapy. Instead it can be agreed that a review takes place at set intervals (for example every two years) provided that medication is maintained.

The advantage for hospitals is that compliance is improved and, hence, the uptake of lithium treatment may be enhanced with consequent budgetary reductions. An instant lithium monitoring service is also available for emergency admissions with suspected overdose or impending toxicity. The response from our clinicians was very positive and we are developing the procedure for wider acceptance (King *et al.*, 1990). We feel that this analytical advance may prove to be one of the most important improvements to be made in the clinical use of lithium in the immediate future.

Distribution/Concentration of Lithium in Tissues

Early studies established that lithium is widely distributed in tissues following oral administration to experimental animals. It appears to be accumulated in bone (Birch and Jenner, 1973; Birch, 1974; Birch *et al.*, 1982) and endocrine glands (Birch, 1982). The earliest studies (Davenport, 1950; Schou, 1958) reported the distribution and rate of uptake in a number of tissues. Thellier, in an elegant neutron activation experiment using the isotope ^{6}Li, provided

TABLE 3   Concentration of lithium in different tissues following different treatment regimens in rats

	Birch and Hullin (1972)	Birch and Jenner (1973)	Ebadi *et al.* (1974)	Stern *et al.* (1977)	
Type	Chronic (18 month)	Chronic (28 day)	Acute	Acute	Chronic (12 day)
Route	Drinking fluid	Drinking fluid	Intravenous	Intraperitoneal	
Daily Li intake (mmol Kg^{-1})	0.8	0.8	7.2	0.6	0.6
Time after last dose (h)	12	12	24	12	12
Tissue (concentrations in mmol Kg^{-1} wet wt)					
Brain	0.18	0.18	1.22		
Muscle	0.30	0.22	1.60		
Bone	1.38	1.15	1.59		
Liver	0.15		0.53		
Kidney	0.53		1.35		
Heart			0.97		
Spleen			0.82		
Serum/ plasma	0.17	0.14	1.22	0.08	0.21
Pituitary				0.26	0.39
Thyroid				0.31	0.43
Adrenal				0.15	0.24

visual localization of lithium in brain tissue (Thellier *et al.*, 1980). A summary of the distribution of lithium is seen in Table 3.

While it is clear that no tissue appears to have a very high lithium content, it is not true to say that higher concentrations do not occur locally. It may be calculated that at a plasma concentration of 1 mmol l^{-1} it is possible that the concentration of lithium in the renal papilla during relative water deprivation may reach a concentration of some 60–65 mmol l^{-1} (K. Thomsen, pers. commun.) and, indeed, urine concentrations determined in subjects whose water intake is low tend to confirm this observation. Similarly, local concentrations in the fluids bathing cells of the gastrointestinal tract during absorption of a single tablet of lithium may be high. Most cells of the body, however, are exposed to lithium concentrations of less than 2 mmol l^{-1}, even at the highest lithium doses.

Intracellular Lithium

The concentration of lithium in the cytoplasm of cells has not received much attention because of the difficulties of determining cytosolic lithium. Previous studies have relied on the experience of manipulating erythrocytes in a variety of lithium solutions; this suggests that approximately 50% of the external lithium concentration is seen in the cellular compartment after equilibration for a number of hours. Recently we have investigated the concentration of lithium intracellularly using ^7Li NMR techniques. These studies are described in detail by Hughes (Chapter 10, this volume), but in general we observe that the intracellular concentration is much lower than hitherto reported.

Methods for Determination of Intracellular Lithium and Lithium Fluxes at Membranes Using Isotopes of Lithium

There are five isotopes of lithium, of which three are radioactive with half-lives of 0.8 s, 0.2 s and 10^{-21} s. The radioactive isotopes are not very useful in the biological study of lithium because of their extremely short half-lives, and this lack of radioisotopes has been an impediment to the understanding of the distribution and metabolic effects of lithium. The two stable isotopes ^6Li and ^7Li, however, may be determined using a number of techniques, and these are described by Thellier and Wissocq (Chapter 4, this volume). We have exploited two methods: an atomic absorption spectroscopy technique and nuclear magnetic resonance spectroscopy. Each of these techniques can measure simultaneously both isotopes present and allow two-way flux measurements to be made of transport occurring into and out of an identified compartment. This contrasts with other studies when only one of the isotopes is present in the experimental system and its concentration is determined by methods which are unable to discriminate between the two isotopes.

Atomic Absorption Methods for the Isotopes of Lithium

The two stable isotopes of lithium have absorption spectra which are doublets, the two lines being separated by 0.015 nm. By coincidence, the separation of the two isotopes is also 0.015 nm and thus 'natural lithium', which comprises 93% ^7Li and 7% ^6Li, is apparently a triplet (Fig. 1).

The separation of the various lines is much below the level of resolution of conventional absorption spectroscopy. However, we have found that by having two atomic absorption hollow cathode lamps made of the two separate isotopes it is possible to distinguish between the two because the atoms of each isotope absorb light most strongly from the hollow cathode lamp made

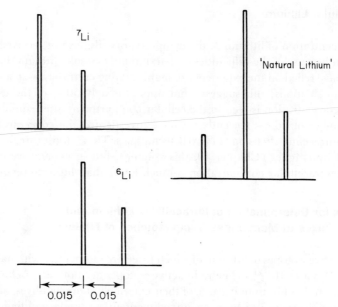

FIG. 1 Schematic diagram of the absorption lines at 670.8 nm of 'natural' lithium and its two stable isotopes ^{6}Li, ^{7}Li. The separation of the lines is 0.015 nm and the isotopic shift is also 0.015 nm.

of the same isotope. Figure 2 shows the absorbance for the same concentration of each isotope determined on each of the two lamps. It is possible, therefore, to set up calibration curves for the absorbance ratio (A_6/A_7) versus the ratio of the number of atoms of each isotope ($[^6Li]/[^7Li]$ atom ratio) at different concentrations. A family of curves is produced as in Fig. 3.

The exact concentration of each isotope can now be determined by solving the exponential equation ($y = ae^{bx}$) once the **total lithium** ($[^6Li] + [^7Li]$) concentration has been determined by atomic emission spectroscopy. The original technique (Birch *et al.*, 1978), which was based on that of Wheat (1971), involved the independent determination of the absorbance of an unknown solution against a ^{6}Li lamp, a ^{7}Li lamp and determination by atomic emission spectroscopy of the total lithium present. Using these three unknowns the absorbance ratio A_6/A_7 was plotted against the atom ratio $[^6Li]/[^7Li]$ for the given concentration.

We have now modified the technique to incorporate the improved precision available by the use of a dual channel atomic absorption spectrometer. We use a Jarrell–Ash Smith–Heiftje Video 22 spectrometer. The advantage of such a system is that the ^{6}Li and ^{7}Li lamps are in use simultaneously so that any variation in flame condition or sampling rate is reflected in both the ^{6}Li and ^{7}Li measurements, and hence the precision of each estimation is greater.

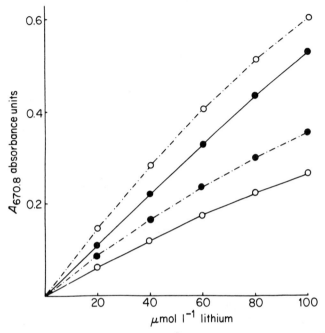

FIG. 2 Absorbance of ⁶Li (—·—·—·—) and ⁷Li (————————) as determined by hollow cathode lamps manufactured with cathode material ⁶Li (○) and ⁷Li (●).

Using this system only two determinations are required, one to determine the A_6/A_7 ratio and the second to determine the total lithium concentration by flame emission.

Using these techniques we have carried out a number of experiments to determine transport rates in different systems. First, we have determined the plasma lithium concentration following a single dose of ⁶Li, the subject having been previously equilibrated on a repeated dose of ⁷Li. In a preliminary report we showed that a single dose of ⁶Li showed an identical plasma profile in a previously equilibrated subject to that of a naive subject receiving ⁷Li for the first time (Birch *et al.*, 1978). We conclude that previous loading with lithium does not affect the kinetics of a subsequent lithium dose. Recent studies have confirmed this result (Fig. 4).

The atomic absorption method for stable isotopes has been used for investigation of the kinetics of lithium in pre-loaded erythrocytes (see Hughes, Chapter 10, and Riddell, Chapter 5, this volume). Atomic absorption was also used to determine the two-way fluxes of lithium in the isolated jejunal mucosa (Karim, 1985) using a technique similar to that described by Davie (Chapter 13, this volume). ⁶Li and ⁷Li were tested on opposite sides of the mucosa and the flux in both directions determined. Karim was unable to

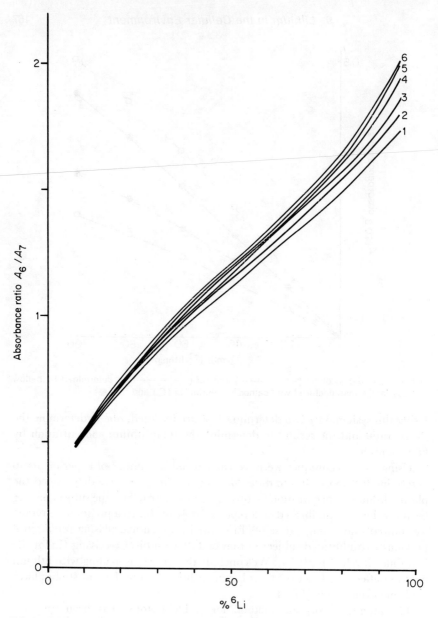

FIG. 3 Calibration curves of A_6/A_7 for a range of isotopic atom ratios [^{6}Li]/[^{7}Li].

Curve 1 =   10 μmol l^{-1} total [Li]; $y = 0.5273$ exp $0.0131x$
Curve 2 =   20 μmol l^{-1} total [Li]; $y = 0.5310$ exp $0.139x$
Curve 3 =   40 μmol l^{-1} total [Li]; $y = 0.5122$ exp $0.0147x$
Curve 4 =   60 μmol l^{-1} total [Li]; $y = 0.4991$ exp $0.0154x$
Curve 5 =   80 μmol l^{-1} total [Li]; $y = 0.4834$ exp $0.0159x$
Curve 6 = 100 μmol l^{-1} total [Li]; $y = 0.4726$ exp $0.0167x$

From Hughes (1989).

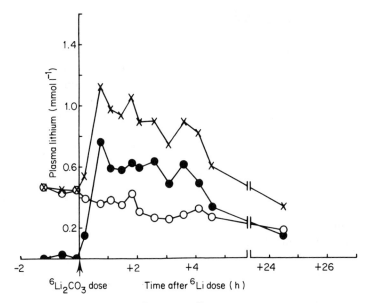

FIG. 4 Plasma concentrations of ^{6}Li($\bullet$) and ^{7}Li($\bigcirc$) following a single dose (400 mg) of ^{6}Li$_2$CO$_3$ at zero time in a normal subject previously equilibrated on ^{7}Li$_2$CO$_3$ for three days (400 mg t.d.s.). ×, total lithium.

measure any difference between the flux rate in either direction, thus indicating that asymmetry did not exist and it was likely that transport processes were passive (Karim, 1985).

Nuclear Magnetic Resonance Spectroscopy

The NMR spectra of ^{7}Li and ^{6}Li are markedly different, but much of the work in biological systems has been carried out with ^{7}Li because the acquisition time for ^{6}Li spectra is excessively long and not very useful for biological experiments. In our group we have studied transport across isolated cells using NMR and some of this work is reported in detail in a later chapter (Hughes, Chapter 10, this volume). A detailed discussion of NMR in relation to lithium is presented by Riddell (Chapter 5, this volume) and a general review of cellular transport measurement by NMR is given by Kirk (1990). We have also studied the transport of lithium in cell types other than erythrocytes and these results have been reported by Thomas *et al.* (1988), Thomas and Olufunwa (1988), Partridge *et al.* (1988), Hughes *et al.* (1988) and Hughes (1988). Essentially the results confirm those findings reported by Hughes that lithium apparently does not penetrate into the cell cytoplasm as readily as previously has been assumed.

Intracellular and Extracellular Lithium Concentrations

Most theories of lithium action make the assumption that the ion occurs in significant concentration in the fluid compartment of cells susceptible to its action. That this is true has not been questioned, and indeed the concentration of lithium at which experiments have been carried out frequently has ignored even the best current estimates of relevant cell lithium concentration. Much of the early lithium 'pharmacology', still cited, was carried out in experiments in which sodium in a physiological solution was replaced to an equimolar concentration (often 150 mmol l^{-1}).

Studies from our laboratory initially questioned the accepted concentrations while studying the cellular concentrations in intestinal cells during lithium absorption in the gut (Davie *et al.*, 1988; Phillips *et al.*, 1988). We showed that lithium is transported via a paracellular route into the blood stream and the concentration in intestinal cells was very low: the majority of 'tissue' lithium was accounted for by trapped extracellular lithium. We progressed from this position to question whether lithium did, in fact, easily enter other cells.

Studies with erythrocytes, hepatocytes, 3T3 fibroblast culture cells and liposome models have brought us to the view that intracellular lithium concentrations are much lower than hitherto imagined and, indeed, the apparent cellular uptake rate of lithium is very low (Phillips *et al.*, 1988). Following long-term exposure cells accumulate more of the ion and this may be a consequence of long-term changes to cellular transport processes (Phillips and Birch, 1990). It appears that there is a gradual reduction in the Na–Li countertransport mechanism which is the major path of efflux of lithium in erythrocytes (Rybakowski, 1990). We have yet to study excitable cells and, of course, it may be that these cells are functionally different.

The significance of these findings is that we should reconsider the postulated sites of action, perhaps to emphasize the extracellular sites, such as the glycoproteins and lipoproteins of the cell surface. In collaboration with Professor Crow (see Chapter 3, this volume) we are carrying out experimental and theoretical studies to investigate lithium ion solvation and interaction with likely pericellular ligands.

Our clinical studies using ion-selective electrodes confirm that in blood there is the expected ionic *activity* which corresponds to the spectroscopically determined ion *concentration*. This indicates, as might be expected, an insignificant binding of lithium to covalent species. In erythrocytes we have shown by isotope studies in collaboration with Dr Riddell (see Chapter 5, this volume) that the exchange rate between cell and surrounding buffer is much faster than the rate of uptake of naive cells. This suggests a membrane carrier mechanism for exchange which is not necessarily available for import of the free ion (Riddell *et al.*, 1990).

We conclude that care must be taken to avoid unquestioning acceptance of plausible mechanisms of action of lithium which have not been shown to be relevant at lithium concentrations of perhaps 20% of the clinically useful plasma concentration, that is at below 0.1 mmol l^{-1} Li$^+$. Many enzyme systems, confidently cited, have K_m values of 10 or 100 times larger than this concentration and Li may indeed prove to be ineffective on these systems at the level actually occurring in cell cytoplasm.

References

Abou Saleh, M. and Coppen, A. (1989). The efficacy of low-dose lithium: Clinical, psychological and biological correlates. *J. Psychiat. Res.* **23**, 157–162.

Birch, N.J. (1970). The effects of lithium on plasma magnesium. *Br. J. Psychiat.* **116**, 461.

Birch, N.J. (1973a). Biological effects of lithium salts. *Lancet* ii, 46.

Birch, N.J. (1973b). The role of magnesium and calcium in the pharmacology of lithium. *Biol. Psychiat.* **7**, 269–272.

Birch, N.J. (1974). Lithium accumulation in bone following its oral administration in the rat and in man. *Clin. Sci. Mol. Med.* **46**, 409–413.

Birch, N.J. (1976). Possible mechanism for the biological action of lithium. *Nature* **264**, 681.

Birch, N.J. (1978a). Metabolic effects of lithium. *In* "Lithium in Medical Practice" (ed. F.N. Johnson and S. Johnson), pp. 89–114. MTP Press, Lancaster.

Birch, N.J. (1978b). Lithium in medicine. *In* "New Trends in Bioinorganic Chemistry" (ed. R.J.P. Williams and J.J.R.E. da Silva), Ch. 11, pp. 389–435. Academic Press, London.

Birch, N.J. (1982). Lithium in psychiatry. *In* "Metal Ions in Biological Systems", Vol. 14 (ed. H. Sigel), Ch. 11, pp. 257–313. Marcel Dekker, New York.

Birch, N.J. and Hullin, R.P. (1972). The distribution and binding of lithium following its long-term administration. *Life Sci. (II)* **11**, 1095–1099.

Birch, N.J. and Jenner, F.A. (1973). The distribution of lithium and its effects on the distribution and excretion of other ions in the rat. *Br. J. Pharmacol.* **47**, 586–594.

Birch, N.J., Goodwin, J.C., Hullin, R.P. and Tyrer, S.P. (1974). Absorption and excretion of lithium following administration of slow-release and conventional preparations. *Br. J. Clin. Pharmacol.* **1**, 339 P.

Birch, N.J., Robinson, D., Inie, R.A. and Hullin, R.P. (1978). ^{6}Li, a stable isotope of lithium, determination by atomic absorption spectroscopy and use in human pharmacokinetic studies. *J. Pharm. Pharmacol.* **30**, 683–685.

Birch, N.J., Horsman, A. and Hullin, R.P. (1982). Lithium, bone and body weight studies in long-term Lithium treated patients and in the rat. *Neuropsychobiology* **8**, 86–92.

Davenport, V.A. (1950). Distribution of parenterally administered lithium administered lithium in plasma, brain and muscle of rats. *Amer. J. Physiol.* **163**, 633–641.

Davie, R.J., Coleman, I.P.L. and Partridge, S. (1988). Lithium transport in isolated epithelial preparations. *In* "Lithium: Inorganic Pharmacology and Psychiatric Use" (ed. N.J. Birch), pp. 107–111. IRL Press, Oxford.

Ebadi, M.S., Simmons, V.J., Hendrickson, M.J. and Lacy, P.S. (1974). Pharmacokinetics of lithium and its regional distribution in rat brain. *Eur. J. Pharmacol.* **27**, 324–329.

Hughes, M.S. (1988). Multinuclear NMR studies in lithium pharmacology. *In* "Lithium: Inorganic Pharmacology and Psychiatric Use" (ed. N.J. Birch), pp. 285–288. IRL Press, Oxford.

Hughes, M.S. (1989). "Lithium interactions with erythrocytes studied by NMR and AAS". PhD Thesis (CNAA), Wolverhampton Polytechnic. 182 pp.

Hughes, M.S., Thomas, G.M.H., Partridge, S. and Birch, N.J. (1988). An investigation into the use of dysprosium shift reagent in the nuclear resonance spectroscopy of biological systems. *Biochem. Soc. Trans.* **16**, 207–208.

Hullin, R.P. (1979). Minimum effective plasma lithium levels for long term treatment. *In* "Proceedings of 1st International Lithium Congress", New York, June 1978 (ed. S. Gershon, N.S. Kline and M. Schou), p. 333. Excerpta Medica, International Congress Series, Amsterdam.

Hullin, R.P. and Birch, N.J. (1979). Effects on renal and thyroid function and bone metabolism in long-term maintenance treatment with lithium salts. *In* "Proceedings of 1st International Lithium Congress", New York, June 1978 (ed. S. Gershon, N.S. Kline and M. Schou), pp. 584–611. Excerpta Medica International Congress Series, Amsterdam.

Jerram, T.C. and McDonald, R. (1978). Plasma lithium control with particular reference to maximum effective levels. *In* "Lithium in Medical Practice" (ed. F.N. Johnson and S. Johnson), pp. 407–408. MTP Press, Lancaster.

Karim, A.R. (1985). "Transport of lithium ion across rodent small intestine". PhD Thesis (CNAA), Wolverhampton Polytechnic.

Kirk, K. (1990). NMR methods for measuring membrane transport rates. *NMR in Biomedicine* **3**, 1–16.

King, J.R., Phillips, J.D., Armond, A., Corbett, J.R. and Birch, N.J. (1990). Clinical use of lithium ion selective electrode. *Psychiat. Bull.*, in press.

Linder, J., Fyro, B., Pettersson, U. and Werner, S. (1989). Acute antidepressant effect of lithium is associated with fluctuation of calcium and magnesium in plasma. *Acta. Psychiatr. Scand.* **80**, 27–36.

Partridge, S., Hughes, M.S., Thomas, G.M.H. and Birch, N.J. (1988). Lithium transport in erythrocytes. *Biochem. Soc. Trans.* **16**, 205–206.

Phillips, J.D. and Birch, N.J. (1990). Lithium in medicine. *In* "Monovalent Cations in Biological Systems" (ed. C.A. Pasternak), pp. 339–355. CRC Press, Boca Raton, FL.

Phillips, J.D., Davie, R.J. and Birch, N.J. (1988). Low cellular lithium concentrations: a consequence of paracellular transport in acute studies? *Br. J. Pharmacol.* **95**, 836 P.

Phillips, J.D., King, J.R., Myers, D.H. and Birch, N.J. (1989). Lithium monitoring close to the patient. *Lancet* **ii**, 1461.

Riddell, F.G., Patel, A. and Hughes, M.S. (1990). Lithium uptake rate of lithium:lithium exchange rate in human erythrocytes at a nearly pharmacologically normal level monitored by Li^7 NMR. *J. Inorg. Biochem.* **39**, 187–192.

Rybakowski, J.K. (1990). Lithium in erythrocytes:pathogenetic and clinical significance. *Lithium* **1**, 75–85.

Seely, E.W., Moore, T.J., LeBoff, M.S. and Brown, E.M. (1989). A single dose of lithium carbonate acutely elevates intact parathyroid hormone levels in humans. *Acta. Endocrinol.* **121**, 174–176.

Schou, M. 1958). Lithium studies. 3. Distribution between serum and tissues. *Acta. Pharmacol. Toxicol.* **15**, 115–124.

Schou, M. (1968). Lithium in psychiatric therapy and prophylaxis. *J. Psychiat. Res.* **6**, 67–95.

Stern, S., Frazer, A., Mendels, J. and Frustaci, C. (1977). Distribution of the lithium ion in endocrine organs of the rat. *Life Sci.* **20**, 1669–1674.

Thellier, M., Wissocq, J.C. and Heurteaux, C. (1980). Quantitative microlocation of lithium in the brain by a (*n*, *α*) nuclear reaction. *Nature* **283**, 299–302.

Thomas, G.M.H. and Olufunwa, R. (1988). Studies of lithium ion transport in isolated rat hepatocytes using 7-Li NMR. *In* "Lithium: Inorganic Pharmacology and Psychiatric Use" (ed. N.J. Birch), pp. 289–291. IRL Press, Oxford.

Thomas, G.M.H., Hughes, M.S., Partridge, S., Olufunwa, R.I., Marr, G. and Birch, N.J. (1988). NMR studies of lithium transport in hepatocytes. *Biochem. Soc. Trans.* **16**, 208.

Tyrer, S.P., Hullin, R.P., Birch, N.J. and Goodwin, J.C. (1976). Absorption of lithium following slow-release and conventional preparations. *Psychol. Med.* **6**, 51–58.

Wheat, J.A. (1971). Isotopic analysis of lithium by atomic absorption spectrophotometry. *App. Spectroscopy*, 25, 328–330.

Williams, P.J., Browne, J.L. and Patel, R.A. (1989). Bayesian forecasting of serum lithium concentrations. Comparison with traditional methods. *Clin. Pharmacokinet.* **17**, 45–52.

Xie, R.Y. and Christian, G.D. (1986). Serum lithium analysis by coated wire lithium ion selective electrodes in a flow injection analysis system. *Anal. Chem.* **58**, 1806.

10 Intracellular Concentrations of Lithium as Studied by Nuclear Magnetic Resonance Spectroscopy

MARK S. HUGHES
Biomedical Research Laboratory, School of Health Sciences,
Wolverhampton Polytechnic, Wolverhampton WV1 1DJ, UK

Introduction

Human red blood cells have been used in lithium research as a model system for nerve cells. It has been suggested that erythrocyte lithium concentrations more accurately reflect brain lithium concentration than plasma lithium (Mallinger *et al.*, 1975) and that erythrocytes possess many transport systems similar to those found in neurones (Mendels *et al.*, 1976; Richelson *et al.*, 1986).

Nuclear magnetic resonance spectroscopy (NMR) is a technique that can be used to investigate the magnetic properties exhibited by many atomic nuclei and their interactions, within their environment, with an externally applied magnetic field. The sample is placed inside a magnetic field where resonance is induced by pulsing with radio frequency energy. One of the main advantages of using NMR is that it is non-invasive, so that metabolic and ionic changes can be observed as they take place in the cellular environment. Thus the technique has proved both popular and successful in the study of the dynamic state of inorganic ions and organic metabolites in isolated cells, tissues and organs as well as in intact organisms (Roberts and Jardetzky, 1981; Balschi *et al.*, 1982; Gupta *et al.*, 1984; Pettergrew *et al.*, 1984, 1987; Brauer *et al.*, 1985; Ogino *et al.*, 1985; Avison *et al.*, 1986; Adam *et al.*, 1987; Espanol and Mota de Freitas, 1987; Höfeler *et al.*, 1987; Springer, 1987).

Many intracellular ions, such as Na^+, K^+ and Li^+, can be observed directly by NMR, whilst others such as Mg^{2+} and Ca^{2+}, are presently indirectly observed through their effects on the NMR spectra of either ^{31}P (Mg^{2+}) or ^{19}F (Ca^{2+}). The information available from interpretation of the NMR spectrum can often complement data from other techniques. Unlike atomic absorption and atomic emission techniques that can determine the total concentration of various ions in a cell, NMR may, depending on the

LITHIUM AND THE CELL
ISBN 0-12-099300-7

chemical characteristics of the nuclei under observation, provide information on the concentration of free ions.

Nuclear magnetic resonance is a relatively modern technique that allows nuclear specific investigation of a wide variety of elements and elemental isotopes. Several characteristics of NMR spectra are useful as an aid to analysis. The position of the absorption signal within the NMR spectrum, the chemical shift, depends on the amount of electron shielding about the nucleus under observation. The longitudinal (T_1) and transverse (T_2) relaxation times of the nuclei are respectively dependent on changes of enthalpy and entropy of the system under investigation. Finally, the multiplicity of the absorption peaks gives a coupling constant value (J) that is, in effect, a measure of the degree of interaction between coupled nuclei. NMR is, therefore, an invaluable technique in the investigation of the molecular environment surrounding the observed species.

NMR Observation of Intracellular Li$^+$ Ions

Observation of lithium ions in cellular systems is hindered due to the very small range of shift values it occupies. The chemical shift of the lithium ion is virtually anion and solvent independent. As lithium ions are believed to exist primarily as free hydrated ions in both the intra- and extracellular medium, detection of separate NMR resonances from these two lithium pools is only made possible by the addition of a chemical shift reagent.

Aqueous shift reagents for biological cations (including lithium) are generally composed of a paramagnetic lanthanide cation and a highly charged anion. The large anion binds the cation under investigation and due to its negative charge has little tendency to bind to macromolecules or to cross biomembranes (Pike *et al.*, 1983). The paramagnetic cation renders the resonance frequency of the cations bound to the anion different from the resonance frequency of the same cationic species present in an environment where the shift reagent is not present. Because the large shift reagent complex does not penetrate the cell membrane, the extracellular environment is magnetically non-equivalent to the intracellular environment and hence the cations present in each of these environments produce signals that are at different positions in the NMR spectrum (Fig. 1).

Dysprosium tripolyphosphate has been shown to be a very good hyperfine shift reagent for use in ^{7}Li NMR (Hughes *et al.*, 1988a). By increasing the molar ratio of shift reagent to lithium it can be seen that the relative separation between the intracellular and extracellular peaks increases linearly (Fig. 2).

The addition of a shift reagent to a sample induces some line-broadening in the shifted resonance peak. This broadening is independent of the magnetic

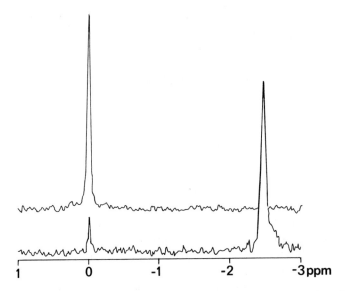

FIG. 1 ^{7}Li NMR spectrum of 10 m$_M$ lithium in phosphate-buffered saline without (top) and with (bottom) 2.5 m$_M$ dysprosium tripolyphosphate. The reference signal positioned at 0 ppm is 2 m$_M$ lithium contained in a coaxial tube.

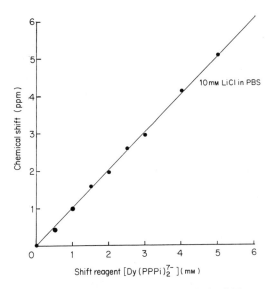

FIG. 2   The separation between intracellular and extracellular lithium as a function of the molar ratio of the shift reagent dysprosium tripolyphosphate to lithium. PBS, phosphate-buffered saline.

field strength employed and is small compared with the shift produced (Pike and Springer, 1982).

The Use of Dysprosium Tripolyphosphate in Human Erythrocytes

In some tissue types it has been reported that an active pyrophosphatase catalyses the decomposition of the tripolyphosphate component of the shift reagent to orthophosphate, a potentially damaging agent to the tissue under investigation. In human erythrocytes this decomposition does not take place. This indicates that either the pyrophosphatase is inactive in erythrocytes, or there is no intracellular transport of the shift reagent (Matwiyoff *et al.*, 1986). Exposure of erythrocytes to dysprosium tripolyphosphate does not affect the adenosine triphosphate, diphosphoglycerate or free magnesium levels in the human erythrocyte, even over a period of several hours (Fabry and San George, 1983). Thus this shift reagent would appear to be non-toxic to erythrocyte cellular energy metabolism.

In tissues that do contain an active pyrophosphatase, other anionic ligands can be used in the formation of the shift reagent. Triethylenetetraminehexa-acetic acid ($TTHA^{6-}$) complexed with dysprosium provides a shift reagent that is capable of resolving separate lithium resonances, although it does not produce shifts that are as large as those with dysprosium tripolyphosphate. However, unlike the tripolyphosphate reagent the shifts produced by $Dy(TTHA)^{3-}$ are virtually pH independent (Hughes, 1989).

Figure 3 shows a ^7Li NMR spectrum of packed human erythrocytes in phosphate buffered saline containing 2.5 mM dysprosium tripolyphosphate. The upfield resonance peak (right of spectrum) corresponds to the extra-cellular lithium, whilst the downfield resonance peak (left of spectrum) arises from intracellular lithium ions. Even when lithium is present in the external solution at therapeutic levels (approximately 1 mM), which is at a much lower concentration than other cations, studies have shown that the intra- and extracellular lithium NMR signals are clearly resolved using only small concentrations (3 mM) of the shift reagent (Hughes, 1989).

To obtain NMR spectra that can be analysed quantitatively, it is important to know the exact value of both the intra- and extracellular transverse relaxation (T_1) time. This is so that a sufficient delay can be included during spectral acquisition in order to allow complete relaxation of the lithium nuclei. The delay time is usually five times the length of the longest transverse relaxation time.

The addition of dysprosium tripolyphosphate to erythrocyte suspensions will cause a reduction in the external, but not the internal, lithium relaxation time. The T_1 time of lithium in water is in the region of 15–20 s

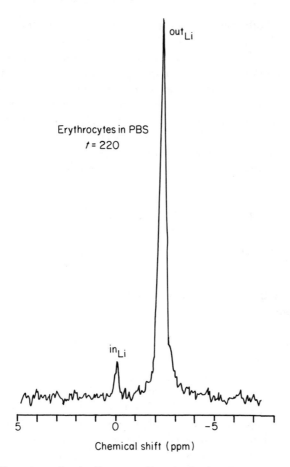

FIG. 3 ^{7}Li NMR spectrum of packed human erythrocytes in phosphate-buffered saline (PBS) containing 2.5 mM dysprosium tripolyphosphate.

(Mota de Freitas and Espanol, 1987). However, in a physiological buffer containing 3 mM dysprosium tripolyphosphate, the extracellular T_1 time has been determined to be 0.17 s, whilst the intracellular lithium T_1 time is in the region of 4.7 s (Hughes, 1989).

Quantitation of Intracellular Li$^+$ Ions by NMR

Several methods are available for the quantitation of intracellular lithium ions. The production of calibration curves using a range of non-cellular standards (Pike *et al.*, 1984) or a reference standard (Espanol and Mota de

Freitas, 1987) both give internal lithium concentrations that are consistent with atomic absorption analyses (Hughes, 1989). These data confirm the work of Pettergrew *et al.* (1987) which showed that, unlike sodium, no intracellular lithium was NMR invisible due to membrane binding.

Alternatively the intracellular concentration can be determined by comparison of the intensity of the resonance from the extracellular ions (A_{out}) with that from a non-cellular control (A_0) which contains the same concentration of lithium ions as the external medium ($[Li_{out}]$). The concentration of intracellular lithium ($[Li_{in}]$) can be calculated from the intensity of the intracellular $^7Li^+$ (A_{in}) using the following equations (Gupta and Gupta, 1982).

$$[Li_{in}] = \left\{ \left(\frac{A_{in}}{A_{out}} \right) \times \frac{S_{out}}{(1 - S_{out})} \right\} \times [Li_{out}]$$

where $S_{out} = A_{out}/A_0$

Determining Intracellular Lithium Concentrations Without the Use of a Shift Reagent

Intracellular lithium can also be observed by NMR without the use of a shift reagent. A modified inversion recovery technique (MIR) is employed which exploits the large difference in the spin-lattice relaxation times of the intracellular (approximately 4.7 s) and extracellular (approximately 16.5 s) lithium ions (Espanol *et al.*, 1989). The basis of the method is that a delay is incorporated between the 180° and 90° pulses that is long enough to allow complete detection of the intracellular signal but short enough to provide no signal from the extracellular ions (Seo *et al.*, 1987). This modified inversion recovery technique allows analysis of the system under observation without the addition of any compounds, such as shift reagents, that would not be normally present *in vivo*.

Intracellular Concentrations of Lithium

Experiments in this laboratory have been performed to investigate intracellular lithium levels by NMR. Incubations of erythrocytes in a 40 mM lithium-containing buffer at 37°C for 3 h gave an intracellular lithium concentration of 3.1 mM, which is just under 8% of the external value. Uptake over this time period was observed to be linear although an intracellular lithium signal was not clearly differentiated from the noise until after a period of 30 min (Partridge *et al.*, 1988; Hughes, 1989). Detection of intracellular lithium by

NMR prior to this 30 min period can be achieved by employing a spectrometer with a greater field strength and/or increasing the period of acquisition.

These experiments gave an uptake rate constant of lithium into human erythrocytes of 10.05×10^{-4} min^{-1}, which is in excellent agreement with a value of 10.00×10^{-4} min^{-1} determined by atomic absorption spectroscopy (Hughes, 1989). These rate constant values can be converted to permeability coefficients of 8.9×10^{-10} cm s^{-1} and 8.896×10^{-10} cm s^{-1} for the NMR and AAS determinations, respectively, by using the simple equation:

$$P = (\text{rate constant})(\text{erythrocyte volume})(\text{surface area})^{-1}$$

(Pike *et al.*, 1982).

Extended loading of erythrocytes at 37°C for approximately $17\frac{1}{2}$ h at an external lithium concentration of 40 mM gave an intracellular lithium concentration of 12.96 ± 0.36 mM ($n = 71$, $\times \pm$ SEM). This gives an intracellular lithium concentration of approximately $32\frac{1}{2}\%$ of the external value (Hughes, 1989).

However, using such high external values of lithium *in vitro* is unlikely to give an accurate indication of the events that are occurring *in vivo*. The pharmacological level of lithium in the plasma has conventionally been maintained at a level of between 0.6 and 1.2 mM, determinations being taken 12 h after the last oral dose of lithium (Abou-Saleh, 1987). A lower target therapeutic range of 0.4–0.8 mM 12 h after last dose has been reported to reduce side effects (Jerram and McDonald, 1978) and is increasingly accepted as good practice (Phillips and Birch, 1990). Obviously at these lower external levels the percentage of lithium entering the cell, and the rate at which it does so, may be significantly different from those determined with much higher lithium-containing incubating mediums.

The strength of an NMR signal is proportional to the square of the concentration of analyte and hence there is an analytical 'trade-off' between concentration and time taken to acquire sufficient data for accurate measurement. NMR experiments using a near pharmacological level of 2 mM lithium have given a lithium rate constant for uptake by erythrocytes of 3.33×10^{-4} min^{-1}. This can be converted to a permeability coefficient of 2.96×10^{-10} cm s^{-1}. Again these results were confirmed using atomic absorption spectroscopy where a rate constant of 2.90×10^{-4} min^{-1} and a permeability coefficient of 2.58×10^{-10} cm s^{-1} were determined.

Incubations of erythrocytes with 2 mM lithium for periods up to 32 h at 37°C gave intracellular lithium concentrations of 21.75% (0.435 mM) (NMR) and 24.1% (0.48 mM) (AAS), which are significantly lower percentage values than those observed using a 40 mM lithium loading medium. These values were determined using erythrocytes from normal healthy volunteers whose cells were naive with respect to lithium therapy. An initial study on two patients currently receiving lithium therapy by Riddell has shown, using NMR, intracellular lithium values of 27.1% and 29.4% of the external plasma lithium level (1.35 mM and 1.5 mM, respectively) (F.G. Riddell, pers. commun.). Thus the actual internal concentration of lithium in these patients is 0.365 mM and 0.441 mM.

In addition to this, a study by Pettergrew *et al.* (1987), also using a dysprosium shift reagent, showed even lower concentrations of internal lithium. After incubating at 25°C for 30 h the intracellular lithium levels were only just over 11% of the external, 50 mM lithium incubating medium. However, incubation at a higher physiological temperature of 37°C does enhance lithium uptake by increasing the rate of passive lithium diffusion, which is the main *in vivo* uptake mechanism of lithium in human erythrocytes (Phillips, 1989).

Erythrocyte Uptake of Stable Lithium Isotopes

Naturally occurring lithium is composed of two stable isotopes, ^6Li and ^7Li. Both isotopes are NMR visible, although ^6Li nuclei are not commonly observed by this method due to the very long periods of data accumulation required (see Chapter 5 by Riddell, this volume). Recent experiments using these isotopes have shown that lithium uptake is a very dynamic process with a fast acting Li/Li exchange reaction taking place (Riddell *et al.*, 1990; see also Chapter 5, this volume). However, the rate at which this Li/Li exchange reaction progressed was observed to depend on the intracellular isotopic pool of lithium with ^7Li–^6Li (intracellular–extracellular) exchange being approximately three times as fast as ^6Li–^7Li exchange (Hughes, 1989).

In red blood cells it has been observed that lithium can partially displace magnesium from ATP (Ramasamy and Mota de Freitas, 1989). The addition of lithium to ATP has also been observed, by NMR methods, to give complexes similar to those formed by magnesium (Hughes, 1988; Hughes *et al.*, 1988b). Although these experiments were either performed *in vitro* or in lithium-loaded cells, it could be that a mechanism for the biological action of lithium involving competition between lithium and magnesium is feasible.

In conclusion, it can be stated that NMR methods provide a non-invasive technique for the investigation of intracellular lithium uptake and transport across biomembranes that yields results that are directly comparable to more conventional, invasive methods such as atomic absorption spectroscopy.

References

Abou-Saleh, M.T. (1987). The dosage regimen. *In* "Depression and Mania: Modern Lithium Therapy" (ed. F.N. Johnson), Ch. 26, pp. 99–105. IRL Press, Oxford.

Adam, W.R., Koretsky, A.P. and Weiner, M.W. (1987). Measurement of tissue potassium *in vivo* using ^{39}K NMR *Biophys. J.* **52**, 265–271.

Avison, M.J., Hetherington, H.P. and Shulman, R.G. (1986). Applications of NMR to studies of tissue metabolism. *Ann. Rev. Biophys. Biophys. Chem.* **15**, 377–402.

Balschi, J.A., Cirillo, V.P. and Springer, C.S. (1982). Direct high resolution NMR studies of cation transport *in vivo*. *Biophys. J.* **38**, 323–326.

Brauer, M., Spread, C.Y., Reithmeier, R.A.F. and Sykes, B.D. (1985). ^{31}P and ^{35}Cl nuclear magnetic resonance measurements of anion transport in human erythrocytes. *J. Biol. Chem.* **260**, 11 643–11 650.

Espanol, M.C. and Mota De Freitas, D. (1987). ^{7}Li NMR studies of lithium transport in human erythrocytes. *Inorg. Chem.* **26**, 4356–4359.

Espanol, M.T., Ramasamy, R. and Mota De Freitas, D. (1989). Measurement of lithium transport across human erythrocyte membranes by ^{7}Li NMR spectroscopy. *In* "Biological and Synthetic Membranes" (ed. D.A. Butterfield). A.R. Liss, New York.

Fabry, M.E. and San George, P.C. (1983). Effect of magnetic susceptibilities on nuclear magnetic resonance signals arising from red cells. A warning! *Biochemistry* **22**, 4119–4125.

Gupta, R.K. and Gupta, P. (1982). Direct observation of resolved resonances from intra- and extracellular sodium-23 ions in NMR studies of intact cells and tissues using dysprosium (III) tripolyphosphate as a paramagnetic shift reagent. *J. Magn. Reson.* **47**, 344–350.

Gupta, R.K., Gupta, P. and Moore, R.D. (1984). NMR studies of intracellular metal ions in intact cells and tissues. *Ann. Rev. Biophys. Bioeng.* **13**, 221–246.

Höfeler, H., Jensen, D., Pike, M.M., Delayre, J.L., Cirillo, V.P., Springer, C.S., Fossel, E.T. and Balschi, J.A. (1987). Sodium transport and phosphorus metabolism in sodium loaded yeast: simultaneous observation with sodium-23 and phosphorus-31 NMR spectroscopy *in vivo*. *Biochemistry* **26**, 4953–4962.

Hughes, M.S. (1988). Multinuclear NMR studies on lithium pharmacology. *In* "Lithium: Inorganic Pharmacology and Psychiatric Use" (ed. N.J. Birch), pp. 285–288. IRL Press, Oxford.

Hughes, M.S. (1989). "Lithium interactions with erythrocytes studied by NMR and AAS". PhD Thesis, Wolverhampton Polytechnic.

Hughes, M.S., Thomas, G.M.H., Partridge, S. and Birch, N.J. (1988a). An investigation into the use of dysprosium shift reagents in the nuclear magnetic resonance spectroscopy of biological systems. *Biochem. Soc. Trans.* **16**, 207–208.

Hughes, M.S., Partridge, S., Marr, G. and Birch, N.J. (1988b). Magnesium interactions with lithium and sodium salts of adenosine triphosphate. *Magnesium Res.* **1**, 35–38.

Jerram, T.C. and McDonald, R. (1978). *In* "Lithium in Medical Practice" (ed. F.N. Johnson). MTP Press, Lancaster.

Mallinger, A.G., Kupfer, D.J., Poust, R.I. and Hanin, J. (1975). *In vitro* and *in vivo* transport of lithium by human erythrocytes. *Clin. Pharmacol. Ther.* **18**, 467–474.

Matwiyoff, N.A., Gasparovic, C., Wenk, R., Wicks, J.D. and Rath, A. (1986). ^{31}P and ^{23}Na NMR studies of the structure and liability of the sodium shift reagent Bis(tripolyphosphate)dysprosium (III) ($[Dy(P_3O_{10})]^{7-}$) ion, and its decomposition in the presence of rat muscle. *Magn. Reson. Med.* **3**, 164–168.

Mendels, J., Frazer, A., Baron, J., Kukopulos, A., Reginaldi, D., Tondo, L. and Caliari, B. (1976). Intra erythrocyte lithium ion concentration and long term maintenance treatment. *Lancet* **ii**, 966.

Mota de Freitas, D. and Espanol, M. (1987). Characteristics of ^{7}Li NMR shift reagents suitable for the study of lithium transport in human erythrocytes. *Biophys. J.* **51**, 73c.

Ogino, T., Shulman, G.I., Aulson, M.J., Gullans, S.R., Den Hollander, J.A. and Shulman, R.G. (1985). ^{23}Na and [39]K NMR studies of ion transport in human erythrocytes. *Proc. Natl. Acad. Sci. USA* **82**, 1099–1103.

Partridge, S., Hughes, M.S., Thomas, G.M.H. and Birch, N.J. (1988). Lithium transport in erythrocytes. *Biochem. Soc. Trans.* **16**, 205–206.

Pettegrew, J.W., Woessner, D.E., Minshew, N.J. and Glonek, T. (1984). ^{23}Na NMR analysis of human whole blood, erythrocytes and plasma. Chemical shift, spin relaxation and intracellular sodium concentration studies. *J. Magn. Reson.* **57**, 185–196.

Pettergrew, J.W., Post, J.F.M., Panchalingam, K., Withers, G. and Woessner, D.E. (1987). ^{7}Li NMR study of normal human erythrocytes. *J. Magn. Reson.* **71**, 519–540.

Phillips, J.D.(1989). "The transport and inorganic biochemistry of lithium and magnesium". PhD Thesis, Wolverhampton Polytechnic.

Phillips, J.D. and Birch, N.J. (1990). Lithium in medicine. *In* "Monovalent Cations in Biological Systems" (ed. C.A. Pasternak), pp. 339–356. CRC Press, Boca Raton, FL.

Pike, M.M. and Springer, C.S. (1982). Aqueous shift reagents for high resolution cationic nuclear magnetic resonance. *J. Magn. Reson.* **46**, 348–353.

Pike, M.M., Simon, S.R., Balschi, J.A. and Springer, C.S. (1982). High resolution NMR studies of transmembrane cation transport: Use of an aqueous shift reagent for [23]Na. *Proc. Nat. Acad. Sci. USA* **79**, 810–814.

Pike, M.M., Yarmush, D.M., Balschi, J.A., Leninski, R.E. and Springer, C.S. (1983). Aqueous shift reagents for high resolution cationic nuclear magnetic resonance. 2. [25]Mg, [39]K and [23]Na resonances shifted by chelidamate complexes of dysprosium (III) and thulium (III). *Inorg. Chem.* **22**, 2388–2392.

Pike, M.M., Fossel, E.T., Smith, T.W. and Springer, C.S. (1984). High resolution ^{23}Na NMR studies of human erythrocytes: Use of aqueous shift reagents. *Amer. J. Phys.* **246**, c528–536.

Ramasamy, R. and Mota de Freitas, D. (1989). Competition between Li^+ and Mg^{2+} for ATP in human erythrocytes. A ^{31}P NMR and optical spectroscopy study. *FEBS Lett.* **244**, 223–226.

Richelson, E., Snyder, K., Carlson, J., Johnson, M., Turner, S., Lumry, A., Boerwinkle, E. and Sing, C.F. (1986). Lithium ion transport in erythrocytes of randomly selected blood donors and manic depressive patients: lack of association with affective illness. *Amer. J. Psychiat.* **143**, 457–462.

Riddell, F.G., Patel, A. and Hughes, M.S. (1990). Lithium uptake rate and lithium:lithium exchange rate in human erythrocytes at a nearly pharmacologically normal level monitored by ^{7}Li NMR. *J. Inorg. Biochem.* **39**, 187–192.

Roberts, J.K.M. and Jardetzky, O. (1981). Monitoring of cellular metabolism by NMR. *Biochim. Biophys. Acta* **639**, 53–76.

Seo, Y., Murakami, M., Suzuki, E. and Watori, H. (1987). A new method to discriminate intracellular and extracellular potassium by [39]K NMR without chemical shift reagents. *J. Mag. Reson.* **78**, 529–533.

Springer, C.S. (1987). Measurement of metal cation compartmentalization in tissue by high resolution metal cation NMR. *Ann. Rev. Biophys. Biophys. Chem.* **16**, 379–399.

11 Effect of Lithium on Blood Cells and the Function of Granulocytes

VINCENT S. GALLICCHIO
Lucille P. Markey Cancer Center, University of Kentucky Medical Center, Lexington, KY 40536-0084, USA

Introduction

Lithium is the lightest metal known. In nature it exists only as lithium ions or salts and is not considered to be an essential trace element for man (Schou, 1957; Mertz, 1981). Lithium carbonate is now widely used for psychiatric purposes (Gershon and Yuwiler, 1960). Following therapeutic doses of lithium, fairly reproducible haematopoietic effects have been documented, especially on the granulocyte leukocytes. This report will review our current understanding of the effects of lithium on blood cell development as it pertains to early haematopoietic progenitors and their differentiation into mature granulocytes, erythrocytes and megakaryocytes.

Although not considered to be an important trace element for biological systems in man, lithium has been postulated to play an important physiological role in haematopoiesis (Barr *et al.*, 1987). This observation has been supported by experimental data demonstrating that lithium influences various cellular lines of differentiation without any particular cell lineage specificity, and that *in vivo* and *in vitro* effects are observed at similar ionic concentrations. These *in vitro* observations have been documented to involve not only lithium, but also other monovalent alkali metal ions such as rubidium and caesium. Because those ions exist either at micromolar or nanomolar concentrations in human systems, they may play an important role in the mechanisms that control the formation of blood cells. The effects of lithium on blood cell formation have been reviewed by Barr and Galbraith (1983), Boggs and Joyce (1983) and Gallicchio (1988).

The well-documented effects of lithium on blood cell formation have suggested that the element might be of value in treating various haematopoietic disorders, e.g. following the use of anti-cancer therapy to ameliorate the bone marrow toxicity associated with chemotherapeutic drugs and/or radiation.

LITHIUM AND THE CELL
ISBN 0-12-099300-7

Effect of Lithium on Peripheral Blood Cells

Attention was first focused on the ability of lithium to induce leukocytosis in manic-depressive patients receiving lithium as therapy (Shopsin et al., 1971). Patients or normal subjects receiving lithium, in doses sufficient to produce a systemic lithium concentration of 0.3 mmol l^{-1} or more, usually produced a sustained elevation in the blood neutrophil concentration (Mayfield and Brown, 1966; Perez-Cruet et al., 1977; Ricci et al., 1981). For therapeutic administration, lithium is given orally. The vast majority of human studies designed to study the haematopoietic effects of lithium have reported a dosage of 300 mg three times a day, producing a blood systemic level of 0.5–1.5 mmol l^{-1}. Serum or plasma lithium levels are usually checked at regular intervals and are usually obtained 2 h before the morning dose is administered. In experimental animals studies, either performed in vivo or in vitro, lithium chloride (LiCl) is widely used and is preferred over lithium carbonate (Li_2CO_3) because Li_2CO_3 is associated with a high pH. For the vast majority of studies on eosinophils, their concentration is usually elevated (Murphy et al., 1971; Bille et al., 1975), but not always. Monocytes have been variously reported to be increased, decreased or not affected (Pointud et al., 1976; Stein et al., 1981; Barr, et al., 1987). The basophil has not been reported to be influenced following lithium administration. The ability of lithium to influence the total number of neutrophilic granulocytes has been shown by numerous studies. In addition, lithium is also an effective inducer of neutrophil function.

Blood platelet concentrations following lithium administration have been reported to be increased in some studies on humans (Steinherz et al., 1980), and mice (Gallicchio et al., 1986a), but other human studies (Ricci et al., 1981) did not show this. The functional activity of platelets as measured by bleeding time or aggregation has been reported to be unaffected by lithium administration (Friedenberg and Marx, 1980).

There is evidence to indicate that lithium, in non-toxic doses, does not significantly influence either blood erythrocyte or reticulocyte levels in man; prolonged administration (> 10 weeks) in mice did not produce a significant reduction in blood erythrocyte levels (Gallicchio and Watts, 1985).

In humans receiving lithium, the neutrophil concentration has been reported not to exceed 1.5 times baseline for any individual. Neutrophilia usually develops within the first week of lithium administration where it persists usually on a stable plateau. It has been difficult to determine the exact lithium dose that results in a blood lithium level producing neutrophilia (Shopsin et al., 1971). A progressive increase in mean neutrophil concentration has been observed in 56 adult patients who received either 600, 900 or 1200 mg lithium day^{-1}. This regimen produced an increase in the mean neutrophil concentration as follows: 3.7×10^9 cells/kg in controls, 4.0×10^9 cells/kg in the 600 mg group; 4.8×10^9 cells/kg in the 900 mg group; and 5.8×10^9 cells/kg in the 1200 mg group (Ricci *et al.*, 1981). In a study performed on normal subjects there was a weak correlation between the neutrophil concentration and the blood lithium levels between 0.2 and $0.9 \, \text{mmol} \, l^{-1}$, but no correlation at levels above $0.9 \, \text{mmol} \, l^{-1}$ (Stein *et al.*, 1981). There have been no reports of the presence of immature neutrophil forms following lithium administration.

Effect of Lithium on Haematopoietic Stem Cells

The observation that lithium stimulation of leukocytosis involved a true proliferative response rather than just a shift of cell populations from the marginating to the circulatory pool of cells led investigators to examine the bone marrow for changes in rates of mitotic cell proliferation (Tisman *et al.*, 1973; Joyce and Chervernick, 1980).

Pluripotential (CFU-S)

The pluripotential haematopoietic stem cell, as identified by the *in vivo* bioassay originally described by Till and McCulloch (1961), is the haematopoietic progenitor responsible for producing all of the differentiated lineage progeny that constitute the haematopoietic system. Lithium has been demonstrated to increase the number of CFU-S harvested from bone marrow following *in vivo* administration to mice (Gallicchio and Chen, 1980), incubation with normal bone marrow cells *in vitro* (Gallicchio and Chen, 1981), obtained from normal murine long-term bone marrow cultures *in vitro* (Levitt and Quesenberry, 1980), and following administration in the diet. CFU-S reconstitution is accelerated following lithium administration to protect haematopoietic tissues from the toxicity induced either by chemotherapeutic drugs (Friedenberg and Marx, 1980; Gallicchio, 1986b, 1987, 1988a) or ionizing radiation (Vacek *et al.*, 1982; Gallicchio *et al.*, 1983, 1984a; Moroz *et al.*, 1988). These results imply that lithium influences both steady-state and regenerating haematopoiesis following the use of agents known to suppress haematopoiesis by modulating pluripotential haematopoietic progenitors (Joyce, 1984).

Granulocyte-macrophage (CFU-GM)

Lithium has been demonstrated to increase the number of granulocyte-macrophage progenitors, CFU-GM, from many species, e.g. man (Greco and

Brereton, 1977; Morley and Galbraith, 1978; Spitzer *et al.*, 1979; Joyce and Chervernick, 1980), dog (Rossof and Fehir, 1979; Hammond and Dale, 1980), and mouse (Harker *et al.*, 1978; Levitt and Quesenberry, 1980; Gallicchio and Chen, 1982). These results indicate that lithium effectively increases the granulocyte-macrophage progenitor stem cell pool and therefore is capable of producing large numbers of granulocytes. These effects would explain the appearance of neutrophilia following lithium administration.

Megakaryocyte (CFU-Meg)

Lithium has been demonstrated to increase megakaryocyte progenitors cultured from murine (Friedenberg and Marx, 1980b; Gamba-Vitalo *et al.*, 1983) and human (Chatelain *et al.*, 1983) marrow. This stimulatory effect would allow for the thrombocytosis observed following lithium administration *in vivo*.

Erythroid (CFU-E and BFU-E)

Lithium has been reported to influence erythroid progenitors by reducing their number, in contrast to its ability to increase CFU-S, CFU-GM and CFU-Meg. CFU-E and BFU-E reduction as the result of lithium exposure has been demonstrated from human (Chan *et al.*, 1980) and murine haematopoietic tissues (Levitt and Quesenberry, 1980; Gallicchio and Chen, 1981; Gallicchio and Murphy, 1983).

Effect of Lithium on Growth Factor Production

Although lithium has been shown to influence many classes of haematopoietic progenitors, several investigators have also indicated that lithium was capable of stimulating increased production of the growth factor(s) responsible for the sustained proliferation of these haematopoietic progenitors. A family of growth-promoting substances, collectively termed 'colony stimulating factors', has been recognized to be obligatory for the growth of haematopoietic precursors, e.g. GM-CSF for CFU-GM, Meg-CSF for CFU-Meg (Metcalf, 1989). These growth-promoting factors are known to be produced by cells of bone marrow origin and are collectively termed 'accessory cells'. Lithium has been shown to increase GM-CSF and Meg-CSF production from a variety of tissues: the cell responsible usually has been the macrophage (Harker *et al.*, 1977; Ramsey and Hays, 1979; Greenberg *et al.*, 1980; Chatelain *et al.*, 1983) or T-lymphocytes (Gallicchio *et al.*, 1984b). Not all lithium-induced responses have been associated with increased levels of haematopoietic growth factors. For example increased GM-CSF present in serum could not be detected from mice given lithium following exposure to sublethal whole body irradiation, even though increased granulopoietic recovery as the result of lithium, compared with irradiation controls, was demonstrated (Gallicchio et al., 1985).

Investigations using diffusion chambers (DC) have attempted to address the question regarding the role of lithium and humoral mediators (Doukas et al., 1985, 1989). Diffusion chambers only allow the passage of diffusible growth factors, enabling their detection as haematopoietic growth-stimulating factors. Studies utilizing DCs have indicated that lithium stimulation of granulopoiesis is indeed partially mediated by growth factor elaboration, the exact nature of which has not yet been clearly eludicated; however, GM-CSF and CSF-1 were not detectable in chamber fluid following lithium administration (Doukas et al., 1989). These studies imply that other candidate molecules, possibly the interleukins or other growth factors not yet identified, may be involved in the mechanism that allows lithium to influence haematopoiesis.

Stromal cells

Several studies have identified the capacity of lithium to influence the support matrix of the haematopoietic microenvironment or stroma under normal steady-state conditions (Gallicchio et al., 1986), in short-term (Doukas et al., 1986) or long-term marrow culture systems (Quesenberry et al., 1984), and following acute radiation exposure (Gallicchio et al., 1983). In fact, stromal cell recovery precedes any stem cell response observed following lithium-radiation exposure (Vacek et al., 1982). These results indicate that lithium can effectively increase numbers of stromal cells and therefore can modulate the microenvironment, either to enhance progenitor cell support capacity or increase stromal cell-derived humoral/growth factor production that would produce or promote local cell–cell interactions.

Lithium Potentiation of Haematopoiesis via Transport-related Mechanisms

Recent studies have focused on the role of sodium transport processes in the mechanism(s) whereby lithium influences granulopoiesis and megakaryocyto-poiesis. Studies have demonstrated that in the presence of sodium transport ionophores, but not potassium transport ionophores, stimulation by lithium was promoted (Gallicchio, 1986c). In the presence of ouabain, a Na/K ATPase inhibitor, lithium potentiation of in vitro CFU-GM growth was inhibited and irreversible. Furthermore, studies conducted in the presence of the calcium ionophore A23187 blocked the effect of lithium (Gallicchio, 1986c). These observations demonstrate that in the presence of activated calcium transport, the ability of lithium to influence granulopoiesis is restricted. In addition, studies conducted in the presence of sodium transport inhibitors were effective in blocking the effect of lithium on both granulopoietic and megakaryocytic progenitors (Gallicchio, 1990). These studies are consistent with the recent observation that changes in sodium influx induced with haematopoietic growth factors may be important for their target cell interactions (Imamure and Kufe, 1988).

Lithium-induced Inhibition of Suppressor Lymphocytes and Prostaglandin Production

Another alternative mechanism for the effects mediated by lithium on haematopoiesis may be attributable in part to the ability of lithium to inhibit the activity of suppressor T-lymphocytes. These cells are thymus-derived lymphocytes that are known to limit haematopoiesis (Barr, 1979; Barr and Stevens, 1982). The therapeutic levels of lithium observed clinically are within the concentration range where lithium effectively reduces the levels of cyclic-AMP (Gelfand et al., 1979).

However, lithium can stimulate or enhance the responsiveness of lymphocytes to mitogens such as phytohaemagglutinin (Stobo et al.,1979), which can be inhibited in the presence of prostaglandin E. The effect of prostaglandin is mediated in part by increasing the activity of cyclic-AMP (Goodwin et al., 1979). Therefore lithium and prostaglandins have opposing effects, apparently on the same target cells utilizing similar pathways. These opposing actions have been demonstrated to influence in vitro haematopoiesis (Chan et al., 1980), and in the presence of a prostaglandin inhibitor, indomethacin, lithium augments the production of CSAs from mitogen-stimulated lymphocytes in vitro (Gallicchio et al., 1984b). Lithium stimulates granulopoiesis while at the same time inhibiting erythropoiesis, where prostaglandins are capable of the opposite effects (Rossi et al., 1980). These effects with respect to cyclic-AMP activity and lithium have recently been demonstrated utilizing long-term marrow cultures (Gualtieri et al., 1986).

Granulocyte Function in Patients Receiving Lithium Carbonate

The importance of polymorphonuclear leukocytes (PMNs) is that they provide protection against a wide variety of microbial pathogens. A reduction in the number of neutrophils (Bodey *et al.*, 1966) or an abnormality in their cellular function (Quie, 1975) is most often associated with infection. Studies reported evaluating random migration, chemotaxis, phagocytosis and measurement of bactericidal ability have been evaluated from patients receiving lithium (Cohen *et al.*, 1980). These studies observed no defects among neutrophils examined either from patients receiving lithium or from cells exposed to lithium *in vitro*. Cohen *et al.* concluded that the cells are fully capable of responding and contributing fully to the phagocytic host defence system of the body. This observation confirmed earlier studies that cells and neutrophils harvested from lithium-treated patients were capable of ingesting yeast to the same degree as controls (Stein *et al.*, 1978).

Lithium has been reported to decrease granulocyte adherence (MacGregor, 1977; MacGregor and Dyson, 1980), although in one study this response was not due to lithium but to a plasma factor, removal of which by patient dialysis caused granulocyte adhesiveness to return (MacGregor and Dyson, 1978). The adherence of neutrophils to blood vessel walls is critically important in the inflammatory response *in vivo* (Marchesi and Florey, 1960). Another study reported normal to increased migration of neutrophils into inflammatory sites as the result of lithium therapy (Rothstein *et al.*, 1978). These results implied that the ratio of cyclic-AMP to cyclic-GMP may affect neutrophil chemotaxis. Together with the observation that lithium is known to decrease cyclic-AMP concentrations from several tissues (Gelfand *et al.*, 1979), these results have suggested that lithium activity on granulocyte adhesiveness may involve alterations in cyclic-AMP concentrations.

Other studies that focused on assessing granulocyte function from neutrophils taken from patients who had been on long-term lithium (Friedenberg and Marx, 1980a,b) demonstrated a decrease in granulocyte function tests such as nitroblue tetrazolium (NBT) reduction, chemotaxis in response to bacteria-derived factors and zymosan-induced C5a, and phagocytic and bactericidal capabilities. However, in another study assessing bactericidal capacity, five patients exposed to long-term lithium were not found to have decreased activity (Cohen *et al.*, 1979).

In Vitro Studies on Granulocyte Function Involving Lithium

Previous investigations had identified that lithium was capable of influencing a number of activities that involved blood mononuclear cells. These included increasing thymidine incorporation by phytohaemagglutinin (Shenluman *et al.*, 1978); stimulating phagocytosis of polystyrene latex particles by human monocytes; and reversing prostaglandin E and theophylline induced inhibition of mitogen responsiveness by mononuclear cells (Daniel-Perez *et al.*, 1980).

Similar results found using β-adrenergic agonists indicated that cyclic-AMP was involved (Gelfand et al., 1979). Further investigations demonstrated both that neutrophil chemotaxis can be inhibited by agents that elevate cellular levels of cyclic-AMP, and that lithium is an effective stimulator of neutrophil chemotaxis by inhibiting cyclic-AMP induced inhibition by all agents that are capable of elevating cyclic-AMP, except cyclic-AMP itself. Lithium does not prevent the inhibition of chemotaxis of cells when they are directly exposed to cyclic-AMP (Simchowitz, 1988). More recent experiments have studied the effect on neutrophils of the chemotactic factor N-formyl-methionyl-leucyl-phenylalanine (FMLP), which activates an amiloride-sensitive alkali metal cation-H^+ counter transport system that exhibits a 1:1 stoichiometry. Lithium can effectively utilize this carrier system, as well as sodium, to promote chemotaxis, while potassium, rubidium and caesium are not effective (Reiser et al., 1982). These results indicate that there is a cation-exchange mechanism operating on neutrophils, similar to mechanisms present in a wide variety of other cell types, that when activated can promote chemotaxis.

Studies have also examined the ability of lithium to influence the metabolic function of granulocytes. Various reports investigating a variety of homogeneous cell populations have shown that lithium can increase the adhesion of nervous system cells and interfere with the effects of colcemid in vitro (Imandt et al., 1977), increase the intensity of platelet aggregation and prolong the duration of disaggregation in vitro (Imandt et al., 1977), and enhance neutrophil skin window migration in vivo (Rothstein et al., 1978). Studies (Joyce and Chervenick, 1980) have indicated that, within the therapeutic concentration range $(0.5-1.5\,\text{mmol}\,l^{-1})$, lithium was neither toxic nor stimulating to resting neutrophils in any oxidative or membrane function measured, such as O_2 generation, as an indicator of the respiratory burst of stimulated neutrophils. Glucose-1-^{14}C oxidation of resting neutrophils was not affected by the presence of lithium. However, from phagocytic activated neutrophils $^{14}CO_2$ production was increased. Lithium is effective in elevating lysosomal enzyme release of unstimulated and phagocytic neutrophils. Also, lithium increased neutrophil cell migration as measured by in vivo skin chambers (Siegel et al., 1980).

In general, within the therapeutic concentration range, lithium has no effect on O_2 generation, chemiluminescence and candidacidal activity of optimally stimulated neutrophils. Energy- and membrane-dependent activities such as aggregation and degranulation are unaffected by lithium. Lithium has been suggested in certain systems to act like calcium and can amplify calcium-mediated actions or, as was described earlier, reverse reactions dependent upon activated adenylcyclase activity, a calcium cofactor-dependent reaction (Rothstein et al., 1978).

Areas of Future Promise

Within the field of haematopoiesis the use of lithium may hold future promise in the areas of AIDS treatment and bone marrow transplantation. In AIDS patients lithium may reverse or minimize faulty lymphocyte production (Sztein *et al.*, 1987) and also combat the burden of increased febrile episodes by augmenting the production of interleukin-2 and neutrophil activation. An additional use in AIDS patients is to reverse the profound myelosuppression and bone marrow toxicity associated with the use of 3'-azido-3'-deoxythymidine (AZT), the only FDA approved drug in the treatment of AIDS. A recent clinical study demonstrates its efficacy in this area (Roberts *et al.*, 1988).

In bone marrow transplantation lithium increases the effectiveness of transplantation by minimizing the degree and extent of suppressed haematopoiesis when administered to the donor prior to harvesting of the transplant. This has been successfully accomplished using an animal model system for syngeneic transplantation (Gallicchio *et al.*, 1990).

Conclusion

Blood cell formation, particularly the production of granulocytes and platelets, can be influenced by the administration of lithium salts. Lithium effectively enhances both granulopoiesis and megakaryocytopoiesis *in vivo* and *in vitro* at the same concentrations at which it reduces erythropoiesis. Lithium can effectively increase the production of colony-stimulating factors, the obligatory molecules for the sustained proliferation and differentiation of various classes of haematopoietic progenitors. Undoubtedly, with the recent knowledge on the synergistic activity of various cytokines such as the interleukins, lithium may be found to alter the production of this class of accessory growth factors. More recent studies have demonstrated that an important component of the ability of lithium to influence blood cells involves the transport of cations across the cell membrane and suggest that such processes play an important role in normal cell proliferation and differentiation. Further studies will undoubtedly more clearly eludicate the role of cation flux in the mechanism of haematopoietic progenitor cell proliferation and differentiation.

The administration of lithium produces a neutrophilia. This increased production of granulocytes has been associated with influences on the functional activities of these cells. As a general rule, the administration of lithium is associated with an increased activity of neutrophils to combat infections, therefore lithium is an effective agent to increase not only the number of phagocytes but also their function. Mechanistically, it appears that lithium promotes granulocyte function by its ability to inhibit adenylcyclase activity. Activation of the enzyme increases cyclic-AMP which limits granulocyte function; this effect is reversed by lithium. These studies demonstrate that lithium can increase granulocyte numbers not only where their production is faulty or inadequate, but it is also effective in conditions where neutrophil function may be inadequate.

References

Barr, R.D. (1979). The role of lymphocytes in haematopoiesis. *Scott. Med. J.* **24**, 267–272.
Barr, R.D. and Galbraith, P.R. (1983). Lithium and hematopoiesis. *Can. Med. Assoc. J.* **128**, 123–138.
Barr, R.D. and Stevens, C.A. (1982). The role of autologous helper and suppressor T-cells in the regulation of human granulopoiesis. *Amer. J. Hematol.* **12**, 323–326.
Barr, R.D., Koekebakker, M., Brown, E.A. and Falbo, M.C. (1987). Putative role for lithium in human hematopoiesis. *J. Lab. Clin. Med.* **109**, 159–163.
Bille, P.E., Jensen, M.K., Kaakind-Jensen, J.P. and Paulsen, J.C. (1975). Studies on the hematological and cytogenic effect of lithium. *Acta Med. Scand.* **198**, 281–286.
Bodey, G.P., Buckley, M., Sathe, Y.S. and Feireich, E.J. (1966). Quantitative relationships between circulating leukocytes and infection in patients with acute leukemia. *Ann. Int. Med.* **64**, 328–331.
Boggs, D.R. and Joyce, R.A. (1983). The hematopoietic effects of lithium. *Sem. Hematol.* **20**, 129–138.
Chan, H.S.L., Saunders, E.F. and Freedman, M.H. (1980). Modulation of human hematopoiesis by prostaglandins and lithium. *J. Lab. Clin. Med.* **95**, 125–132.
Chatelain, C., Burstein, S.E., Samuel, E. and Harker, L.A. (1983). Lithium enhancement of megakaryocytopoiesis in culture: mediation via accessory marrow cells. *Blood* **62**, 172–176.
Cohen, M.S., Zakhiveh, B., Metcalf, J.A. and Root, R.K. (1979). Granulocyte function during lithium therapy. *Blood* **53**, 913–917.
Cohen, M.S., Zahkiveh, B., Metcalf, J.A. and Root, R.K. (1980). Granulocyte function in patients receiving lithium carbonate. *Adv. Exp. Med. Biol.* **127**, 335–346.
Daniel-Perez, H., Kaplan, H.B., Goldstein, I.M., Shenkman, L. and Borkowsky, W. (1980). Effects of lithium on polymorphonuclear leukocyte chemotaxis. *Adv. Exp. Med. Biol.* **127**, 357–369.
Doukas, M.A., Niskanen, E.O. and Quesenberry, P.J. (1985). Lithium stimulation of granulopoiesis in diffusion chambers—a model of a humoral, indirect stimulation of stem cell proliferation. *Blood* **65**, 163–168.
Doukas, M.A., Niskanen, E.O. and Quesenberry, P.J. (1986). Effect of lithium on stem cell and stromal cell proliferation *in vitro. Exp. Hematol.* **14**, 215–221.
Doukas, M.A., Shadduck, R.K., Waheed, A. and Gass, C. (1989). Lithium stimulation of diffusion chamber colony growth is mediated by factors other than colony stimulating factor. *Int. J. Cell Clon.* **7**, 168–178.
Friedenberg, W.R. and Marx, J.J. (1980). The bactericidal defect of neutrophil function with lithium therapy. *Adv. Exp. Biol. Med.* **127**, 389–399.
Friedenberg, W.R. and Marx, J.J. (1981). Effect of lithium carbonate on lymphocyte, granulocyte, and platelet function. *Cancer* **45**, 91–97.

Gallicchio, V.S. (1986a). Lithium and hematopoietic toxicity I. Recovery in vivo of murine hematopoietic stem cells (CFU-S and CFU-MIX) following single-dose administration of cyclophosphamide. *Exp. Hematol.* **14**, 395–400.

Gallicchio, V.S. (1986b). Lithium stimulation of *in vitro* granulopoiesis: evidence for mediation via sodium transport pathways. *Br. J. Haematol.* **62**, 455–466.

Gallicchio, V.S. (1987). Lithium and hematopoietic toxicity II. Acceleration in vivo of murine hematopoietic stem cell (CFU-GM and CFU-Meg) following treatment with vinblastine sulfate. *Int. J. Cell Clon.* **5**, 122–133.

Gallicchio, V.S. (1988). Lithium and hematopoietic toxicity III. *In vivo* recovery of hematopoiesis following single-dose administration of cyclophosphamide. *Acta Haematol.* **79**, 192–197.

Gallicchio, V.S. (1988a). Lithium and granulopoiesis: mechanisms of lithium action. *In* "Lithium: Inorganic Pharmacology and Psychiatric Use" (ed. N.J. Birch), p. 93. IRL Press, Oxford.

Gallicchio, V.S. (1990). Lithium enhanced granulopoiesis and megakaryocytopoiesis *in vitro*: potentiation with agents that accelerate sodium transport. *Lithium* **1**, 93–100.

Gallicchio, V.S. and Chen, M.G. (1980). Modulation of pluripotential stem cell proliferation *in vivo* by lithium carbonate. *Blood* **56**, 804–811.

Gallicchio, V.S. and Chen, M.G. (1981). Lithium influences the proliferation of hematopoietic stem cells. *Exp. Hematol.* **9**, 804–811.

Gallicchio, V.S. and Chen, M.G. (1982). Cell kinetics of lithium induced granulopoiesis. *Cell Tiss. Kinet.* **15**, 179–186.

Gallicchio, V.S. and Hulette, B.C. (1989). *In vitro* effect of lithium on carbamezapine-induced inhibition of murine and human bone marrow derived granulocyte-macrophage, erythroid, and megakaryocyte progenitor stem cells. *Proc. Soc. Exp. Biol. Med.* **190**, 1109–1116.

Gallicchio, V.S. and Murphy, M.J. Jr (1983). Cation influences on the *in vitro* growth of erythroid stem cells (CFU-E and BFU-E). *Cell Tiss. Res.* **233**, 175–181.

Gallicchio, V.S. and Watts, T.D. (1985). Sustained elevation of murine granulopoiesis *in vivo* with lithium carbonate. *IRCS Med. Sci.* **13**, 1050–1051.

Gallicchio, V.S., Chen, M.G., Watts, T.D. and Gamba-Vitalo, C. (1983). Lithium stimulates the recovery of granulopoiesis following acute radiation. *Exp. Hematol.* **11**, 553–563.

Gallicchio, V.S., Chen, M.G. and Watts, T.D. (1984a). Ability of lithium to accelerate the recovery of granulopoiesis after sub-acute radiation injury. *Acta Radiol. Oncol.* **23**, 361–366.

Gallicchio, V.S., Chen, M.G. and Watts, T.D. (1984b). Specificity of lithium (Li^+) to enhance the production of colony stimulating factor (GM-CSF) from mitogen stimulated lymphocytes *in vitro*. *Cell. Immunol.* **85**, 58–66.

Gallicchio, V.S., Chen, M.G. and Watts, T.D. (1985). Lithium stimulated recovery of granulopoiesis after sub-lethal irradiation is not mediated via increased levels of colony stimulating factor. *Int. J. Rad. Biol.* **47**, 581–590.

Gallicchio, V.S., Gamba-Vitalo, C., Watts, T.D. and Chen, M.G. (1986). *In vivo* and *in vitro* modulation of megakaryocytopoiesis and stromal colony formation by lithium. *J. Lab. Clin. Med.* **108**, 199–205.

Gallicchio, V.S., Messino, M.J., Hulette, B.C. and Hughes, N.K. (1990). Accelerated bone marrow engraftment in syngeneic bone marrow transplantation following lithium treatment to the donor. *Lithium*, in press.

Gamba-Vitalo, C., Gallicchio, V.S., Watts, T.D. and Chen, M.G. (1983). Lithium stimulated *in vitro* megakaryocytopoiesis. *Exp. Hematol.* **11**, 153–155.

Gelfand, E.W., Dosch, H.M., Hastings, B. and Shore, A. (1979). Lithium: a modulator of cyclic AMP-dependent events in lymphocytes? *Science* **203**, 365–367.

Goodwin, J.S., Kaszubowski, P.A. and Williams, R.C. Jr (1979). Cyclic adenosine monophosphate response to prostaglandin E2 in populations of human lymphocytes. *J. Exp. Med.* **150**, 1260–1264.

Greco, F.A. and Brereton, H.D. (1977). Effect of lithium carbonate on the neutropenia caused by chemotherapy: a preliminary clinical trial. *Oncology* **34**, 153–155.

Greenberg, P.L., Packard, B. and Steed, S.M. (1980). Effects of lithium chloride on human and murine marrow myeloid colony formation and colony stimulating activity. *Adv. Exp. Biol. Med.* **127**, 137–144.

Gualtieri, R.J., Berne, R.M., McGrath, H.E., Huster, W.J. and Quesenberry, P.J. (1986). Effect of adenosine nucleotides on granulopoiesis and lithium-induced granulocytosis in long-term bone marrow cultures. *Exp. Hematol.* **14**, 689–695.

Hammond, W.P. and Dale, D.C. (1980). Lithium therapy of canine cyclic hematopoiesis. *Blood* **55**, 26–28.

Harker, W.G., Rothstein, G., Clarkson, D., Athens, J.W. and MacFarlane, J.L. (1977). Enhancement of colony stimulating factor by lithium. *Blood* **49**, 263–267.

Harker, W.G., Rothstein, G., Clarkson, D.R., Larsen, W., Grossner, B.D. and Athens, J.W. (1978). Effect of lithium on neutrophil mass and production. *New Engl. J. Med.* **298**, 178–180.

Imamura, K. and Kufe, D. (1988). Colony stimulating factor-induced Na$^+$ influx into human monocyte involves activation of a pertussis toxin-sensitive GTP-binding protein. *J. Biol. Chem.* **263**, 14 093–14 095.

Imandt, L., Genders, T., Wessesls, H. and Haanen, C. (1977). The effect of lithium on platelet aggregation and platelet release reaction. *Thromb. Res.* **11**, 297–300.

Joyce, R.A. (1984). Sequential effects of lithium on haematopoiesis. *Br. J. Haematol.* **56**, 307–321.

Joyce, R.A. and Chervenick, P.A. (1980). Lithium effects on granulopoiesis in mice following cytotoxic chemotherapy. *Adv. Exp. Med. Biol.* **127**, 145–154.

Levitt, L.J. and Quesenberry, P.J. (1980). The effect of lithium on murine hematopoiesis in a liquid culture system. *New Engl. J. Med.* **302**, 713–719.

MacGregor, R.R. (1977). Granulocyte adherence changes induced by hemodialysis, endotoxin, epinephrine, and glucocorticoids. *Ann. Int. Med.* **86**, 35–40.

MacGregor, R.R. and Dyson, W.L. (1978). Effect of lithium on granulocyte adherence. *Clin. Res.* **26**, 352 (abstract).

MacGregor, R.R. and Dyson, W.L. (1980). Inhibition of granulocyte adherence by lithium: possible relationship to lithium-induced leukocytosis. *Adv. Exp. Med. Biol.* **127**, 347–355.

Marchesi, J.T. and Florey, H.W. (1960). Electron micrographic observations on the emigration of leukocytes. *Q. J. Exp. Physiol.* **45**, 343–347.

Mayfield, O. and Brown, R.G. (1966). The clinical laboratory and electroencephalographic effects of lithium. *J. Psychiat. Res.* **4**, 207–219.

Metcalf, D. (1989). The role of stem cell self-renewal and autocrine growth factor production in the biology of myeloid leukemias. *Cancer Res.* **49**, 2305–2311.

Mertz, W. (1981). The essential trace elements. *Science* **213**, 1332–1338.

Morley, D.C. and Galbraith, P.R. (1978). Effect of lithium on granulopoiesis in culture. *Can. Med. Assoc. J.* **118**, 288–290.

Moroz, B.B., Deshevoi, I.B., Tsybaner, O.A. and Adiushlan, A.I. (1988). Effect of lithium carbonate on hematopoietic stem cells (CFU-S) in acute radiation lesions. *Pat. Fiz. Eksp. Ter. (Rus.)* **1**, 41–43.

Murphy, D.L., Goodwin, F.K. and Burney, W.E. (1971). Leukocytosis during lithium treatment. *Am. J. Psychiat.* **127**, 135–137.

Perez-Cruet, J., Dancey, J.T. and Waite, J. (1977). Lithium effects on leukocytosis and lymphopenia. *In* "Lithium in Medical Practice" (ed. F.N. Johnson and S. Johnson) MTP Press, Lancaster.

Pointud, C., Clerc, C.A. and Manegand, G. (1976). Essai de traitment du syndrome de Felty par le lithium. *Sem. Hop. Paris* **52**, 1719–1723.

Quesenberry, P.J., Coppola, M.A. and Gualtieri, P.J. (1984). Lithium stimulation of murine hematopoiesis in liquid culture: an effect mediated by marrow stromal cells. *Blood* **63**, 121–127.

Quie, P.G. (1975). Pathology of bactericidal power of neutrophils. *Sem. Hematol.* **12**, 143–146.

Ramsey, R. and Hays, E.F. (1979). Factors promoting colony stimulating activity (CSA) production in macrophages and epithelial cells. *Exp. Hematol.* **7**, 245–254.

Reiser, G., Lautenschlager, E. and Hamprecht, B. (1982). Effects of colemid and lithium ions on processes of cultured cells derived from the nervous system. *In* "Microtubules and Microtubule Inhibitors" (ed. M. Borgers and M. deBradander) North-Holland, Amsterdam.

Roberts, D.E., Berman, S.M., Nakasato, S., Wyle, F.A., Wishnow, R.M. and Segal, G.P. (1988). Effect of lithium carbonate on zidovudine-associated neutropenia in the acquired immunodeficiency syndrome. *Am. J. Med.* **85**, 428–241.

Rossi, G.B., Migliaccio, A.R., Migliaccio, G., Lettieri, F., DiRosa, M., Peschle, C. and Mastroberardino, G. (1980). *In vitro* interactions of PGE and cyclic AMP with murine and human erythroid precursors. *Blood* **56**, 74–79.

Rossof, A.H. and Fehir, K.M. (1979). Lithium carbonate increases marrow granulocyte committed colony forming units and peripheral blood granulocytes in a canine model. *Exp. Hematol.* **7**, 255–258.

Rothstein, G., Clarkson, D.R., Larsen, W., Grossen, B.I. and Athens, J.W. (1978). Effect of lithium on neutrophil mass and production. *New Engl. J. Med.* **298**, 178–180.

Schou, M. (1957). Biology and pharmacology of the lithium ion. *Pharmacol. Rev.* **9**, 17–58.

Shenluman, L., Borkowsky, W., Holzman, R.S. and Shopsin, B. (1978). Enhancement of lymphocyte and macrophage function *in vitro* by lithium chloride. *Clin. Immunol. Immunopathol.* **10**, 187–191.

Shopsin, B., Friedman, R. and Gershon, S. (1971). Lithium and leukocytosis. *Clin. Pharmacol. Ther.* **12**, 923–928.

Siegel, J.N., Johnston, R.B., Lowe, R.S., Epstein, P.S. and Rossof, A.H. (1980). Effects of lithium on neutrophil metabolism *in vitro* and on neutrophil function during therapy. *Adv. Exp. Biol. Med.* **127**, 371–388.

Simchowitz, L. (1988). Lithium movements in resting and chemotactic factor activated human neutrophils. *Am. J. Physiol.* **254**, C526–C534.

Spitzer, G., Verma, D.S., Barlogi, B., Beran, N.A. and Dicke, K.A. (1979). Possible mechanisms of action for lithium on augmentation of *in vitro* spontaneous myeloid colony formation. *Cancer Res.* **39**, 3215–3220.

Stein, R.S., Hanson, G., Koethe, S. and Hansen, R. (1978). Lithium induced granulocytosis. *Ann. Intern. Med.* **88**, 809–810.

Stein, R.S., Howard, C.A., Brennan, M. and Czorniak, M. (1981). Lithium carbonate and granulocyte production: dose optimization. *Cancer* **48**, 2696–2701.

Steinherz, P.G., Rosen, G., Ghavimi, F., Wollner, Y. and Miller, D. (1980). The effect of lithium carbonate on leukemia after chemotherapy. *Ped. Pharm. Ther.* **96**, 923–927.

Stobo, J.D., Kennedy, M.S. and Goldyne, M.E. (1979). Prostaglandin E modulation of the mitogenic response of human T cells. Differential response of T-cell subpopulations. *J. Clin. Invest.* **64**, 1188–1203.

Sztein, M.B., Simon, G.L., Parenti, D.M., Scheib, R., Goldstein, A.L., Goodman, R., Digioia, R., Paxton, H., Skotuicki, A.B. and Schulof, R.S. (1987). *In vitro* effects of thymosin and lithium on lymphoproliferative responses of normal donors and HIV seropositive male homosexuals with AIDS-related complex. *Clin. Immunol. Immunopathol.* **44**, 51–62.

Tisman, G., Herbert, V. and Rosenblatt, S. (1973). Evidence that lithium induces human granulocyte proliferation: elevated serum vitamin B12 binding capacity *in vivo* and granulocyte colony formation *in vitro. Br. J. Haematol.* **24**, 767–771.

Vacek, A., Sikulova, J. and Bartonickova, A. (1982). Radiation resistance in mice increased following chronic application of lithium carbonate. *Acta Radiol. Oncol.* **21**, 325–330.

12 Mechanism of Action of Lithium: Comparison and Contrast with Carbamazepine

ROBERT M. POST and DE-MAW CHUANG
Biological Psychiatry Branch, National Institute of Mental Health, Bethesda, MD 20892, USA

Introduction

Lithium carbonate and carbamazepine share some similarities in their clinical profile of action in manic-depressive illness but diverge in others. The convergences and divergences provide a critical backdrop to the interpretation of their biochemical effects pertinent to their potential mechanisms of action in manic-depressive illness. In addition to its effects in manic-depressive illness, which take time to develop, carbamazepine works rapidly in the treatment of seizures and paroxysmal pain syndromes such as trigeminal neuralgia. Given the different time courses of onset of action of the therapeutic effects of carbamazepine in seizure, pain and affective disorders, it is likely that different biological effects of carbamazepine are related to its efficacy in these different syndromes. This complicates the problem of the comparison and contrast with lithium carbonate, as it becomes critical to separate the biological mechanisms of carbamazepine related to its anticonvulsant and antinociceptive effects (that should be different from those of lithium) from those related to its psychotropic effects (that may overlap with lithium).

In addition, the analysis is further complicated by the fact that even in the areas of similarities of lithium and carbamazepine in their effects on mood disorders, there is some preliminary evidence that subgroups of responsive patients may exist, suggesting that these two drugs may target differential mechanisms even for their psychotropic effects. As summarized in Table 1, it is widely acknowledged that lithium shows better acute antimanic than antidepressant properties. A similar profile appears to be emerging for carbamazepine, where 19 double-blind studies exist to document its acute antimanic efficacy in some 50–60% of patients (Post *et al.*, 1987; Post, 1990a), while only one extensive study exists for acute antidepressant properties where substantial improvement was observed only in approximately one-third of the

TABLE 1   Comparative spectrum of therapeutic effects.

	Lithium	Carbamazepine
Mania	+ +	+ +
Depression	+	+
Antidepressant potentiation	+ +	?
Prophylaxis of:		
Mania	+ +	+ +
Depression	+ +	+ +
Trigeminal neuralgia	− −	+ +
Complex partial and amygdala-kindled seizures	− −	+ +
Cocaine-kindled seizures[a]		+ +
Behavioural sensitization to cocaine		
Motor activating effects	(+)	− −
Sensitization to stereotypy		(±)

[a] Requires chronic administration.

subjects (Post et al., 1986d). Both drugs have been reported to exert prophylactic effects on the recurrence of both manic and depressive episodes. This is clearly well documented for lithium carbonate using a variety of controlled methodologies and designs, but has been less well studied for carbamazepine. Preliminary evidence exists from eight controlled studies (either using double-blind, randomization, or cross-over designs) which supplement a more extensive open literature supporting the proposition that carbamazepine, like lithium, can prevent the recurrences of manic and depressive episodes (Okuma et al., 1981; Post et al., 1983; Post, 1990b).

However, within this general convergence of similar acute and prophylactic profiles in bipolar illness, subtle differences are becoming evident. Lithium has been reported to be less effective in patients with initially more psychotic or aggressive manic presentations, those with dysphoric mania, rapid cycling, and without a family history of primary affective illness in first-degree relatives, while carbamazepine may treat some of these patients if not differentially target these subgroups (Post et al., 1987; Okuma, 1984). In the treatment of acute depression, we have observed better responses to carbamazepine in those who were initially more severely depressed at the

TABLE 2   Comparative biological effects of lithium and carbamazepine.

		Lithium	Carbamazepine	
I.	CSF: VIP	↓	ND	
	Somatostatin	—	↓	
	HVA-probenecid	?	↓	
	NE	ND	↓	
II.	Plasma			
	Tryptophan	(↑)	↑, —	
	Calcium	(↑)	↓	Opposite
	AVP (hypertonic saline)	↑	↓	Opposite
	T$_4$	↓	↓	Similar
	TSH, basal	↑	—	
	TSH (TRH)	↑	(↓)	Opposite
	Prolactin (tryptophan)	↑	↑	Similar
	GH (apomorphine)	—	↑	
III.	RBC: Choline	↑	ND	
IV.	Brain			
	A. Levels and turnover			
	ACh			
	Levels	—	↑ (striatum)	
	Turnover	↓ then ↑	ND	
	Choline	ND	↓	
	GABA			
	Levels	↑	±	
	Turnover	↓	↓	Similar
	Substance P			
	(striatum and SN)	↑	↑	Similar
	sensitivity	ND	↑	
	Neurotensin	↑	ND	
	Proenkephalin	↑	ND	
	Dynorphin A	↑	ND	
	B. Presynaptic activity			
	Biosynthesis			
	Tryptophan	↑ then ↓	(↓)	
	hydroxylase			
	5HT synthesis	↑	↑ then, No ▲	
	DA synthesis	ND	↓	
	(frontal cortex)			
	Release			
	5HT (chronic),	↑	↑	Similar
	hippocampus			
	DA	—	(↑)	
	NE	↓,↑	↓	
	Re-uptake			
	NE	↑	↓	Opposite
	Tryptophan	↑	ND	

TABLE 2 (continued).

	Lithium	Carbamazepine	
C. Receptor activity			
DA autoreceptor	↓[1]	↓[2]	
supersensitivity			
[1] Unit firing			
[2] Response to			
apomorphine			
Response to agonists			
$5HT_{1A}$	↑,↓	—	
$5HT_2$	↓	↑	Opposite
α_2	↓	ND	
Haloperidol-induced	↓	—	
supersensitivity			
(behavioural)			
$GABA_B$ receptors in	↑	↑	Similar
hippocampus			
(chronic)			
$GABA_A$ (chronic)	↓	ND	
Adenosine receptor	—	↑	
(chronic)			
'Peripheral-type'	ND	(↑)	
benzodiazepine			
receptors			
Vasopressin	↓	↑	Opposite
β-adrenergic (chronic)	—	—	Similar
$5HT_2$ (chronic)	—	—	Similar
D. Ion Channels			
Compete with:			
Na (counter	↓	ND	
transport)			
Batrachotoxin	ND	↓	
binding			
E. Second messenger			
Cyclic AMP NE	↓	↓	Similar
(acute) DA	↓	↓	Similar
Adenosine	ND	↓	
Ouabain	ND	↓	
(chronic) NE	↓	ND	
Cyclic GMP (acute)	↓	↓	Similar
PI Turnover[a] (acute)			
Diacylglycerol	↑	ND	
IP_3 (chronic)	↓	—,↓	
IP_3	↓	—,↓	

[a] See Table 4 for details.
ND, not determined; ▲, change.

outset (Post *et al.*, 1986c), while this was not found to be a predictor for acutely depressed patients responsive to lithium.

Thus, it is suggested by several authors (Okuma *et al.*, 1973; Okuma, 1984; Post *et al.*, 1987) that it is possible that lithium and carbamazepine, while sharing a common spectrum of clinical efficacy in bipolar disorder, may target different subpopulations of patients. Therefore, one might be induced to look for subtle differences in the mechanism of action of these two agents within their broader context of common effects capable of stabilizing both phases of the mood disorder.

The behavioural pharmacology of these two agents also shows interesting differences. While lithium is ineffective as an anticonvulsant and does not block amygdala or hippocampal seizures acutely or their development, carbamazepine is a highly effective anticonvulsant for a variety of seizure mechanisms and is highly effective in blocking completed amygdala-kindled seizures (although it will not block the development of amygdala-kindling in the rat) (Weiss and Post, 1987). Conversely, carbamazepine is ineffective in blocking the development or expression of cocaine-induced behavioural sensitization (Weiss *et al.*, 1990) while preliminary evidence exists that lithium can block cocaine-induced behavioural sensitization (Post *et al.*, 1984). An extensive series of studies summarized by Bunney and Garland-Bunney (1987) documents that lithium is capable of inhibiting behavioural manifestation of dopamine supersensitivity following chronic haloperidol administration, while preliminary evidence suggests that carbamazepine does not share this effect (Elphick, 1989).

At the current time, we can only surmise which of these differences in behavioural pharmacology is reflective of biochemical differences in actions that are relevant to effects in manic-depressive illness (as opposed to being related to the other mechanisms of action of carbamazepine in seizure and pain disorders that may not be relevant to effects in the mood disorders) (Table 2). Nonetheless, categorization of the similarities and differences in lithium and carbamazepine's actions ultimately may assist in the definition of their convergent and divergent psychotropic properties in patients with manic-depressive illness.

This chapter focuses on strategies and types of approaches pertinent to the study of the mechanisms of lithium and carbamazepine. Thus, examples in several areas of interest are presented in an illustrative fashion, not attempting to be inclusive or comprehensive. Interesting candidate mechanisms for lithium's actions are discussed in the next section, while carbamazepine and its similarities to and differences from lithium are discussed in pp. 217ff. Examination of the temporal characteristics of drug effects, particularly those emerging after chronic administration, may be important in discerning which effects are likely to be relevant to effects in manic-depressive illness.

The Psychobiology of Lithium

Any theory of the mechanism of action of lithium must account for its acute and prophylactic efficacy in both mania and depression. These effects are of slow onset, with the exception of the finding of lithium-induced potentiation of a variety of antidepressant modalities, which can occur rapidly over a period of 24–48 h (de Montigny *et al.*, 1981; Kramlinger and Post, 1989a) although maximum effects may require three weeks or longer (Heninger *et al.*, 1983; Thase *et al.*, 1989) (Fig. 1). There is considerable evidence that

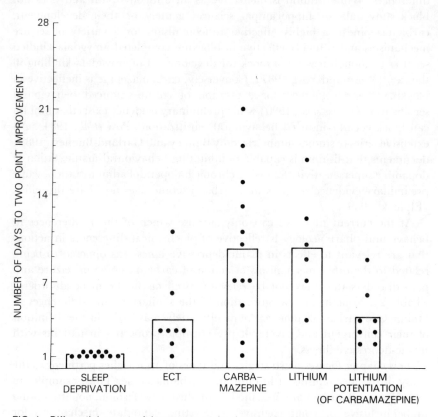

FIG. 1 Differential onset of improvement in depression after sleep deprivation, ECT, carbamazepine and lithium. Improvement induced by sleep deprivation occurs literally overnight but tends to be transient and the patient often relapses following one night's recovery sleep. Onset of clinically important improvement (2 point change on Global Bunney–Hamburg depression scale) occurs more slowly in responders to lithium, carbamazepine, and traditional heterocyclic and MAOI antidepressants. Improvement during this potentiation can occur rapidly, in a matter of 24–28 h in some patients.

in long-term prophylaxis, patients demonstrating a pattern of mania followed by depression and then a well interval (M-D-I) respond considerably better than those showing the opposite pattern; i.e. a depression followed by the switch into mania and then a well interval (D-M-I) (Grof et al., 1987; Haag et al., 1987; Maj et al., 1989; Kukopulos et al., 1980). Lithium also prevents a variety of rebound phenomena both in preclinical models as well as in the clinic. For example, Baxter et al.(1986) have reported that patients who typically only demonstrate acute response to one night's sleep deprivation treatment for their depression, when maintained on lithium co-administration, may have a more sustained antidepressant response, presumably by blocking the sleep-induced rebound exacerbation of the effects of sleep deprivation. Lithium also appears to attenuate a variety of circadian rhythms which may be pertinent to its effects on the inherent mechanism of rhythmicity and cyclicity in manic-depressive illness (Kripke and Wyborney, 1980; McEachron et al., 1982).

A critical focus is the definition of how a simple ion such as lithium alters biological machinery on a cellular and molecular basis to produce its acute and long-term effects in manic-depressive illness. The effects of lithium appear to be selective in some cases and opposite to the effects of other ions such as rubidium. For example, in opiate-mediated excitement, lithium is specifically capable of decreasing motor activity, while rubidium enhances this activity (Carroll and Sharp, 1971). Is this ion-selective effect related to its ability to dampen the motor excitement of mania?

A basic conundrum of most biochemical effects of the mechanism of action of lithium has been how any single biochemical mechanism of this ion could be translated into bimodal effects on both manic and depressive phases of the illness. One approach to the problem is the conceptualization of manic and depressive swings as pathological exacerbations of normal adaptive mechanisms. One might then consider lithium's ability to dampen overactive processes (in both excitatory and inhibitory neural systems) as potentially pertinent to its bimodal effects. Yet, in the consideration of this process, it is apparent that lithium may be more effective in mania and in patients showing the pattern of manic-depressive illness (M-D-I) rather than in depression and the D-M-I pattern, as noted above.

Prien and colleagues (1972) present data that are representative of a variety of comparative studies of neuroleptics, indicating that the onset of lithium's acute effects in mania are slower than those of chlorpromazine, which begins to demonstrate a significant efficacy even in the first week of treatment. These data are also convergent with our observations that pimozide shows a more rapid onset of acute antimanic effects than lithium carbonate in our series of refractory affectively ill patients (Post et al., 1980). Parenthetically, most studies comparing the antimanic efficacy of carbamazepine with that of neuroleptics show an approximately equal time-course, a finding that may be pertinent ultimately to establish a differential time-course of lithium and carbamazepine.

While it may be argued that slow onset in action of lithium is related to pharmacokinetic parameters, attempts to speed the onset of action of lithium using 'loading' or digitalizing doses have not substantially hastened the onset of clinical response. Thus, it appears that some mechanism of lithium's pharmacodynamics relates to this lag in onset and efficacy, similar to that observed with even such aggressive treatment as electroconvulsive therapy (Post *et al.*, 1986a and Fig. 1).

Thus, one of the focuses of consideration of lithium's psychotropic effects must be this lag in onset of action, as one might establish the general principle that effects of lithium, which require chronic administration in order to be observed, may be closer to psychotropic efficacy in manic-depressive illness than effects that are apparent acutely. For example, Jope (1979) reports differences in effects of lithium on acetylcholine synthesis with decreases acutely but increases chronically. Moreover, lithium blocks the compensatory increase in acetylcholine receptors following denervation (Pestronk and Drachman, 1980, 1987) and alters effects of acetylcholine on phosphoinositide (PI) turnover (see below). Similarly, in the clinical realm, lithium has been reported to increase red blood cell choline to a marked extent (Jope *et al.*, 1980; Shea *et al.*, 1981; Gutterman *et al.*, 1985). In one study, better lithium prophylaxis was achieved in those patients who showed a higher red blood cell : plasma choline ratio (Haag *et al.*, 1984). Thus, chronic changes in cholinergic mechanisms remain candidates for lithium's psychotropic effects. A variety of other biological effects, several of which are discussed below, obviously also remain important candidates for this action of lithium.

Serotonergic mechanisms have been postulated for the relatively acute effects of lithium on potentiation of inadequate antidepressant modalities. As noted above, a rich literature exists which documents lithium potentiation of the effects of a variety of antidepressant modalities including tricyclics, heterocyclics (HCAs), monoamine oxidase inhibitors (MAOIs), and other agents in some 167 to 243 patients reported in the literature (68% response rate) (Kramlinger and Post, 1989a). The onset of action of lithium potentiation may be rapid in some studies. In our own data, illustrated in Fig. 1, not only is the onset of lithium potentiation more rapid than the onset of antidepressant effects of lithium alone, but lithium potentiation of carbamazepine has a faster onset in depression than in mania (Kramlinger and Post, 1989a,b) (Fig. 2). These data suggest that, in contrast to other effects of lithium, lithium potentiation of antidepressant modalities may occur relatively rapidly. De Montigny and Aghajanian (1978) and Blier and de Montigny (1985a,b) have postulated that tricyclics may sensitize

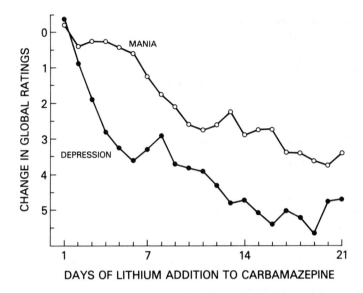

FIG. 2 Lithium potentiation of carbamazepine: faster onset in depression than in mania. Although antimanic response to carbamazepine occurs more rapidly than antidepressant response, the response to lithium potentiation shows the opposite pattern.

post-synaptic serotonin receptors, particularly in the CA_3 region of the hippocampus, that could provide a substrate for acute lithium potentiation by increasing CA_3 response by a pre-synaptic mechanism with a relatively rapid onset.

While one might continue to consider a variety of individual neurotransmitter mechanisms for lithium's action in mania or depression, it becomes difficult to conceptualize how a single effect of lithium could impact on mood disorders of essentially opposite poles. As mentioned above, one might circumvent this difficulty by postulating a common effect of lithium in overactive processes that mediates the opposite moods of mania and depression. For example, an increased dopaminergic or noradrenergic tone (which is generally considered inhibitory in electrophysiological terms) has been postulated to be associated with mania. Conversely, an increase in acetylcholinergic tone (which has generally been associated with excitatory effects electrophysiologically) has been postulated as crucial in depression (Janowsky *et al.*, 1974). One might then focus the search on potential mechanisms of action of lithium that are able to dampen overactive effects in either of these systems.

In this fashion, focus of attention could shift from considering pre- and post-synaptic effects of lithium, which are likely to be linked to unimodal actions of lithium, to subreceptor and second-messenger effects of lithium, which have the potential for acting on several neurotransmitter systems and, thus, could mediate the bimodal actions of lithium. Just as lithium has been postulated to exert a panoply of effects on pre- and post-synaptic neurotransmitter systems, it has recently been postulated to be involved in a variety of second-messenger systems as well.

Adenylate and guanylate cyclases

Lithium has been reported to inhibit brain adenylate cyclase with little or no effect on the basal activity. At therapeutically relevant concentration (e.g. 2 mM), lithium inhibits norepinephrine (NE)- and isoproterenol-induced cyclic AMP accumulation in rat brain preparations (Ebstein *et al.*, 1980) and intact pineal glands (Zatz, 1979). This effect of lithium is likely due to competition with magnesium, which is required for the binding of GTP to the stimulatory G protein. *In vitro*, lithium is also reported to inhibit forskolin- and calcium calmodulin-stimulated activity by displacing magnesium from its allosteric site on this protein (Geisler *et al.*, 1987).

Three-week treatment of rats with lithium in the therapeutic range markedly inhibits NE-sensitive adenylate cyclase activity in cerebral cortex (Ebstein *et al.*, 1980). This effect may be distal to the receptor recognition site (Maggi and Enna, 1980) and involves uncoupling of receptors to the G protein. Indeed, it has been reported that repeated lithium injections result in an inhibition of isoproterenol-induced binding of GTP to the G protein in rat cortex (Avissar *et al.*, 1988; Avissar and Schreiber, 1989). However, a more recent report shows that three-day treatment with lithium elicits no change in isoproterenol-sensitive adenylate cyclase activity in rat cortical slices, despite a significant decrease in α_1-adrenergic receptor-mediated phosphoinositide (PI) turnover (Godfrey *et al.*, 1989). These conflicting results could be due to differences in the regimens of lithium administration, particularly the time duration of the drug treatment. However, since NE-sensitive cyclic AMP generation has been reported to be influenced by α_1-receptor stimulation (Johnson and Minneman, 1986), it is also possible that the decrease of NE-sensitive adenylate cyclase following chronic lithium treatment is in part secondary to the loss of NE-induced (α_1-mediated) PI turnover (see below). Lithium *in vivo* and *ex vivo* after chronic treatment also inhibits forskolin-stimulated adenylate cyclase activity (Mork and Geisler, 1989). The *ex vivo* effect of lithium is not reversed by magnesium and is additive to the *in vitro* effect, suggesting that distinct mechanisms are involved.

Regardless of the mechanism involved, inhibition of stimulated adenylate cyclase activity by lithium *in vivo* could provide a hypothesis for the therapeutic and/or adverse effect of this ion. In fact, Belmaker and associates

(Ebstein *et al.*, 1976) have presented clinical data consistent with preclinical observations indicating that lithium blunts the plasma cyclic AMP response to epinephrine in psychiatric patients. Moreover, using another compound, dimethylchortetracycline (DMC), which also inhibits adenylate cyclase activity, they observed in preliminary studies that DMC tended to exert significantly greater decrements in Brief Psychiatric Rating Scale (BPRS) scores in patients with excited psychoses compared with placebo (Belmaker and Roitman, 1988). In this regard it is also of interest that carbamazepine is reported acutely to inhibit adenylate cyclase activity stimulated by NE, dopamine, adenosine, veratridine, and ouabain (Lewin and Bleck, 1977; Palmer, 1979; Palmer *et al.*, 1979, 1981; Ferrendelli and Kinscherf, 1979). Further preclinical studies are necessary to ascertain whether the effects of carbamazepine and lithium on adenylate cyclase activity are achieved by a common or differential locus of action.

At this time there is no compelling evidence that receptors negatively coupled to adenylate cyclase are affected by lithium. Thus, in CA_3 pyramidal neurons of guinea pig hippocampus, lithium applied extracellularly (1 mM) or intracellularly blocked M_1- but not M_2-muscarinic receptor-mediated electric current (Muller *et al.*, 1989), suggesting that M_2-receptor-mediated adenylate cyclase inhibition is unaffected. Prior chronic lithium treatment alters neither dopamine D_1 receptor-mediated cyclic AMP accumulation nor the inhibitory effect of carbachol on the D_1-receptor-coupled cyclase activity (Whitworth and Kendall, 1988). Chronic administration of lithium has been shown to abolish the upregulation of rat brain muscarinic receptor sites induced by atropine (Levy *et al.*, 1982) and the guanylyl nucleotide's effect on the binding of oxotremorine to muscarinic receptors (Avissar *et al.*, 1988). However, the subtype of muscarinic receptors and the second messenger involved remain to be elucidated.

Lithium has been reported to reduce cyclic GMP levels in the pineal gland stimulated by NE and depolarizing concentration of KCl (Zatz, 1979). In neuroblastoma N1E-115 cells, lithium at relatively high doses inhibits cyclic GMP synthesis mediated by muscarinic (low affinity), neurotensin, angiotensin II and bradykinin receptors (Kanba and Richelson, 1984; Kanba *et al.*, 1986). Since all these agents are also agonists of calcium-mobilizing receptors coupled to phospholipase C and guanylate cyclase is a cytosolic enzyme activated by intracellular calcium, it cannot be excluded that lithium's effect on cyclic GMP synthesis may occur secondary to the dampening of the receptor-mediated PI hydrolysis and calcium mobilization (see below). However, lithium, but not carbamazepine, has been reported to increase dark adaptation threshold, consistent with a subsensitivity to light possibly mediated by lithium's effects on adenylate cyclase activity (Carney *et al.*, 1988; Kaschka *et al.*, 1988; Seggie *et al.*, 1989).

Phosphoinoside hydrolysis by phospholipase C

The possibility that lithium can interfere with inositol lipid metabolism was first suggested by Allison and Stewart (1971), who reported that the level of brain inositol is reduced in lithium-treated rats. This was later found to be associated with a rise of inositol monophosphate (Allison *et al.*, 1976) due to non-competitive inhibition by lithium of inositol monophosphatase (Hallcher and Sherman, 1980) which dephosphorylates various forms of inositol monophosphates to free *myo*-inositol. Levels of inositol monophosphate, *myo*-inositol and lithium in rat cerebral cortex are significantly correlated over the course of chronic treatment with lithium (Sherman *et al.*, 1985). Thus, lithium's effect can occur in the clinically efficacious serum concentration range (0.5–1.0 mM). It has been proposed that the therapeutic effect of lithium is due to its ability to interfere with PI metabolism (Berridge *et al.*, 1982). The idea is that manic-depressive illness is the consequence of uncontrollable hyperactivity of receptor-mediated PI turnover in neurons and that lithium, by decreasing the level of free *myo*-inositol, would eventually reduce the

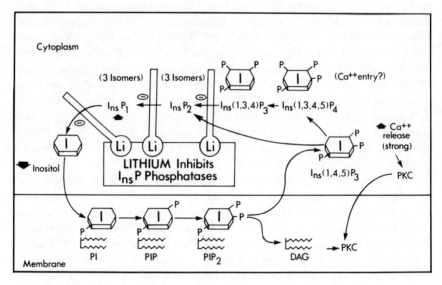

FIG. 3 Schematic illustration of phosphoinositide cycle. PIP_2 is hydrolysed by phospholipase C to form dural second messengers, DAG and Ins (1,4,5) P_3. DAG is an activator of PKC translocated from the cytosol to plasma membranes following stimulation of phospholipase C-coupled cell-surface receptors. Ins (1,4,5) P_3 is either dephosphorylated to form Ins P_2, Ins P_1 and, ultimately, free inositol or phosphorylated to form Ins (1,3,4,5) P_4 which is then dephosphorylated by distinct pathways. Lithium blocks the dephosphorylation of Ins (1,3,4) P_3, certain forms of IP_2 and all forms of IP_1 (for details, see Fig. 4). Therefore, the presence of lithium reduces the level of inositol and may attenuate the synthesis of phosphoinositides. Modified from Worley *et al.* (1987).

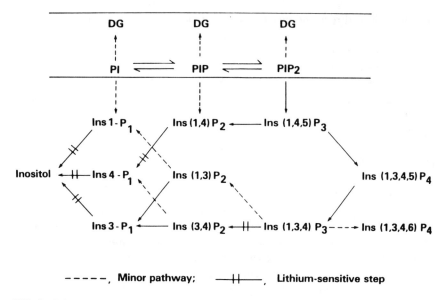

—————, **Minor pathway;** ——||——, **Lithium-sensitive step**

FIG. 4 Schematic illustration of the metabolism of inositol phosphates and the pathways sensitive to lithium. Ins (1,4,5) P_3 generated by PIP_2 hydrolysis is metabolized by two major pathways: a stepwise dephosphorylation to form first Ins (1,4) P_2 and then Ins 4-P_1 and a phosphorylation by the 3-kinase to form Ins (1,3,4,5) P_3 and various forms of bis- and monophosphates. Inositol produced by the action of inositol monophosphatase is utilized for the synthesis of PI, the precursor of PIP_2. Lithium at therapeutically effective concentrations inhibits the dephosphorylation of Ins (1,3,4) P_3, Ins (1,4) P_2 and all three forms of inositol monophosphates due to lithium's effects on their respective phosphatases. Modified from Chuang (1989).

biosynthesis of phosphatidylinositol 4,5 biphosphate (PIP_2), thereby limiting the generation of dural second messengers, Ins (1,4,5) P_3 and diacylglycerol by phospholipase C (Figs 3 and 4). Ins (1,4,5) P_3 mobilizes calcium from non-mitochondrial stores, while diacylglycerol is an activator of protein kinase C. These two messengers often act synergistically to trigger a spectrum of synaptic events (for review, see Chuang. 1989).

Myo-inositol does not readily pass the blood–brain barrier; in fact, it has been shown that only about 2% of *myo*-inositol in the CSF originates from the plasma (Barkai, 1981) with the rest being *de novo* synthesized from glucose. Thus, lithium administration would selectively reduce *myo*-inositol level in the brain, notably in neurons that are over-active in receptor-stimulated PI metabolism, as they have much greater demand of *myo*-inositol for PI resynthesis. This would explain the lack of significant effect of lithium in people not suffering from manic-depressive illness whose PI metabolism is normal and whose lithium sensitivity is, therefore, minimal. However, at this time there is no direct evidence that the brain level of PIP_2 is reduced following chronic lithium treatment (Sherman *et al.*, 1985; Honchar *et al.*, 1989). This finding does not necessarily negate the inositol depletion hypothesis because there are compartmentations of the PI pool with less than 20% being stimulus-sensitive. Moreover, as suggested by Berridge (1989), there may be only a small population of neurons that are over-stimulated and involved in the etiology of manic-depressive illness.

Lithium has additional sites of action in the PI cycle. Beside affecting inositol monophosphatase, this ion also effectively blocks inositol poly-phosphatase that dephosphorylates Ins $(1,3,4)$ P_3 and Ins $(1,4)$ P_2. Detailed effects of lithium on inositol phosphosphate metabolism are shown in the scheme in Fig. 4. Lithium has been reported to selectively inhibit muscarinic receptor-stimulated Ins $(1,3,4,5)$ P_4 accumulation in cerebral cortical slices by unknown mechanisms (Batty and Nahorski, 1987; Whitworth and Kendall, 1988). In therapeutic concentration ranges, lithium dampens the relaxation of smooth muscle mediated by carbachol or histamine-induced PI turnover (Menkes *et al.*, 1986) and diminishes the blockade of PI-mediated cholinergic responses in hippocampal slices (Worley *et al.*, 1988). The latter effect may result from reduced production of Ins $(1,4,5)$ P_3, as shown recently in cerebral cortical slices following carbachol stimulation (Kennedy *et al.*, 1989). The relationship between these *in vitro* effects of lithium and its clinical efficacy remains to be explored.

Increasing evidence from preclinical studies indicates that lithium *in vivo* attenuates stimulus-coupled but not basal PI metabolism in the brain. Following acute or chronic lithium treatment, carbachol-, histamine (H_1)-, serotonin $(5\text{-}HT_2)$- and depolarization-induced PI breakdowns in rat cerebral cortex are significantly reduced, while NE (α_1)-induced inositol phosphate production is decreased only after chronic treatment (Kendall and Nahorski, 1987; Casebolt and Jope, 1987). Furthermore, subchronic lithium administration results in inhibition of fluoride-sensitive inositol phosphate release (Godfrey *et al.*, 1989). Several lines of evidence support the notion that these *ex vivo* effects are not the consequence of depletion of PI pool or loss of cellular receptor sites, but may be due to interference of the receptor-phospholipase C coupling to the putative Gp protein (Kendall and Nahorski, 1987; Godfrey *et al.*, 1989). It is also unclear whether these *in vivo* effects are secondary to persistent blockade by this ion of inositol phosphatase activity or other metabolic events such as protein kinase C activation.

Muscarinic receptor-mediated PI turnover seems to be a major site of action exerted by lithium *in vivo*. It has been demonstrated that the decrease of brain *myo*-inositol content by a single injection of lithium is blocked by the cholinergic antagonists atropine and scopolamine (Allison and Blisner, 1976). In this context, it is interesting to note that exposure of cultured cerebellar granule cells to lithium (> 2 mM) for more than three days induces a marked attenuation of M_3-muscarinic receptor-mediated PI turnover, while basal and other receptor-mediated PI responses are much less affected (X.M. Gao and D.-M. Chuang, unpubl. obsv.). This effect of lithium is associated with a significant loss of surface muscarinic receptor sites assessed by binding of labelled *n*-methylscopolamine to intact cells. Preliminary observation of F. Fukamauchi in our Branch indicates that treatment of cultured cerebellar granule cells with 2 mM lithium for more than three days results in downregulation of the level of M_3—but not M_2—muscarinic receptor mRNA, suggesting that lithium may modulate the transcription or alter the stability of certain receptor mRNA species. The finding of lithium's ability to antagonize the effects of physostigmine on methylphenidate-induced gnawing activity are also of considerable interest. Janowsky *et al.* (1973) reported that physostigmine itself inhibited methylphenidate-induced stereotypic gnawing activity. Lithium alone had no effect on this behaviour. However, when used in combination, lithium was able partially to inhibit the effects of physostigmine on this behaviour, suggesting that it was capable of countering some of the effects of this cholinomimetic. This inhibition could lead to a decrement in cholinergic tone which, on a clinical level, could be translated into less depression. Yet, the overall effect of lithium on cholinergic tone is complex, as lithium has been reported to *potentiate* the effects of carbachol on the induction of the proto-oncogene c-fos in cultured rat astrocytes (Arenander *et al.*, 1989).

As illustrated in Figs 5(a), (b), lithium could be modulating intercellular processes at a variety of second-messenger loci. Thus, in addition to effects on PI turnover and effects on adenylate cyclase (reviewed above), the effects of lithium directly on G proteins (Avissar *et al.*, 1988), calcium flux (Aldenhoff and Lux, 1985; Meltzer, 1986; Meltzer *et al.*, 1988) and even guanylate cyclase (Kanba *et al.*, 1986), have been postulated.

In addition, lithium competes with magnesium ions which are essential for binding of GTP to G proteins (Worley *et al.*, 1986; Avissar *et al.*, 1988), block voltage-gated ion channels (Mayer and Westbrook, 1987), affect ATPase function (Naylor *et al.*, 1974), and tonically affect neural excitability. Therefore, competition of lithium with normal cellular functions occupied by calcium and magnesium could remarkably alter intercellular second-messenger systems.

Avissar and co-workers (1988) have postulated an effect of lithium on G proteins that could also explain the ability of lithium to dampen overactivity of excitatory and inhibitory systems. They have reported that chronic lithium treatment attenuates the inhibitory effects of guanylyl nucleotides on agonist binding to G protein-coupled receptors in the brain. Although this phenomenon is not replicated by some investigators (R.H. Lenox, pers.

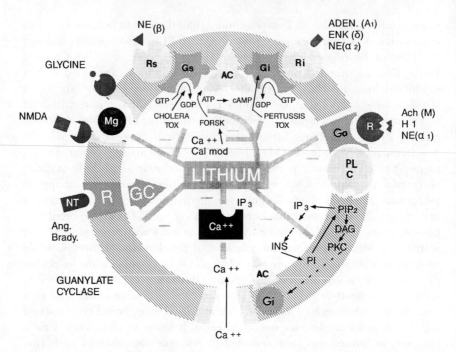

FIG. 5 (a) Schematic illustration of intracellular sites of actions of lithium on the signal transduction systems. Lithium at or near therapeutically efficacious concentration ranges inhibits a variety of receptor second-messenger systems. Thus, lithium reduces the production of cyclic-AMP by adenylate cyclase (AC), the generation of cyclic-GMP by guanylate cyclase (CG), the dephosphorylation of inositol phosphates to form free inositol for the resynthesis of phosphoinositides (PI), and the calcium influx mediated by magnesium-sensitive NMDA receptors. In addition, lithium inhibits the activity of Ca^{2+}/calmodulin for the stimulation of adenylate cyclase.

commun., March 1990), there is a general consensus that lithium affects the coupling of receptors to their effectors via G proteins. A variety of G proteins have been recently discovered that appear to couple receptors to second-messenger and to ion channel systems. To the extent that lithium can interfere with G protein activity, it could uncouple overactive pathways in a variety of neurotransmitter systems. Thus, the mechanism of lithium's bimodal action could arise from a common action on opposing systems such as G proteins, with lithium altering effects mediated by inhibitory or stimulatory G proteins. Its effects on adenylate versus phospholipase C could likewise provide for common actions on opposing systems as well as its effects on other effector systems. However, it should be stressed that the G protein hypothesis, although attractive, cannot easily explain why lithium has little effect on metabolic

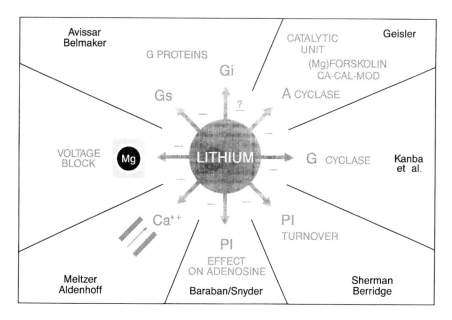

FIG. 5 (b) Schematic illustration of lithium's effects on various receptor-mediated signal transduction systems. Lithium has been reported to reduce the function of guanylate cyclase, phospholipase C in catalysing PI hydrolysis, adenosine receptors dependent on PI metabolism, the influx of calcium and Mg^{2+}-sensitive NMDA receptors. In addition, lithium inhibits the activity of adenylate cyclase stimulated by G_s and calmodulin; however, it is less certain whether G_i-dependent cyclase inhibition is affected by this ion.

activities in areas other than the brain. Moreover, it is difficult to rationalize why lithium has virtually no effect on normal individuals or on patients not suffering from manic-depressive illness.

The postulated ability of lithium to dampen overactivated systems in a variety of neural pathways could similarly be translated to the mechanism involved in the timing of circadian and ultradian clocks. That is, lithium may be able to dampen overactive oscillations and slow rhythmic functions. This has been demonstrated in a variety of systems such as the ability of lithium to slow and split the wheel-running circadian rhythms in the rat (Kripke and Wyborney, 1980) and to change hamster rhythms to light pulses (Han, 1984). Lithium appears capable of delaying biochemical circadian rhythms in rats (McEachron *et al.*, 1982); of plasma prolactin, parathyroid hormone, corticosterone and aldosterone; of serum calcium and magnesium; and of cerebellar content of calcium and magnesium. Interestingly, chronic lithium treatment was shown to shift the melatonin peak to an earlier time in the 24-h cycle (Seggie *et al.*, 1983). Lithium has also been reported to affect markedly a variety of circadian rhythms in neurotransmitter receptor number (Kafka *et al.*, 1982). It abolished rhythms of α- and β-adrenergic receptors, as well as benzodiazepine receptors in cortex. It delayed rhythms of acetylcholine, opiate benzodiazepine receptors in the striatum and it produced bimodal effects on rhythms of dopamine in α-MSH receptors.

Overview

In this brief overview on the mechanisms of action of lithium potentially pertinent to manic-depressive illness, we have emphasized how differences in time-course of clinical efficacy may imply different mechanisms of action. Antidepressant potentiation may occur more rapidly than antimanic or antidepressant effects, which in turn may be present more rapidly than prophylactic effects. We have suggested that it may be parsimonious to conceptualize both manic and depressive processes as overactivated neuro-transmitter affective systems which, in turn, might provide a bimodal basis for the mechanism of action of lithium. Effects on G proteins, adenylate cyclase, phospholipase C and other effectors could each in turn provide the basis for this bimodal effectiveness. Since lithium in this case would be postulated to dampen overexcitable systems and even block compensatory adaptations, it may, in a similar fashion, slow rhythmic oscillations and provide a link between lithium's effects on acute episodes and on the cyclicity of manic-depressive illness.

These series of mechanistic vignettes are obviously highly preliminary and superficially presented, but the interested reader is referred to the more detailed reviews of a variety of workers in the field who have dedicated their careers to the further understanding of the mechanism of action of lithium. These might include all of those listed in Fig. 5(b), as well as many others not dealt with in this brief overview. Nonetheless, it is hoped that this synopsis will now provide a template for considering mechanisms of action of carbamazepine, which can be compared and contrasted with those of lithium.

While the hope that the discovery of the mechanism of action of lithium would provide the insight necessary to uncover the basic defects in manic-depressive illness has not yet been realized despite many decades of active research, we hope that it is also not too naive to hope that some progress can be gained in the field by comparing and contrasting the mechanisms of action of lithium and carbamazepine. The process is obviously made difficult by the panoply of effects of both of these agents on a variety of neurotransmitter, neuropeptide, and second-messenger systems. Nonetheless, additional leverage may be gleaned from the comparative analysis of lithium and carbamazepine.

Mechanisms of Action of Carbamazepine in Affective Illness

Consideration of the time-course of onset of carbamazepine's clinical effects also provides a possible key to relating specific mechanisms of the drug to clinical efficacy in different neuropsychiatric syndromes. As illustrated schematically in Fig. 6, there is considerable evidence that the effects of carbamazepine on seizure and pain disorders can be manifest acutely. In some seizure models, the anticonvulsant effects of carbamazepine can be elicited by a single injection. Antinociceptive effects of carbamazepine have been

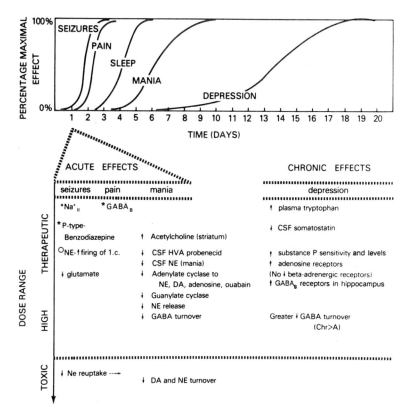

FIG. 6 Time-course of clinical and biochemical effects of carbamazepine. Seizure and paroxysmal pain syndromes respond rapidly to carbamazepine, while maximal effects in mania and depression are delayed two to three weeks or more, respectively. The acute effects observed in animals and man are probably related to anticonvulsant and antinociceptive efficacy, while chronic effects are more likely to be related to antidepressant mechanisms. Candidate mechanisms for each clinical effect are listed according to their occurrence at clinically relevant concentrations.

reported to occur in 24–48 h; i.e. essentially the time it takes to reach therapeutic blood levels. Thus, the effects on seizures and pain are likely to involve mechanisms of carbamazepine that are present acutely. Similarly, we have observed a rapid onset of improved sleep in our patients with manic and depressive illness, with improvement in sleep occurring within the first week of administration and independently of subsequent antidepressant effects. In contrast, acute antimanic effects, while rapid in onset and comparable to those of neuroleptics, appear to require two to three weeks before they become maximal (Post *et al.*, 1987). In addition, the effects of carbamazepine in depression often require four to six weeks in order to become maximal (Post *et al.*, 1986d). Thus, the psychotropic effects of carbamazepine (those related to its antimanic and antidepressant effects) are likely to involve mechanisms that take time to develop (Post, 1987, 1988).

Anticonvulsant mechanisms

As illustrated in Fig. 6, acute effects of carbamazepine on four different neurotransmitter systems have been most closely linked to its anticonvulsant efficacy. In addition, each of these effects has been reported to occur within the therapeutic dose range and achieved blood levels that are pertinent to the clinical situation. Considerable evidence links the effects of carbamazepine on amygdala-kindled seizures to 'peripheral-type' benzodiazepine mechanisms, α-2 noradrenergic mechanisms, voltage-sensitive sodium channels (as marked by the ligand batrachotoxinin α-benzoate) and lesser evidence that carbamazepine could interact indirectly with glutaminergic mechanisms. It is of interest, in light of the foregoing discussion, that lithium does not appear to interact directly with any of these putative mechanisms for carbamazepine's acute anticonvulsant effects, which is consistent with lithium's lack of anticonvulsant efficacy.

'Peripheral-type' benzodiazepine mechanism

In contrast with diazepam, which acts relatively selectively, and clonazepam, which acts exclusively at central-type benzodiazepine receptors, carbamazepine is ineffective in displacing binding at this receptor and is more potent in displacing binding at the so-called 'peripheral-type' benzodiazepine receptor in brain (Marangos *et al.*, 1983). This receptor is marked by the binding of [^{3}H]Ro5-4864 or PK-11195 in contrast with the central-type receptor which binds [^{3}H]diazepam, clonazepam or β-carboline. The biochemical potency at the 'peripheral-type' receptor is consistent with the biochemical pharmacology of carbamazepine's anticonvulsant actions mediated by this so-called 'peripheral-type' receptor. It is important to note that the 'peripheral-type' benzodiazepine receptor, while discovered initially in the periphery, is located in the brain and has a different regional distribution from the central-type ligands.

Blockers or inverse agonists at the central-type site, RO15-1788 or the β-carboline (BCCM), respectively, interfere with the anticonvulsant effects of diazepam but not with those of carbamazepine on amygdala-kindled seizures. In contrast, a ligand for the 'peripheral-type' benzodiazepine receptor, Ro5-4864, does not reverse the anticonvulsant effects of diazepam, but does reverse the anticonvulsant effects of carbamazepine (Weiss *et al.*, 1985, 1986). This double dissociation implies that the 'peripheral-type' site and not the central-type site is important to carbamazepine's actions. A variety of controlled studies have been conducted which support this analysis, including the finding that the effects of Ro5-4864 are themselves blockable by PK-11195 (Weiss *et al.*, 1986).

While the central-type benzodiazepine receptor has been linked to the GABA receptor and the chloride ionophore, the 'peripheral-type' receptor appears to be closely associated with calcium channels (Le Fur *et al.*, 1985; Mestre *et al.*, 1985; Snyder *et al.*, 1987), although recent data from Puia and colleagues (1989) suggest that 'peripheral-type' receptors may also modulate chloride channels in some instances. Most recently, the 'peripheral-type' receptor has been localized to the outer membrane of mitochondria, where it also interacts with protoporphyrins (Snyder *et al.*, 1987; Verma *et al.*, 1987; Slobodyansky *et al.*, 1989). Interestingly, carbamazepine affects haem biosynthesis, increasing 5-aminolaevulinic acid (ALA) synthesis and reducing ALA dehydratase activity, as well as increasing urinary excretion of porphobilinogen and total porphyrins (McGuire *et al.*, 1988).

The potential involvement of 'peripheral-type' benzodiazepine receptors to calcium channel modulation, as well as effects on calcium flux mediated by carbamazepine's effects on adenosine receptors (Gasser *et al.*, 1988) and GABA$_B$ mechanisms (discussed below) are of particular interest in relationship to the emerging data that calcium channel blockers may also be effective in the treatment of acute mania and to the postulated effects of lithium on calcium metabolism (see above). While the acute time frame does not fit with a putative mechanism for carbamazepine's psychotropic effects, chronic administration has been reported to upregulate 'peripheral-type' benzodiazepine receptors (Weizman *et al.*, 1987) and repeated carbamazepine induces tolerance to anticonvulsant effects mediated by 'peripheral-type' but not central-type receptors (Weiss and Post, 1990a,b). Papart *et al.* (1988) also reported that the 'peripheral-type' benzodiazepine ligand PK-11195 reduced symptomatology in anxious depressed patients.

Carbamazepine and phenytoin both inhibit the binding of [^{3}H]batracho-toxinin-A20-α-benzoate, which is a ligand that binds to and regulates type-2 sodium channels (Willow *et al.*, 1984; McLean and MacDonald, 1986). The actions of these agents on sodium currents are prominent under conditions thought to be relevant to a seizure focus; i.e. they show use dependency and effects are increased during depolarization and rapid firing (Willow *et al.*, 1985; McLean and MacDonald, 1986). Similar to the arguments that have been put forward for the 'ideal' putative mechanism of lithium carbonate on PI turnover (i.e. it provides a mechanism for inhibition of pathological function while leaving normal baseline function unharmed), the effects of carbamazepine on sodium channels could provide a parallel process that could be important not only to anticonvulsant effects but to other systems that are paroxysmally activated in the fulminant pain syndromes and severe affective disorders as well.

While lithium has been reported to facilitate uptake of norepinephrine (NE), carbamazepine has only a weak ability to block NE re-uptake, an effect that is not thought to be related to its mechanism of action (Purdy *et al.*, 1977). Carbamazepine is unable to inhibit the uptake of 6-hydroxydopamine (6-OHDA) into noradrenergic nerve terminals and protect them, as does desmethylimipramine, for example (Quattrone *et al.*, 1981). Yet, the anticonvulsant effects of carbamazepine are thought to involve noradrenergic mechanisms, as selective depletion of NE with 6-OHDA and DMI impairs the anticonvulsant efficacy of carbamazepine (Quattrone *et al.*, 1978), and the α-2 antagonist yohimbine reverses the anticonvulsant effects of carbamazepine on amygdala-kindled seizures (S.R.B. Weiss and R.M. Post, unpubl. obsv., 1987). Indirect effects of carbamazepine on glutaminergic mechanisms potentially relevant to anticonvulsant effects are reviewed elsewhere (Post, 1988). It is noteworthy that low concentrations (10 μM) of carbamazepine decrease [^{3}H]L-glutamate release (Olpe *et al.*, 1985), suggesting that carbamazepine's anticonvulsant actions could also involve decreases in excitatory amino acid function.

Antinociceptive effects

The antinociceptive effects of carbamazepine have been most closely linked to $GABA_B$ mechanisms. Terrence *et al.* (1983) suggested a similarity in the structure of carbamazepine and baclofen and further physiological studies reveal that carbamazepine and baclofen shared putative antinociceptive effects in a model of trigeminal neuralgia in the cat. These effects of both *l*-baclofen and carbamazepine were, in turn, blocked by the inactive isomer *d*-baclofen, further supporting a $GABA_B$ mechanism for this effect. Moreover, Foong and Sato (1985) confirmed a GABAergic mechanism for carbamazepine's antinociceptive effects in a different pain model (bradykinin-induced rat tooth pulp pain). In addition, Fromm *et al.* (1984) demonstrated that *l*-baclofen was effective in the treatment of trigeminal neuralgia. Given this convincing evidence linking the antinociceptive effects of carbamazepine to $GABA_B$ mechanisms, it was important to establish whether $GABA_B$ mechanisms were also important to carbamazepine's anticonvulsant or psychotropic efficacy. In preliminary studies, we observed that l-baclofen was not a potent anticonvulsant on amygdala-kindled seizures and the inactive isomer d-baclofen was insufficient to reverse the acute anticonvulsant effects of carbamazepine on amygdala-kindled seizures. Thus, it appeared that the anticonvulsant effects of carbamazepine were not mediated by $GABA_B$ mechanisms (S.R.B. Weiss *et al.*, unpublished observations).

Psychotropic effects

This still left open the possibility that the psychotropic effects of carbamazepine could be mediated by a $GABA_B$ mechanism, similar to its antinociceptive effects. Thus, we initiated a clinical trial of l-baclofen in affective illness, bolstered by the preclinical data of Lloyd and associates (1986) linking upregulation of $GABA_B$ receptors to the putative mechanism of action of a variety of antidepressant modalities including chronic, but not acute, heterocyclic antidepressant treatment, MAOI treatment, and ECT. Recently, Motohashi *et al.* (1989) found that chronic, but not acute, lithium and carbamazepine upregulated $GABA_B$ receptors in hippocampus but not cortex.

None of the first five patients treated with l-baclofen showed antidepressant effects and three of the five showed some pattern of clinical deterioration during treatment, with improvement upon withdrawal, suggesting the possibility that increases in $GABA_B$ tone could be counter-therapeutic (Post *et al.*, 1990). These data are of considerable interest from the perspective that chronic treatment with lithium, carbamazepine, propranolol, and valproate have been reported to decrease GABA turnover (Bernasconi and Martin, 1979; Bernasconi, 1982). It is noteworthy that the effects of these antimanic and mood-stabilizing agents appeared to be more prominent following chronic than acute administration. Thus, the common ability of carbamazepine and lithium to decrease GABA turnover is convergent with their common spectrum of psychotropic activity and meets the appropriate time frame analysis for consideration of effects that take time to develop.

These effects are also of interest in relationship to the observations of the efficacy of valproate in manic and in long-term prophylaxis and to the postulated role of GABA in mood disorders (Emrich *et al.*, 1980; Berrettini and Post, 1984). It is possible that the lack of efficacy of l-baclofen in affective illness is not inconsistent with this formulation. Recent data suggest that $GABA_B$ mechanisms may, in some instances, inhibit GABA release and therefore functionally decrease GABAergic tone. It is possible that a decrement in $GABA_B$ tone is actually important to the antidepressant effects of a variety of agents, as reflected in their ability to upregulate $GABA_B$ receptors (Lloyd et al., 1986) or inhibit GABA turnover (Bernasconi, 1982). Further study of the comparative effects of lithium and carbamazepine and other bimodal mood-stabilizing agents, including ECT and valproate, on $GABA_A$ and $GABA_B$ mechanisms would appear indicated. The GABAergic neuro-transmitter system appears to be a prime candidate for the psychotropic effects of these agents.

In addition to effects on GABA turnover, carbamazepine affects a variety of neurotransmitter and second-messenger systems that have been putatively implicated in the manic process. Thus, carbamazepine increases acetylcholine in the striatum (Consolo et al., 1976), decreases CSF HVA and other indices of dopamine turnover (Post et al., 1986c), decreases CSF NE and inhibits NE release (Post et al., 1985), and decreases adenylate cyclase and guanylate cyclase activity (Post et al., 1982), as reviewed above. The effects of carbamazepine on each of these systems also deserve comparison and contrast with the effects of lithium. For example, the ability of carbamazepine to decrease NE release could provide a biochemical target for carbamazepine's mechanism of action comparable to lithium's ability to facilitate NE re-uptake, both agents making less NE available at the synapse. Similarly, while carbamazepine's mechanism of action on dopamine is poorly understood (Post et al., 1986c), recent evidence suggests the possibility that carbamazepine could make more dopamine available extracellularly and, therefore, down-regulate agonist effects in this system (Kaneko et al., 1981; Kowalik et al., 1984; Barros et al., 1986; Barros and Leite, 1986) although contradictory findings have been reported by Elphick (1989). Perhaps equally mysterious effects on dopaminergic mechanisms have been reported for lithium carbonate, although the data are highly consistent on lithium's ability to block dopamine supersensitivity behaviourally as elicited by chronic neuroleptic administration (Bunney and Garland-Bunney, 1987). Elphick (1989) documents that carbamazepine is not able to parallel this effect of lithium. Thus, it is possible that these two agents have common abilities to modulate indirectly dopaminergic mechanisms, but do so through different molecular targets of action. Waldmeier (1987) also postulates important common effects of lithium and the anticonvulsants.

The effects of carbamazepine, which clearly take time to develop and perhaps require chronic administration, yield interesting candidate systems for carbamazepine's antidepressant effects. As noted above, some of these effects are similar in both carbamazepine and lithium, such as GABA turnover, as well as effects on substance P (discussed below), while some of them are divergent; for example, carbamazepine's ability to increase plasma tryptophan, decrease CSF somatostatin, and increase adenosine receptors after chronic treatment. Lithium also increases the rate of uptake of tryptophan into synaptosomes (Knapp and Mandell, 1973; Price *et al.*, 1990). In contrast to a variety of other antidepressant modalities, carbamazepine has not been reported to downregulate β-adrenergic receptor sites (Marangos *et al.*, 1983; Joffe *et al.*, 1988b), although the effects on β-adrenergic receptor-mediated cyclase activity have not been studied directly.

Chronic carbamazepine appears paradoxically to upregulate adenosine receptors. In light of its opposite behavioural and physiological profile to caffeine, which is an adenosine antagonist and also upregulates adenosine receptors, one might have expected opposite biochemical effects of carbamazepine (Marangos *et al.*, 1985, 1987; Daval *et al.*, 1989). One potential explanation for this apparent inconsistency is given by data of Clark and Post (1989) suggesting that carbamazepine selectively binds to adenosine A_1 receptors while caffeine is non-selective for both adenosine A_1 and A_2 sites. These data are convergent with those of Gasser *et al.* (1988) studying effects of carbamazepine on adenosine-induced calcium fluxes in CA_1 neurons of hippocampal slices, and suggest that this drug acts as an antagonist at A_1 receptors. Recently, Elphick *et al.* (1990b) reported that chronic carbamazepine produced behavioural effects similar to an A_2-selective antagonist (PD 115,199), however. It is of interest that chronic electro-convulsive therapy, but not lithium, also upregulates adenosine receptor mechanisms (Newman *et al.*, 1984). The ability of chronic carbamazepine to alter adenosinergic receptor activity is also of interest in relation to the ability of lithium (at least acutely) also to modulate the effects of adenosine, perhaps indirectly through the effects of muscarinic receptor-mediated PI turnover, as described by Worley and colleagues (1988).

Thus, again, divergent molecular targets of action of lithium and carbamazepine may ultimately yield common functional effects on a given neurotransmitter system. If such proves to be the case, it may provide the basis for combination treatment with both lithium and carbamazepine in order to produce additive or potentiative effects on a given molecular mechanism. It is also possible that there is an entirely different target of each agent and this might also provide a mechanism for additive or potentiative effects. A variety of evidence supports the combined use of lithium and carbamazepine in a subpopulation of patients who do not respond adequately to either agent alone (Kishimoto *et al.*, 1984; Post, 1990a).

The effects of carbamazepine on serotonergic mechanisms have not been adequately elucidated. Pratt and colleagues (1984) reported that chronic administration of this anticonvulsant, in contrast to phenytoin and phenobarbital, was associated with increases in plasma tryptophan. These data have not been replicated by Elphick *et al.* (1990a,c), who found no changes in basal tryptophan levels after short-term treatment in normal volunteers and an increased prolactin response to a tryptophan challenge.

Lithium, at least initially, also increases prolactin response to tryptophan (Price *et al.*, 1989, 1990; Cowen *et al.*, 1989), again suggesting a potential common action of these two mood-stabilizing agents.

Later, nonetheless, given Elphick *et al.*'s (1990a) postulation of a pre-synaptic mechanism for carbamazepine's effects on serotonergic tone, and Blier and de Montigny's similar postulate (1985b) of a pre-synaptic serotonergic mechanism for lithium's effects in potentiation, there may again be a common basis of serotonergic effects through different molecular mechanisms.

The ability of carbamazepine to decrease the CSF somatostatin has been well documented in patients with affective illness and seizure disorders (Rubinow *et al.*, 1985; Steardo *et al.*, 1986). This ability of carbamazepine to decrease CSF somatostatin could be relevant to its mechanism in either seizure, pain, or affective disorders. Considerable data link these decreases in CSF somatostatin function to anticonvulsant properties. Somatostatin has also been intimately linked with pain mechanisms and data of Gerner and Yamada (1982), Rubinow *et al.* (1983), and a number of other investigative groups, as reviewed elsewhere (Rubinow *et al.*, 1990), suggest that CSF somatostatin is transiently decreased in patients during acute phases of depression and returns towards normal with improvement. Lithium does not appear to share carbamazepine's ability to decrease CSF somatostatin (Rubinow *et al.*, 1985).

The effects of carbamazepine and lithium on vasopressin appear to be divergent. While lithium blocks the effects of vasopressin by inhibiting stimulation of adenylate cyclase activity or by reducing its level (Yoshikawa and Hong, 1983) and is associated with the induction of diabetes insipidus, carbamazepine has the ability to act directly or indirectly as a vasopressin agonist (and has been used as a treatment for primary diabetes insipidus). It is likely that these opposing effects of lithium and carbamazepine are pertinent to the side effects profiles of these two agents and not to their primary mechanism of therapeutic action in the affective disorders. Thus, lithium carries the liability of inducing diabetes insipidus (which will not be reversed by the actions of carbamazepine) and carbamazepine appears to carry the liability of inducing hyponatraemia and water intoxication. Should similar opposing effects on vasopressin function be related to the putative effects of vasopressin on learning and memory, one would postulate opposite effects of these agents on this side effects profile. Preliminary data pertinent to this hypothesis have not been supportive of this idea, as both drugs have been reported to have minimal impact on psychological function, as assessed by the tests employed (Joffe *et al.*, 1988a).

Subtle differences in the prolactin response to arginine and thyrotropin-releasing hormone (TRH) are also evident in response to challenges on these two drugs. Carbamazepine significantly increases the prolactin response to arginine while leaving the prolactin response to TRH unchanged (Joffe *et al.*, 1988b). Lithium shows the opposite profile of effects (Joffe *et al.*, 1986).

Substance P represents a peptide that appears to have common and converging effects of carbamazepine and lithium which could be implicated in their mechanism of psychotropic action (Table 3). Jones *et al.* (1985) reported that chronic administration of carbamazepine is associated with an increased responsivity of cortical neurons to substance P. This effect is shared by a variety of other antidepressant modalities following chronic but not acute treatment. These include carbamazepine, desipramine, chlorimipramine, trimipramine, zimelidine, oxyprotaline, and tranylcypromine, although, interestingly, not electroconvulsive treatment. Substance P levels were reported to be increased in dopamine-enriched areas following lithium treatment (Hong *et al.*, 1983; Le Douarin *et al.*, 1983). Mitsushio *et al.* (1988) have reported that chronic administration of both lithium and carbamazepine increase substance P levels (in a haloperidol-reversible fashion) in the striatum but not raphe. These effects were not seen following acute administration and were identical to those observed with lithium. Thus, substance P becomes an interesting putative candidate for the psychotropic effects of carbamazepine, both because of its requirement for chronic administration and its shared effects with lithium carbonate. Clearly, this neuropeptide system which modulates psychomotor stimulant-induced behavioural activity and has an interesting pattern of neurotransmitter coexistence with serotonin and TRH, deserves further exploration.

TABLE 3   Effects of lithium and carbamazepine on substance P (data from Mitsushio *et al.*, 1988).

	Lithium	Carbamazepine
Acute treatment		
↑ In substance P levels	None[a,b]	None[d]
↑ In substance P sensitivity	None[c]	None[c]
Chronic treatment		
Substance P levels	↑ Striatum[a,b]	↑ Striatum[d]
	↑ Substantia nigra	↑ Substantia nigra
	↑ Nucleus accumbens	
	↑ Frontal cortex	
↑ Substance P is haloperidol reversible	Yes[a,b]	Yes[d]
↑ Substance P sensitivity	?	Yes[c]

[a] Hong *et al.*, 1983; [b] Le Dourain *et al.*, 1983; [c] Jones *et al.*, 1985; [d] Mitsushio *et al.*, 1988.
↑ = increases in.

Recent data suggests that chronic administration of lithium also affects the peptide neurotensin. This effect may be mediated by an increased level of neurotensin mRNA, as demonstrated in PC-12 pheochromocytoma cell line (Dobner *et al.*, 1988). Subchronic lithium treatment has also been shown by Sivam *et al.* (1986, 1988, 1989) to increase the levels of mRNA of proenkephalin A, dynorphin A (1-8), and preprotachykinin in the brain. These data are again of interest in light of the requirement for chronic administration and the effects of carbamazepine on these neuropeptides remain to be explored.

Clinical and preclinical tests for a given neurotransmitter hypothesis

Thus, a variety of effects of chronic carbamazepine have been elucidated which deserve further study as potential candidate mechanisms for the psychotropic effects of carbamazepine. In order to prove which is critically important to the effects of carbamazepine, systematic studies will have to be undertaken in a clinical population. For example, recent studies by Delgado *et al.* (1990), Charney (1990), etc., have shown that a diet acutely depleting tryptophan will reverse the antidepressant effects of a variety of compounds. Given the ability of chronic carbamazepine to increase plasma tryptophan in epileptic patients (Pratt *et al.*, 1984), such a study could be performed to ascertain whether carbamazepine's effects on serotonergic mechanisms (either mediated directly by its ability to increase plasma tryptophan [Pratt *et al.*, 1984] or indirectly at a pre-synaptic level [Elphick *et al.*, 1990a]) are critical to its antidepressant efficacy. Lithium's putative cyclic-AMP actions and carbamazepine's putative GABA$_B$ actions are being assessed with clinical trials of demeclomycin and *l*-baclofen, as described above. In other putative neurotransmitter systems, critical studies to assess neurotransmitter hypotheses are not so easily conducted, however. One would have to rely on indirect assessments of changes in these neurotransmitters in body fluids and available tissues such as CSF, and then ascertain whether such changes were related to the degree of clinical efficacy. Obviously this strategy is slow and cumbersome. It would be valuable to have an initial screening mechanism to help prioritize the most important candidate mechanisms.

Recently, Weiss *et al.* have elucidated a seizure model that requires chronic administration of carbamazepine in order to demonstrate efficacy. Thus, in contrast to the effects of carbamazepine on amygdala-kindled seizures, which were evident following a single administration, effects requiring chronic administration of the drug may be closer to its psychotropic properties. We have found that chronic administration of carbamazepine is required in order to block the development of cocaine- and lidocaine-kindled seizures (Fig. 7). Acute administration of doses of carbamazepine that are sufficient to block amygdala-kindled seizures are not only without effect on local anaesthetic

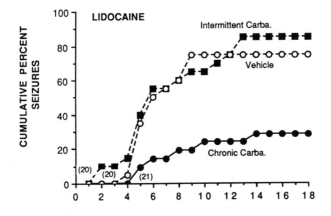

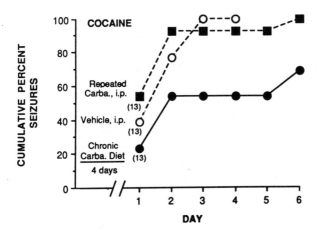

FIG. 7 Pretreatment with chronic carbamazepine (CBZ) in diet (not repeated injection) is required to block both lidocaine- and cocaine-kindled seizures. Rats were treated with chronic CBZ (diet), i.p. injected CBZ (15 mg kg^{-1}) or vehicle for 4 days prior to either lidocaine or cocaine injections and were continued on this regimen throughout the study. The repeated i.p. CBZ had no effect on the development of either lidocaine- or cocaine-kindled seizures, whereas chronic CBZ was highly effective. Increasing the dose of carbamazepine to 50 mg kg^{-1} i.p. exacerbates local anaesthetic kindling. From Weiss *et al.* (1989), with permission.

kindling (Weiss *et al.*, 1989), but, in high doses, will actually exacerbate this kindling process (S.R.B. Weiss *et al.*, unpubl. obsv.).

In this fashion, the inhibition of lidocaine- and cocaine-kindled seizures by carbamazepine may provide an indirect model for further elucidating the mechanisms involved in a process requiring chronic drug administration. For example, preliminary studies that implicated α_2-noradrenergic mechanisms in the acute anticonvulsant effects of carbamazepine on amygdala-kindled seizures have not implicated α_2 mechanisms in the effects of carbamazepine on local-anaesthetic kindling development, since administration of the α_2 antagonist yohimbine does not reverse the ability of carbamazepine to inhibit local anaesthetic kindling (S.R.B. Weiss *et al.*, unpubl. obsv.).

The effects of carbamazepine on corticotropin-releasing hormone (CRH) are ambiguous on this chronic model system. Rivier and Vale (1987) demonstrated that the local anaesthetics release CRF and recent data of Calogero *et al.* (1989) demonstrated that carbamazepine can inhibit local-anaesthetic-induced release of CRF. Dr Weiss examined whether co-administration of CRF would inhibit the ability of carbamazepine to block cocaine kindling. While CRF did achieve this effect, CRF was also capable of increasing cocaine-kindled seizures in the non-drug-treated control condition. Nonetheless, these two sets of studies (with the α_2 antagonist yohimbine and CRF) provide examples of how the cocaine-kindling paradigm can be used to dissect possible mechanisms involved in an effect of carbamazepine requiring chronic administration. Once candidate neuro-transmitter systems are thus identified, they might then be further explored in the appropriate preclinical and clinical studies.

Given lithium's intriguing effects on PI turnover in the brain that have been postulated to be involved in its mechanism of action in bipolar illness, it is of considerable interest to assess the comparative effects of carbamazepine. Different answers appear to emerge, depending on the preparation studied (Table 4). Unlike lithium, carbamazepine is not an inhibitor of inositol phosphatases. Carbamazepine does not amplify agonist-induced accumulation of inositol phosphates in rat brain slices (Elphick *et al.*, 1988) and neurohybrid NCB-20 cells (D.-M. Chuang, unpubl. obsv.), in a manner previously shown with lithium in a variety of systems. In NCB-20 cells, however, carbamazepine dose-dependently elicits an increase in inositol monophosphate accumulation in the presence of lithium. This effect is shared by some tricyclic antidepressants, particularly imipramine and amitriptyline (D.-M. Chuang, unpubl. obsv.). The significance of these findings is uncertain, as their effects in other systems such as cultured cerebellar granule cells are much less significant.

Using an *in vivo* model of electroconvulsive shock-induced PI turnover in rat brain, it has been shown that both lithium and carbamazepine attenuate the PI effect when the drug is administered prior to shock application (Vadnal and Bazan, 1987, 1988). The effect of carbamazepine occurs when mild inhibition of convulsion is observed and is associated with a reduction of $InsP_3$ synthesis in cerebral cortex and hippocampus. In another study, exposure of cultured cerebellar granule cells to carbamazepine was found to

TABLE 4 Comparative effects of lithium and carbamazepine on phosphoinositide turnover in the brain.

	Lithium	Carbamazepine
Inositol phosphatases	↓	—
Acute effects		
Basal PI turnover	—	—,↓
Stimulated PI turnover		
IP$_3$ (CCh-treated slices)		
Cerebral cortex	↓	—
Hippocampus	?	↓
IP$_4$ (cerebral cortex)	↓	ND
IP$_3$ (ECT-treated rats)	↓	↓
DAG (GH$_3$ cells)	↑	ND
Chronic effects		
Basal PI turnover		
Cerebral cortex	—,↓	—
Cerebellar neurons	?	—
Stimulated PI turnover		
Cerebral cortex	↓	—
Cerebellar neurons	↓	↓
(CCh-stimulated)		

ND, not determined; CCh, carbachol.

induce a time- and dose-dependent decrease of muscarinic receptor-mediated PI metabolism (X.M. Gao and D.-M. Chuang, unpubl. obsv.). The inhibitory effect requires at least 2-3 days' treatment and can be observed with a therapeutically related dose (30–50 μM). These results are of interest, as in a parallel study, long-term lithium exposure was found to exert a similar inhibitory effect on muscarinic receptor-stimulated PI hydrolysis in cerebellar granule cells (see previous section). However, Elphick *et al.* (1988) reported that 14 days' treatment of rats with carbamazepine does not affect 5-HT-, carbachol- and NE-induced PI turnover, despite a 25–50% decrease of these responses by 14 days of lithium treatment. Since only cerebral cortex was examined in their studies, it remains possible that receptor-mediated PI turnover in certain brain areas is a common site of action elicited by both chronic lithium and carbamazepine administration. If this were the case, these two drugs might have displayed their effects on PI metabolism by distinct but converging mechanisms.

McDermott and Logan (1989) recently reported that carbamazepine at relatively high doses decreased basal and carbachol-evoked [^{3}H]inositol phosphate accumulation in rat hippocampal slices. The drug also inhibited responses to histamine and veratrine, but not 5HT, NE, or substance P. Their results suggest that carbamazepine may affect membrane inositol phospholipid distribution and the receptor–G protein–phospholipase C complex. Thus, there appear to be some differences in the effects of lithium and carbamazepine on PI turnover and further experimentation is required to establish a direct linkage of these effects to their psychotropic properties. However, effects on the PI system remain a potentially important mechanism for dampening the effects of overactive pathways and for conceptualizing a common site for actions that would be involved in attenuating a variety of neurotransmitter effector systems.

Conclusions

While considerable progress has been made in delineating possible mechanisms of action of lithium carbonate, the field still suffers from an overabundance of excellent hypotheses and an inability to provide convincing experimental and clinical data that narrow the choices. In this regard, perhaps lithium is the prototype of a 'dirty' drug with a panoply of effects on pre- and post-synaptic neurotransmitter functions as well as a host of intriguing effects on intercellular processes that are now providing potentially compelling hypotheses regarding its mechanism of action.

While comparing and contrasting the effects of lithium with carbamazepine and other bimodal agents of use in acute and prophylactic treatment of bipolar disorders provides additional leverage points, this analysis is also not without its difficulties. Further clinical work is required in order to delineate whether lithium and carbamazepine have clearcut differences in the clinical subgroups for which they are targeted, therefore implying differential mechanisms of action, or whether there is considerable overlap in patient responsivity consistent with their similar general clinical profile of acute and prophylactic efficacy. Perhaps what is needed is a variety of clinical and preclinical techniques in order to generate and test the most fruitful hypotheses.

In this chapter, we have attempted to elucidate promising areas for further exploration with a particular focus on the time-course of action of lithium and carbamazepine, recognizing that there are dysjunctions that may be useful in the study of both compounds. It appears that lithium potentiation may have a more rapid onset of action than either lithium treatment alone in depression or mania. Similarly, chronic administration of carbamazepine appears required in order to observe maximum antimanic and antidepressant effects while the ability to treat seizure, pain, and sleep disorders is evident more rapidly. This acute–chronic dysjunction may be useful in focusing on properties of these agents that require chronic administration, as first-line candidates for the psychotropic effects of these agents.

It is hoped that once the more specific molecular mechanisms of action are identified, newer more focused therapies can be directed as these putative target systems with either better clinical efficacy or a cleaner side effects profile. It is also possible, however, that the inherently 'dirty' nature of these agents with multiple mechanisms of action may ultimately be important to their ability to balance several systems and provide mood stabilization and that the hope of finding a specific mechanism may not be realized. Nonetheless, the search should be productive of new therapeutic agents and, ultimately, may be of importance in understanding the basic defects that are critical to the illness.

While this hope may be decades away, it would appear more readily in our grasp to begin to identify possible clinical and biological markers of which patients may respond to which of the available therapeutic agents. This advance alone would be of great importance in considerably reducing potential morbidity and mortality associated with this potentially devastating illness. Clarifying the clinical response data, as well as more subtle questions as to whether different stages in the evolution of manic-depressive illness are differentially pharmaco-therapeutically responsive (Post *et al.*, 1986c), should also provide a critical backdrop to explaining the basic biochemistry of lithium and carbamazepine and linking discrete neurochemical effects to their therapeutic actions.

References

Aldenhoff, J.B. and Lux, H.D. (1985). Lithium slows neuronal calcium regulation in the snail *Helix pomatia*. *Neurosci. Lett.* **54**, 103–108.

Allison, J.H. and Blisner, M.E. (1976). Inhibition of the effect of lithium on brain inositol by atropine and scopolamine. *Biochem. Biophys. Res. Commun.* **68**, 1332–1338.

Allison, J.H. and Stewart, M.A. (1971). Reduced brain inositol in lithium-treated rats. *Nature New Biol.* **233**, 267–268.

Allison, J.H., Blisner, M.E., Holland, W.H., Hipps, P.P. and Sherman, W.R. (1976). Increased brain *myo*-inositol 1-phosphate in lithium-treated rats. *Biochem. Biophys. Res. Commun.* **71**, 664–670.

Arenander, A.T., de Vellis, J. and Herschman, H.R. (1989). Induction of c-fos and TIS genes in cultured rat astrocytes by neurotransmitters. *J. Neurosci. Res.* **24**, 107–114.

Avissar, S. and Schreiber, G. (1989). Muscarinic receptor subclassification and G-proteins: Significance for lithium action in affective disorders and for the treatment of the extrapyramidal side effects of neuroleptics. *Biol. Psychiat.* **26**, 113–130.

Avissar, S., Schreiber, G., Danon, A. and Belmaker, R.H. (1988). Lithium inhibits adrenergic and cholinergic increases in GTP binding in rat cortex. *Nature* **331**, 440–442.

Barkai, A.I. (1981). *Myo*-inositol turnover in the intact rat brain: increased production after *d*-amphetamine. *J. Neurochem.* **36**, 1485–1491.

Barros, H.M. and Leite, J.R. (1986). Effects of acute and chronic carbamazepine administration on apomorphine-elicited stereotypy. *Eur. J. Pharmacol.* **123**, 345–349.

Barros, H.M., Braz, S. and Leite, J.R. (1986). Effect of carbamazepine on dopamine release and reuptake in rat striatal slices. *Epilepsia* **27**, 534–537.

Batty, I. and Nahorski, S.R. (1987). Lithium inhibits muscarinic-receptor-stimulated inositol tetrakisphosphate accumulation in rat cerebral cortex. *Biochem. J.* **247**, 797–800.

Baxter, L.R. Jr., Liston, E.H. and Schwartz, J.M. (1986). Prolongation of the antidepressant response to partial sleep deprivation by lithium. *J. Psychiat. Res.* **19**, 17–23.

Belmaker, R.H., and Roitman, G. (1988). A clinical trial of desmethylchlortetracycline as a lithium-like agent in excited psychoses. *In* "Lithium: Inorganic Pharmacology and Psychiatric Use" (ed. N.J. Birch), pp. 191–193. IRL Press, Oxford.

Bernasconi, R. (1982). The GABA hypothesis of affective illness: influence of clinically effective antimanic drugs on GABA turnover. *In* "Basic Mechanisms in the Action of Lithium—Proceedings of a Symposium held at Schloss Ringberg, Bavaria, 4–6 October 1981", pp. 183–192. Excerpta Medica, Amsterdam.

Bernasconi, R. and Martin, P. (1979). Effects of antileptic drugs on the GABA turnover rate. *Arch. Pharmacol.* **307**, R63–Abstr. 251.

Berrettini, W. and Post, R.M. (1984). GABA in affective illness. *In* "Neurobiology of Mood Disorders" (ed. R.M. Post and J.C. Ballenger), pp. 673–685. Williams and Wilkins, Baltimore, MD.

Berridge, M.J. (1989). Inositol trisphosphate, calcium, lithium, and cell signaling. (The Albert Lasker Medical Awards.) *J. Amer. Med. Assoc.* **262**, 1834–1841.

Berridge, M.J., Downes, C.P. and Hanley, M.R. (1982). Lithium amplifies agonist-dependent phosphatidylinositol responses in brain and salivary glands. *Biochem. J.* **206**, 587–595.

Blier, P. and de Montigny, C. (1985a). Serotonergic but not noradrenergic neurons in rat central nervous system adapt to long-term treatment with monoamine oxidase inhibitors. *Neuroscience* **16**, 949–955.

Blier, P. and de Montigny, C. (1985b). Short-term lithium administration enhances serotonergic neurotransmission: electrophysiological evidence in the rat CNS. *Eur. J. Pharmacol.* **113**, 69–77.

Bunney, W.E., Jr. and Garland-Bunney, B.L. (1987). Mechanisms of action of lithium in affective illness: Basic and clinical implications. *In* "Psychopharmacology: A Third Generation of Progress" (ed. H. Meltzer), pp. 553–565. Raven Press, New York.

Calogero, A.E., Gallucci, W.T., Kling, M.A., Chrousos, G.P. and Gold, P.W. (1989). Cocaine stimulates rat hypothalamic corticotropin-releasing hormone secretion *in vitro*. *Brain Res.* **505**, 7–11.

Carney, P.A., Seggie, J., Vojtechovsky, M., Parker, J., Grof, E. and Grof, P. (1988). Bipolar patients taking lithium have increased dark adaptation threshold compared with controls. *Pharmacopsychiatry* **21**, 117–120.

Carroll, B.J. and Sharp, P.T. (1971). Rubidium and lithium: Opposite effects on amine-mediated excitement. *Science* **172**, 1355–1357.

Casebolt, T.L. and Jope, R.S. (1987). Chronic lithium treatment reduces norepinephrine-stimulated inositol phospholipid hydrolysis in rat cortex. *Eur. J. Pharmacol.* **140**, 245–246.

Charney, D.S. (1990). Serotonin-specific drugs in psychiatry. *Grand Rounds; South Beach Psychiatric Center*, 2 February 1990 (Abstr.).

Chuang, D.M. (1989). Neurotransmitter receptors and phosphoinositide turnover. *Ann. Rev. Pharmacol. Toxicol.* **29**, 71–110.

Clark, M. and Post, R.M. (1989). Carbamazepine, but not caffeine, is highly selective for adenosine A1 binding sites. *Eur. J. Pharmacol.* **164**, 399–401.

Consolo, S., Bianchi, S. and Ladinsky, H. (1976). Effects of carbamazepine on cholinergic parameters in rat brain areas. *Neuropharmacology* **15**, 653–657.

Cowen, P.J., McCance, S.L., Cohen, P.R. and Julier, D.L. (1989). Lithium increases 5-HT-mediated neuroendocrine responses in tricyclic resistant depression. *Psychopharmacology (Berl.)* **99**, 230–232.

Daval, J.L., Deckert, J., Weiss, S.R., Post, R.M. and Marangos, P.J. (1989). Upregulation of adenosine A1 receptors and forskolin binding sites following chronic treatment with caffeine or carbamazepine: a quantitative autoradiographic study. *Epilepsia* **30**, 26–33.

de Montigny, C. and Aghajanian, G.K. (1978). Tricyclic antidepressants: long-term treatment increases responsivity of rat forebrain neurons to serotonin. *Science* **202**, 1303–1306.

de Montigny, C., Grunberg, F., Mayer, A. and Deschenes, J.-P. (1981). Lithium induces rapid relief of depression in tricyclic antidepressant drug non-responders. *Br. J. Psychiat.* **138**, 252–256.

Delgado, P.L., Charney, D.S., Price, L.H., Aghajanian, G.K., Landis, H. and Heninger, G.R. (1990). Serotonin function and the mechanism of antidepressant action. Reversal of antidepressant-induced remission by rapid depletion of plasma tryptophan. *Arch. Gen. Psychiat.* **47**, 411–418.

Dobner, P.R., Tischler, A.S., Lee, Y.C., Bloom, S.R. and Donahue, S.R. (1988). Lithium dramatically potentiates neurotensin/neuromedin N gene expression. *J. Biol. Chem.* **263**, 13 983–13 986.

Ebstein, R., Belmaker, R., Grunhaus, L. and Rimon, R. (1976). Lithium inhibition of adrenaline-stimulated adenylate cyclase in humans. *Nature* **259**, 411–413.

Ebstein, R.P., Hermoni, M. and Belmaker, R.H. (1980). The effect of lithium on noradrenaline-induced cyclic AMP accumulation in rat brain: inhibition after chronic treatment and absence of supersensitivity. *J. Pharmacol. Exp. Ther.* **213**, 161–167.

Elphick, M. (1989). Effects of carbamazepine on dopamine function in rodents. *Psychopharmacology (Berl.)* **99**, 532–536.

Elphick, M., Taghavi, Z., Powell, T. and Godfrey, P.P. (1988). Alteration of inositol phospholipid metabolism in rat cortex by lithium but not carbamazepine. *Eur. J. Pharmacol.* **156**, 411–414.

Elphick, M., Anderson, S.M.P., Hallis, K.F. and Grahame-Smith, D.G. (1990a). Effects of carbamazepine on 5-hydroxytryptamine function in rodent. *Psychopharmacology (Berl.)* **100**, 49–53.

Elphick, M., Taghavi, Z., Powell, T. and Godfrey, P.P. (1990b). Chronic carbamazepine down-regulates adenosine A2 receptors: studies with the putative selective adenosine antagonists PD115,119 and PD116,948. *Psychopharmacology (Berl.)* **100**, 522–529.

Elphick, M., Yang. J.-D. and Cowen, P.J. (1990c). Effects of carbamazepine on dopamine- and serotonin-mediated neuroendocrine responses. *Arch. Gen. Psychiat.* **47**, 135–140.

Emrich, H.M., von Zerssen, D., Kissling, W., Moller, H.-J. and Windorfer, A. (1980). Effect of sodium valproate in mania. The GABA hypothesis of affective disorders.

234   *R.M. Post and D.-M. Chuang*

Arch. Psychiatr. Nervenkr. **229**, 1–16.

Ferrendelli, J.A. and Kinscherf, D.A. (1979). Inhibitory effects of anticonvulsant drugs on cyclic nucleotide accumulation in brain. *Ann. Neurol.* **5**, 533–538.

Foong, F.W. and Satoh, M. (1985). Antinociceptive effect of intracisternal carbamazepine evidenced by the bradykinin-induced biting-like response to rats. *Exp. Neurol.* **90**, 264–267.

Frank, R.A. and Zubrycki, E. (1989). Chronic imipramine does not block cocaine-induced increases in brain stimulation reward. *Pharmacol. Biochem. Behav.* **33**, 725–727.

Fromm, G.H., Terrence, C.F. and Chattha, A.S. (1984). Baclofen in the treatment of trigeminal neuralgia: double-blind study and long-term follow-up. *Ann. Neurol.* **15**, 240–244.

Gasser, T., Reddington, M. and Schubert, P. (1988). Effect of carbamazepine on stimulus-evoked Ca^{2+} fluxes in rat hippocampal slices and its interaction with α_1-adenosine receptors. *Neurosci. Lett.* **91**, 189–193.

Geisler, A., Mork, A. and Klysner, R. (1987). Influence of lithium on adaptability and function of membrane receptors. *In* "Lithium: Inorganic Pharmacology and Psychiatric Use" (ed. N.J. Birch), pp. 195–198. IRL Press, Oxford.

Gerner, R.H. and Yamada, T. (1982). Altered neuropeptide concentrations in cerebrospinal fluid of psychiatric patients. *Brain Res.* **238**, 298–302.

Godfrey, P.P., McClue, S.J., White, A.M., Wood, A.J. and Grahame-Smith, D.G. (1989). Subacute and chronic *in vivo* lithium treatment inhibits agonist- and sodium fluoride-stimulated inositol phosphate production in rat cortex. *J. Neurochem.* **52**, 498–506.

Gold, P.W., Robertson, G.L., Ballenger, J.C., Kaye, W., Chen, J., Rubinow, D.R., Goodwin, F.K. and Post, R.M. (1983). Carbamazepine diminishes sensitivity of the plasma arginine vasopressin response to osmotic stimulation. *J. Clin. Endocrinol. Metab.* **57**, 952–957.

Grof, E., Haag, M., Grof, P. and Haag, H. (1987). Lithium response and the sequence of episode polarities: preliminary report on a Hamilton sample. *Progr. Neuropsychopharmacol. Biol. Psychiat.* **11**, 199–203.

Gutterman, D.F., Correa, E.I., DePaulo, J.R., Jr. and Coyle, J.T. (1985). RBC choline and renal disorders during lithium treatment. *Amer. J. Psychiat.* **142**, 493–495.

Haag, H., Heidorn, A., Haag, M. and Greil, W. (1987). Sequence of affective polarity and lithium response: preliminary report on Munich sample. *Progr. Neuropsychopharmacol. Biol. Psychiat.* **11**, 205–208.

Haag, M., Haag, H., Eisenried, F. and Greil, W. (1984). RBC-choline: changes in lithium and relation to prophylactic response. *Acta Psychiat. Scand.* **70**, 389–399.

Hallcher, L.M. and Sherman, W.R. (1980). The effects of lithium ion and other agents on the activity of *myo*-inositol-1-phosphatase from bovine brain. *J. Biol. Chem.* **255**, 10896–10901.

Han, S.-Z. (1984). Lithium chloride changes sensitivity of hamster rhythm to light pulses. *J. Interdiscipl. Cycle Res.* **15**, 139–145.

Heninger, G.R., Charney, D.S. and Sternberg, D.E. (1983). Lithium carbonate augmentation of antidepressant treatment. An effective prescription for treatment-refractory depression. *Arch. Gen. Psychiat.* **40**, 1335–1342.

Honchar, M.P., Ackermann, K.E. and Sherman, W.R. (1989). Chronically administered lithium alters neither *myo*-inositol monophosphatase activity nor phosphoinositide levels in rat brain. *J. Neurochem.* **53**, 590–594.

Hong, J.S., Tilson, H.A. and Yoshikawa, K. (1983). Effects of lithium and haloperidol administration on the rat brain levels of substance P. *J. Pharmacol. Exp. Ther.* **224**, 590–593.

Janowsky, D.S., El-Yousef, M.K. and Davis, J.M. (1973). Parasympathetic suppression of manic symptoms by physostigmine. *Arch. Gen. Psychiat.* **28**, 542–547.

Janowsky, D.S., El-Yousef, M.K. and Davis, J.M. (1974). Acetylcholine and depression. *Psychosom. Med.* **36**, 248–257.

Joffe, R.T., Post, R.M., Ballenger, J.C., Rebar, R. and Gold, P.W. (1986). The effects of lithium on neuroendocrine function in affectively ill patients. *Acta Psychiat. Scand.* **73**, 524–528.

Joffe, R.T., MacDonald, C. and Kutcher, S.P. (1988a). Lack of differential cognitive effects of lithium and carbamazepine in bipolar affective disorder. *J. Clin. Psychopharmacol.* **8**, 425–428.

Joffe, R.T., Post, R.M., Sulser, F. and Weiss, S.R.B. (1988b). Effects of thyroid alterations and carbamazepine on cortical beta-adrenergic receptors in rat. *Neuropharmacology* **27**, 171–174.

Johnson, R.D. and Minneman, K.P. (1986). Characterization of alpha 1-adrenoceptors which increase cyclic AMP accumulation in rat cerebral cortex. *Eur. J. Pharmacol.* **129**, 293–305.

Jones, R.S., Mondadori, C. and Olpe, H.-R. (1985). Neuronal sensitivity to substance P is increased after repeated treatment with tranylcypromine, carbamazepine or oxaprotaline, but decreased after repeated electroconvulsive shock. *Neuropharmacology* **24**, 627–633.

Jope, R. (1979). Effects of lithium treatment *in vitro* and *in vivo* on acetylcholine metabolism in rat brain. *J. Neurochem.* **33**, 487–495.

Jope, R.S., Jenden, D.J., Ehrlich, B.E., Diamond, J.M. and Gosenfeld, L.F. (1980). Erythrocyte choline concentrations are elevated in manic patients. *Proc. Natl. Acad. Sci. USA* **77**, 6144–6146.

Kafka, M.S., Wirz-Justice, A., Naber, D., Marangos, P.J., O'Donohue, T.L. and Wehr, T.A. (1982). Effect of lithium on circadian neurotransmitter receptor rhythms. *Neuropsychobiology* **8**, 41–50.

Kanba, S. and Richelson, E. (1984). Antimuscarinic effects of lithium [letter]. *New Engl. J. Med.* **310**, 989.

Kanba, S., Pfenning, M., Kanba, K.S. and Richelson, E. (1986). Lithium ions have a potent and selective inhibitory effect on cyclic GMP formation stimulated by neurotensin, angiotensin II and bradykinin. *Eur. J. Pharmacol.* **126**, 111–116.

Kaneko, S., Kurahashi, K. and Mori, A. (1981). The mechanisms of action of carbamazepine. *In* "12th World Congress of Neurology, International Congress Series 548", pp. 318–319. Excerpta Medica, Amsterdam.

Kaschka, W.P., Thurauf, N., Mokrusch, T. and Korth, M. (1988). Electro-oculography and dark adaptation in patients with affective and schizoaffective psychoses: early physiological effects of carbamazepine and lithium. *Pharmacopsychiatry* **21**, 404–406.

Kendall, D.A. and Nahorski, S.R. (1987). Acute and chronic lithium treatments influence agonist and depolarization-stimulated inositol phospholipid hydrolysis in rat cerebral cortex. *J. Pharmacol. Exp. Ther.* **241**, 1023–1027.

Kennedy, E.D., Challiss, R.A. and Nahorski, S.R. (1989). Lithium reduces the accumulation of inositol polyphosphate second messengers following cholinergic stimulation of cerebral cortex slices. *J. Neurochem.* **53**, 1652–1655.

Kishimoto, A., Omura, F., Umezawa, Y. and Fukuma, E. (1984). Combined therapy with lithium and carbamazepine. New Research Abstracts, 137th Annual Meeting, *Amer. Psychiat. Assoc., Abstr.* NR22.

Knapp, S. and Mandell, A.J. (1973). Short- and long-term lithium administration: effects on the brain's serotonergic biosynthetic systems. *Science* **180**, 645–647.

Kowalik, S., Levitt, M. and Barkai, A. (1984). Effects of carbamazepine and antidepressant drugs on endogenous catecholamine levels in the cerebroventricular compartment of the rat. *Psychopharmacology* **83**, 169–171.

Kramlinger, K.G. and Post, R.M. (1989a). The addition of lithium carbonate to carbamazepine: antidepressant efficacy in treatment-resistant depression. *Arch. Gen. Psychiat.* **46**, 794–800.

Kramlinger, K.G. and Post, R.M. (1989b). Adding lithium carbonate to carbamazepine: antimanic efficacy in treatment-resistant mania. *Acta Psychiat. Scand.* **79**, 378–385.

Kripke, D.F. and Wyborney, V.G. (1980). Lithium slows rat circadian activity rhythms. *Life Sci.* **26**, 1319–1321.

Kukopulos, A., Reginaldi, D., Laddomada, P., Floris, G., Serra, G. and Tondo, L. (1980). Course of the manic-depressive cycle and changes caused by treatment. *Pharmakopsychiatria Neuropsychopharmakol.* **13**, 156–167.

Le Douarin, C., Oblin, A., Fage, D. and Scatton, B. (1983). Influence of lithium on biochemical manifestations of striatal dopamine target cell supersensitivity induced by prolonged haloperidol treatment. *Eur. J. Pharmacol.* **93**, 55–62.

Le Fur, G., Mestre, M., Carriot, T., Belin, C., Renault, C., Dubroeucq, M.C., Gueremy, C. and Uzan, A. (1985). Pharmacology of peripheral type benzodiazepine receptors in the heart. *Progr. Clin. Biol. Res.* **192**, 175–186.

Levy, A., Zohar, J. and Belmaker, R.H. (1982). The effect of chronic lithium pretreatment on rat brain muscarinic receptor regulation. *Neuropharmacology* **21**, 1199–1201.

Lewin, E. and Bleck, V. (1977). Cyclic AMP accumulation in cerebral cortical slices: effect of carbamazepine, phenobarbitol, and phenytoin. *Epilepsia* **18**, 237–242.

Lloyd, K.G., Thuret, E.W. and Pilc, A. (1986). GABA and the mechanism of action of antidepressant drugs. In "GABA and Mood Disorders: Experimental and Clinical Research", LERS Monograph Series, Vol. 4, (ed. G. Bartholini, K.G. Lloyd and P.L. Morselli), pp. 33–42. Raven Press, New York.

Maggi, A. and Enna, S.J. (1980). Regional alterations in rat brain neurotransmitter systems following chronic lithium treatment. *J. Neurochem.* **34**, 888–892.

Maj, M., Pirozzi, R. and Starace, F. (1989). Previous pattern of course of the illness as a predictor of response to lithium prophylaxis in bipolar patients. *J. Affective Disord.* **17**, 237–241.

Marangos, P., Post, R.M., Patel, J., Zander, K., Parma, A. and Weiss, S.R.B. (1983). Specific and potent interactions between carbamazepine and brain adenosine receptors. *Eur. J. Pharmacol.* **93**, 175–182.

Marangos, P.J., Weiss, S.R.B., Montgomery, P., Patel, J., Narang, P.K., Cappabianca, A. and Post, R.M. (1985). Chronic carbamazepine treatment increases brain adenosine receptors. *Epilepsia* **26**, 493–498.

Marangos, P.J., Montgomery, P., Weiss, S.R., Patel, J. and Post, R.M. (1987). Persistent upregulation of brain adenosine receptors in response to chronic carbamazepine treatment. *Clin. Neuropharmacol.* **10**, 443–448.

Mayer, M.L. and Westbrook, G.L. (1987). The physiology of excitatory amino acids in the vertebrate central nervous system. *Progr. Neurobiol.* **28**, 197–276.

McDermott, E.E. and Logan, S.D. (1989). Inhibition of agonist-stimulated inositol lipid metabolism by the anticonvulsant carbamazepine in rat hippocampus. *Br. J. Pharmacol.* **98**, 581–589.

McEachron, D.L., Kripke, D.F., Hawkins, R., Haus, E., Pavlinac, D. and Deftos, L. (1982). Lithium delays biochemical circadian rhythms in rats. *Neuropsychobiology* **8**, 12–29.

McGuire, G.M., Macphee, G.J., Thompson, G.G., Park, B.K., Moore, M.R. and Brodie, M.J. (1988). The effects of chronic carbamazepine treatment of haem biosynthesis in man and rat. *Eur. J. Clin. Pharmacol.* **35**, 241–247.

McLean, M.J. and Macdonald, R.L. (1986). Carbamazepine and 10,11-epoxy-carbamazepine produce use- and voltage-dependent limitation of rapidly firing action potentials of mouse central neurons in cell culture. *J. Pharmacol. Exp. Ther.* **238**, 727–738.

Meltzer, H.L. (1986). Lithium mechanisms in bipolar illness and altered intracellular calcium functions. *Biol. Psychiat.* **21**, 492–510.

Meltzer, H.L., Kassir, S., Goodnick, P.J., Fieve, R.R., Chrisomalis, L., Feliciano, M. and Szypula, D. (1988). Calmodulin-activated calcium ATPase in bipolar illness. *Neuropsychobiology* **20**, 169–173.

Menkes, H.A., Baraban, J.M., Freed, A.N. and Snyder, S.H. (1986). Lithium dampens neurotransmitter response in smooth muscle: relevance to action in affective illness. *Proc. Natl. Acad. Sci. USA* **83**, 5727–5730.

Mestre, M., Carriot, T., Berlin, C., Uzan, A., Renault, C., Dubroeucq, M.C., Gueremy, C., Doble, A. and Le Fur, G. (1985). Electrophysiological and pharmacological evidence that peripheral type benzodiazepine receptors are coupled to calcium channels in the heart. *Life Sci.* **36**, 391–400.

Mitsushio, H., Takashima, M., Mataga, N. and Toru, M. (1988). Effects of chronic treatment with trihexyphenidyl and carbamazepine alone or in combination with haloperidol on substance P content in rat brain: a possible implication of substance P in affective disorders. *J. Pharmacol. Exp. Ther.* **245**, 982–989.

Mork, A. and Geisler, A. (1989). The effects of lithium *in vitro* and *ex vivo* on adenylate cyclase in brain are exerted by distinct mechanisms. *Neuropharmacology* **28**, 307–311.

Motohashi, N., Ikawa, K. and Kariya, T. (1989). GABA$_B$ receptors are up-regulated by chronic treatment with lithium or carbamazepine. GABA hypothesis of affective disorders. *Eur. J. Pharmacol.* **166**, 95–99.

Muller, W., Brunner, H. and Misgeld, U. (1989). Lithium discriminates between muscarinic receptor subtypes on guinea pig hippocampal neurons *in vitro*. *Neurosci. Lett.* **100**, 135–140.

Naylor, G.J., Dick, D.A.T., Dick, E.G. and Moody, J.P. (1974). Lithium therapy and erythrocyte membrane cation carrier. *Psychopharmacology* **3**, 81–86.

Newman, M., Zohar, J., Kalian, M. and Belmaker, R.H. (1984). The effects of chronic lithium and ECT on A1 and A2 adenosine receptor systems in rat brain. *Brain Res.* **291**, 188–192.

Okuma, T. (1984). Therapeutic and prophylactic efficacy of carbamazepine in manic-depressive psychosis. *In* "Anticonvulsants in Affective Disorders" (ed. H.M. Emrich, T. Okuma and A.A. Muller), pp. 76–87. Excerpta Medica, Amsterdam.

Okuma, T., Kishimoto, A., Inoue, K., Matsumoto, H., Ogura, A., Matsushita, T., Naklao, T. and Ogura, C. (1973). Anti-manic and prophylactic effects of carbamazepine on manic-depressive psychosis. *Folia Psychiatr. Neurol. Jpn* **27**, 283–297.

Okuma, T., Inanaga, K., Otsuki, S., Sarai, K., Takahashi, R., Hazama, H., Mori, A. and Watanabe, M. (1981). A preliminary double-blind study of the efficacy of carbamazepine in prophylaxis of manic-depressive illness. *Psychopharmacology* **73**, 95–96.

Olpe, H.-R., Baudry, M. and Jones, R.S.G. (1985). Electrophysiological and neurochemical investigations on the action of carbamazepine on the rat hippocampus. *Eur. J. Pharmacol.* **110**, 71–80.

Palmer, G.C. (1979). Interactions of antiepileptic drugs on adenylate cyclase and phosphodiesterases in rat and mouse cerebrum. *Exp. Neurol.* **63**, 322–335.

Palmer, G.C., Jones, D.J., Medina, M.A. and Stavinohka, W.B. (1979). Anticonvulsant drug actions on *in vitro* and *in vivo* levels of cyclic AMP in the mouse brain. *Epilepsia* **20**, 95–104.

Palmer, G.C., Palmer, S.J. and Legendre, J.L. (1981). Guanylate cyclase-cyclic GMP in mouse cerebral cortex and cerebellum: modification by anticonvulsants. *Exp. Neurol.* **71**, 601–614.

Papart, P., Ansseau, M., Cerfontaine, J.L. and von Frenckell, R. (1988). Pilot study of 52028 RP (PK 11195), an antagonist of the peripheral type benzodiazepine binding sites, among inpatients with anxious or depressive symptomatology. Abstracts of the Collegium Internationale Neuro-Psychopharmacologicum (CINP) 219-#28.33.11

Pestronk, A. and Drachman, D.B. (1980). Lithium reduces the number of acetylcholine receptors in skeletal muscle. *Science* **210**, 342–343.

Pestronk, A. and Drachman, D.B. (1987). Mechanism of action of lithium on acetylcholine receptor metabolism in skeletal muscle. *Brain Res.* **412**, 302–310.

Post, R.M. (1987). Mechanisms of action of carbamazepine and related anticonvulsants in affective illness. *In* "Psychopharmacology: A Generation of Progress" (ed. H. Meltzer and W.E. Bunney, Jr), pp. 567–576. Raven Press, New York.

Post, R.M. (1988). Time course of clinical effects of carbamazepine: implications for mechanisms of action. *J. Clin. Psychiat.* **49**, 35–46.

Post, R.M. (1990a). Prophylaxis of bipolar affective disorders. *Int. Rev. of Psychiat.* **2**, 165–208.

Post, R.M. (1990b). Alternatives to lithium for bipolar affective illness. *In* "Review of Psychiatry", Vol. 9, pp. 170–202. (ed. A. Tasman, S. Goldfinger and C. Kaufmann). American Psychiatric Press Inc., Washington, DC.

Post, R.M., Jimerson, D.C., Bunney, W.E. Jr., Goodwin, F.K. and Sharpley, P.H. (1980). Dopamine and mania: behavioral and biochemical effects of the dopamine receptor blocker pimozide. *Psychopharmacology* **67**, 297–305.

Post, R.M., Ballenger, J.C., Uhde, T.W., Smith, C., Rubinow, D.R. and Bunney, W.E., Jr (1982). Effect of carbamazepine on cyclic nucleotides in CSF of patients with affective illness. *Biol. Psychiat.* **17**, 1037–1045.

Post, R.M., Uhde, T.W., Ballenger, J.C. and Squillace, K.M. (1983). Prophylactic efficacy of carbamazepine in manic-depressive illness. *Amer. J. Psychiat.* **140**, 1602–1604.

Post, R.M., Weiss, S.R.B. and Pert, A. (1984). Differential effects of carbamazepine and lithium on sensitization and kindling. *Progr. Neuropsychopharmacol. Biol. Psychiat.* **8**, 425–434.

Post, R.M., Rubinow, D.R., Uhde, T.W., Ballenger, J.C., Lake, C.R., Linnoila, M., Jimerson, D.C. and Reus, V. (1985). Effects of carbamazepine on noradrenergic mechanisms in affectively ill patients. *Psychopharmacology* **87**, 59–63.

Post, R.M., Putnam, F., Uhde, T.W. and Weiss, S.R.B. (1986a). Electroconvulsive therapy: clinical and basic research issues. *In* "Annals of the New York Academy of Sciences", Vol. 462 (ed. S. Malitz and H.A. Sackheim), pp. 376–388. New York Academy of Sciences, New York.

Post, R.M., Rubinow, D.R. and Ballenger, J.C. (1986b). Conditioning and sensitization in the longitudinal course of affective illness. *Br. J. Psychiat.* **149**, 191–201.

Post, R.M., Rubinow, D.R., Uhde, T.W., Ballenger, J.C. and Linnoila, M. (1986c). Dopaminergic effects of carbamazepine: relationship to clinical response in affective illness. *Arch. Gen. Psychiat.* **42**, 392–397.

Post, R.M., Uhde, T.W., Roy-Byrne, P.P. and Joffe, R.T. (1986d). Antidepressant effects of carbamazepine. *Amer. J. Psychiat.* **143**, 29–34.

Post, R.M., Uhde, T.W., Roy-Byrne, P.P. and Joffe, R.T. (1987). Correlates of antimanic response to carbamazepine. *Psychiat. Res.* **21**, 71–83.

Post, R.M., Weiss, S.R.B., Ketter, T., Kramlinger, K.L. and Joffe, R.T. (1990). *L*-baclofen in depression: Lack of relationship of GABA-B agonism to anticonvulsant and psychotropic effects of carbamazepine. (Unpubl. manuscript).

Pratt, J.A., Jenner, P., Johnson, A.L., Shorvon, S.D. and Reynolds, E.H. (1984). Anticonvulsant drugs alter plasma tryptophan concentrations in epileptic patients: implications for antiepileptic action and mental function. *J. Neurol. Neurosurg. Psychiat.* **47**, 1131–1133.

Price, L.H., Charney, D.S., Delgado, P.L. and Heninger, G.R. (1989). Lithium treatment and serotoninergic function. Neuroendocrine and behavioral responses to intravenous tryptophan in affective disorder. *Arch. Gen. Psychiat.* **46**, 13–19.

Price, L.H., Charney, D.S., Delgado, P.L. and Heninger, G.R. (1990). Lithium and serotonin function: implications for the serotonin hypothesis of depression. *Psychopharmacology (Berl.)* **100**, 3–12.

Prien, R.F., Caffey, E.M. and Klett, C.J. (1972). Comparison of lithium carbonate and chlorpromazine in the treatment of mania. *Arch. Gen. Psychiat.* **26**, 146–153.

Puia, G., Santi, M.R., Vicini, S., Pritchett, D.B., Seeburg, P. and Costa, E. (1989). 4′-Chlorodiazepam decreases function of Cl^- channels coupled to native and transiently expressed $GABA_A$ receptor subunits in a manner insensitive to flumazenil. Abstracts. *Soc. Neurosci.* **15**, 641–Abstr. 259.2

Purdy, R.E., Julien, R.M. and Fairhurst, A.S. (1977). Effect of carbamazepine on the *in vitro* uptake and release of norepinephrine in adrenergic nerves of rabbit aorta and in whole brain synaptosomes. *Epilepsia* **18**, 251–257.

Quattrone, A., Crunelli, V. and Samanin, R. (1978). Seizure susceptibility and anticonvulsant activity of carbamazepine, diphenylhydantoin and phenobarbitol in rats with selective depletions of brain monoamines. *Neuropharmacology* **17**, 643–647.

Quattrone, A., Annunziato, L. and Aguglia, K. (1981). Carbamazepine, phenytoin and phenobarbitol do not influence brain catecholamine uptake, *in vivo*, in male rats. *Arch. Int. Pharmacodyn. Ther.* **252**, 180–185.

Rivier, C. and Vale, W. (1987). Cocaine stimulates adrenocorticotropin (ACTH) secretion through a corticotropin-releasing factor (CRF)-mediated mechanism. *Brain Res.* **422**, 403–406.

Rubinow, D.R., Gold, P.W. and Post, R.M. (1983). Somatostatin in affective illness. *Arch. Gen. Psychiat.* **40**, 403–412.

Rubinow, D.R., Post, R.M., Gold, P.W. and Reichlin, S. (1985). Effects of carbamazepine on cerebrospinal fluid somatostatin. *Psychopharmacology* **85**, 210–214.

Rubinow, D.R., Davis, C.L. and Post, R.M. (1990). Somatostatin. *In* "Neuropeptides in Psychiatry" (ed. C. Nemeroff). APA Press, Washington, DC, in press.

Seggie, J., Werstiuk, E., Grota, L. and Brown, G. (1983). Chronic lithium treatment and twenty-four hour rhythm of serum prolactin, growth hormone and melatonin in rats. *Progr. Neuropsychopharmacol. Biol. Psychiat.* **7**, 827–830.

Seggie, J., Carney, P.A., Parker, J., Grof, E. and Grof, P. (1989). Effect of chronic lithium on sensitivity to light in male and female bipolar patients. *Progr. Neuropsychopharmacol. Biol. Psychiat.* **13**, 3–7.

Shea, P.A., Small, J.G. and Hendrie, H.C. (1981). Elevation of choline and glycine in red blood cells of psychiatric patients due to lithium treatment. *Biol. Psychiat.* **16**, 825–830.

Sherman, W.R., Munsell, L.Y., Gish, B.G. and Honchar, M.P. (1985). Effects of systematically administered lithium on phosphoinositide metabolism in rat brain, kidney, and testis. *J. Neurochem.* **44**, 798–807.

Sivam, S.P., Strunk, C., Smith, D.R. and Hong, J.-S. (1986). Proenkephalin-A gene regulation in the rat striatum: influence of lithium and haloperidol. *Mol. Pharmacol.* **30**, 186–191.

Sivam, S.P., Takeuchi, K., Li, S., Douglass, O.C., Calvetta, L., Herbert, E., McGinty, J.F. and Hong, J.S. (1988). Lithium increases dynorphin A(1-8) and prodynorphin mRNA levels in the basal ganglia of rats. *Mol. Brain Res.* **3**, 155–164.

Sivam, S.P., Krause, J.E., Takeuchi, K., Li, S., McGinty, J.F. and Hong, J.S. (1989). Lithium increases rat striatal beta- and gamma-preprotachykinin messenger RNAs. *J. Pharmacol. Exp. Ther.* **248**, 1297–1301.

Slobodyansky, E., Alho, H., Bovolin, P., Guidotti, A. and Costa, E. (1989). The triakontatetraneuropeptide (TTN) a brain processing product of diazepam binding inhibitor (DBI): a putative allosteric modulator of $GABA_{A3}$ receptor. *Abstracts. Soc. Neurosci.* **15**, 641–Abstr. 259.3

Snyder, S.H., Verma, A. and Trifiletti, R.R. (1987). The peripheral-type benzodiazepine receptor: a protein of mitochondrial outer membranes utilizing porphyrins as endogenous ligands. *FASEB J.* **1**, 282–288.

Steardo, L., Barone, P. and Hunnicutt, E. (1986). Carbamazepine lowering effect on CSF somatostatin-like immunoreactivity in temporal lobe epilepsy. *Acta Neurol. Scand.* **74**, 140–144.

Terrence, C.F., Sax, M., Fromm, G.H., Chang, C.H. and Yoo, C.S. (1983). Effect of baclofen enantiomorphs on the spinal trigeminal nucleus and steric similarities of carbamazepine. *Pharmacology* **27**, 85–94.

Thase, M.E., Kupfer, D.J., Frank, E. and Jarrett, D.B. (1989). Treatment of imipramine-resistant recurrent depression: II. An open clinical trial of lithium augmentation. *J. Clin. Psychiat.* **50**, 413–417.

Vadnal, R.E. and Bazan, N.G. (1987). Electroconvulsive shock stimulates polyphosphoinositide degradation and inositol trisphosphate accumulation in rat cerebrum: lithium pretreatment does not potentiate these changes. *Neurosci. Lett.* **80**, 75–79.

Vadnal, R.E. and Bazan, N.G. (1988). Carbamazepine inhibits electroconvulsive shock-induced inositol trisphosphate (IP3) accumulation in rat cerebral cortex and hippocampus. *Biochem. Biophys. Res. Commun.* **153**, 128–134.

Verma, A., Nye, J.S. and Snyder, S.H. (1987). Porphyrins are endogenous ligands for the mitochondrial (peripheral-type) benzodiazepine receptor. *Proc. Natl. Acad. Sci. USA* **84**, 2256–2260.

Waldmeier, P.C. (1987). Is there a common denominator for the antimanic effect of lithium and anticonvulsants. *Pharmacopsychiatry* **20**, 37–47.

Weiss, S.R.B. and Post, R.M. (1987). Carbamazepine and carbamazepine-10,11-epoxide inhibit amygdala kindled seizures in the rat but do not block their development. *Clin. Neuropharmacol.* **10**, 272–279.

Weiss, S.R.B. and Post, R.M. (1990a). Development and reversal of conditioned inefficacy and tolerance to the anticonvulsant effects of carbamazepine. *Epilepsia*, in press.

Weiss, S.R.B. and Post, R.M. (1990b). Contingent tolerance to carbamazepine: A peripheral-type benzodiazepine mechanism. *Eur. J. Pharmacol.*, in press.

Weiss, S.R.B., Post, R.M., Patel, J. and Marangos, P.J. (1985). Differential mediation of the anticonvulsant effects of carbamazepine and diazepam. *Life Sci.* **36**, 2413–2419.

Weiss, S.R.B., Post, R.M., Marangos, P.J. and Patel, J. (1986). Peripheral-type benzodiazepines: behavioral effects and interactions with the anticonvulsant effects of carbamazepine. *In* "Kindling III" (ed. J. Wada), pp. 375–392. Raven Press, New York.

Weiss, S.R.B., Post, R.M., Szele, F., Woodward, R. and Nierenberg, J. (1989). Chronic carbamazepine inhibits the development of local anesthetic seizures kindled by cocaine and lidocaine. *Brain Res.* **497**, 72–79.

Weiss, S.R.B., Post, R.M., Costello, M., Nutt, D.J. and Tandeciarz, S. (1990). Carbamazepine retards the development of cocaine-kindled seizures but not sensitization to cocaine's effects on hyperactivity and stereotypy. *Neuropsychopharmacology* **3**, 273–281.

Weizman, A., Tanne, Z., Karp, L., Martfeld, Y., Tyano, S. and Gavish, M. (1987).Carbamazepine up-regulates the binding of [^{3}H]-PK 11195 to platelets of epileptic patients. *Eur. J. Pharmacol.* **141**, 471–474.

Whitworth, P. and Kendall, D.A. (1988). Lithium selectively inhibits muscarinic receptor-stimulated inositol tetrakisphosphate accumulation in mouse cerebral cortex slices. *J. Neurochem.* **51**, 258–265.

Willow, M. and Catterall, W.A. (1982). Inhibition of binding of [^{3}H]-Batrachotoxinin A 20-alpha-benzoate to sodium channels by the anticonvulsant drugs diphenylhydantoin and carbamazepine. *Mol. Pharmacol.* **22**, 627–635.

Willow, M., Kuenzel, E.A. and Catterall, W.A. (1984). Inhibition of voltage-sensitive sodium channels in neuroblastoma cells and synaptosomes by the anticonvulsant drugs diphenylhydantoin and carbamazepine. *Mol. Pharmacol.* **25**, 228–234.

Willow, M., Gonoi, T. and Catterall, W.A. (1985). Voltage clamp analysis of the inhibitory actions of diphenylhydantoin and carbamazepine on voltage-sensitive sodium channels in neuroblastoma cells. *Mol. Pharmacol.* **27**, 549–558.

Worley, P.F., Baraban, J.M., De Souza, E.B. and Snyder, S.H. (1986). Mapping second messenger systems in the brain: differential localizations of adenylate cyclase and protein kinase C. *Proc. Natl. Acad. Sci. USA* **83**, 4053–4057.

Worley, P.F., Heller, W.A., Snyder, S.H. and Baraban, J.M. (1988). Lithium blocks a phosphoinositide-mediated cholinergic response in hippocampal slices. *Science* **239**, 1428–1429.

Yoshikawa, K. and Hong, J.S. (1983). The enkephalin system in the rat anterior pituitary: regulation by gonadal steroid hormones and psychotropic drugs. *Endocrinology* **113**, 1218–1227.

Zatz, M. (1979). Low concentrations of lithium inhibit the synthesis of cyclic AMP and cyclic GMP in the rat pineal gland. *J. Neurochem.* **32**, 1315–1321.

13 Gastrointestinal Absorption of Lithium

ROBERT J. DAVIE

Biomedical Research Laboratory, School of Health Sciences, Wolverhampton Polytechnic, Wolverhampton WV1 4DJ, UK

Introduction

The characteristics of gastrointestinal absorption of lithium are important because this exclusively is the route of administration of the drug. The two main factors regulating the bioavailability of the preparation are K_{Ab} (absorption coefficient) and K_{El} (elimination constant); the latter is a measure of renal function, the former that of the intestine. Lithium concentrations will be influenced by a number of factors including gut porosity and motility. However, these characteristics are difficult to predict, and the rate of lithium absorption has been shown to exhibit great interindividual variation (Poust *et al.*, 1976). Plasma lithium levels may be modified again by different rates of excretion by the kidney. Both differences in absorption and excretion combine to bring about a wide variation in the lithium plasma levels after administration of the same dose to different subjects.

It is important to be aware that this may result in high plasma lithium levels which are potentially very serious, bringing about increasingly dangerous side effects such as dry mouth, muscle weakness, tremor, nausea, diarrhoea, coma and finally death. Furthermore, the difference in rate of absorption is suggested also to have considerable influence on the extent of associated side effects often persistent in lithium therapy. An understanding of the mechanisms involved in the intestinal absorption of lithium is important therefore in achieving a safe lithium level consistent with its physiological and pharmacological actions.

Gastrointestinal Absorption of Lithium

Few studies have been conducted to investigate the mechanisms of lithium absorption from the gastrointestinal tract. Peak plasma lithium concentrations

LITHIUM AND THE CELL
ISBN 0-12-099300-7

are usually achieved within 1–2 h following the administration of a standard lithium carbonate tablet, indicating that lithium is rapidly absorbed. However, absorption is critically related to tablet design. In an effort to reduce the plasma lithium peak, and therefore associated side effects, a range of 'slow' or 'sustained' release formulations have been developed. In these cases the peak may be delayed to be between 3–6 h. Using 'standard' lithium tablets, Tyrer (1978) showed that virtually 100% of a dose is absorbed. This confirmed earlier work by Hullin et al. (1968), who determined that less than 1% of lithium given as standard lithium carbonate leaves the body in the faeces. This value may rise substantially following administration of a 'slow' release formulation, with a greater percentage of lithium remaining in the faeces. Tyrer et al. (1976) showed also that the amount of lithium remaining in the faeces was inversely proportional to the plasma lithium concentration. Under 'normal' conditions, however, the majority of lithium absorption, between 70 and 90%, occurs in the upper and middle small intestine. Lithium absorption in the mouth is considered to be negligible though relatively high amounts of lithium are secreted in saliva. Ramsey et al. (1979) showed that very little lithium was absorbed in the stomach, although this will depend on the nature of the lithium tablet formulation. Up to 20% of a dose may be absorbed in the stomach from a conventional preparation, while a 'slow release' tablet will carry much of the lithium through the stomach into the small intestine. There is some evidence of intestinal lithium secretion. Kersten et al. (1986) administered lithium intravenously to rats. They found that after ligation of both kidneys and bile duct they obtained an increase in lithium content of the faeces. It is likely that lithium excreted in the upper intestine will be reabsorbed more distally.

In human intestine, the site of lithium absorption had not been studied prior to the work of Diamond et al. (1983). Using healthy volunteers this group employed a standard gut perfusion technique to study lithium absorption at three sites along the intestine: jejunum, ileum and colon. Their results showed lithium absorption to be negligible from the colon but substantial in human ileum and jejunum. They suggest lithium movement to be passive in leaky epithelia such as the jejunum and ileum. This is in agreement with the predominant transepithelial route of passage for small ions in the small intestine. In contrast, the negligible lithium transport from the colon demonstrates the restrictive nature of colonic tight epithelia to lithium movement. The passive nature of lithium movement in human ileum has been confirmed in vitro by Phillips et al. (1988a). Using tissue obtained at surgery this group could not saturate lithium movement up to a mucosal lithium concentration of 50 mM. Their work with metabolic inhibitors and extra-cellular markers suggest that lithium movement is dependent on the nature of the paracellular pathways.

A problem with experiments utilizing human gut perfusion is that they do not provide any direct information on the mode or mechanism of lithium absorption. Our laboratory has been investigating aspects of gastrointestinal absorption of lithium since 1983 (Birch *et al.*, 1983). Early studies using rat everted small intestinal gut sacs suggested that lithium movement was both time and concentration dependent and that lithium tissue kinetics were passive (Birch *et al.*, 1983, 1984). Similar findings were reported for isolated epithelial cells from guinea pig small intestine (Birch *et al.*, 1985).

These studies have continued using a more sophisticated isolated epithelial preparation in guinea pigs (Davie *et al.*, 1988). This technique, derived from that of Lauterbach (1977), involves the removal of the kinetically obstructive muscle and connective tissue layers from the absorptive mucosa. The epithelial sheet obtained is mounted in a small flux chamber which allows directional substrate kinetics to be determined. The use of a radiolabelled extracellular space marker, tritiated polyethylene glycol (M_r 900, PEG-900) provides a measure of membrane permeability. It allows a more accurate assessment of events occurring at the mucosal surface to be obtained. These studies have confirmed lithium kinetics to be a passive process, with no evidence of saturation with increasing luminal lithium concentration. Furthermore, the use of inhibitors (2,4-DNP; amiloride) or a 10°C reduction in temperature was shown to have no significant effect on lithium movement. A simultaneous measurement of glucose kinetics, however, as an indicator of active transport, showed significant metabolic inhibition by a reduction in glucose transport in all cases (Davie *et al.*, 1988). Earlier work by Davie *et al.* (1987) again demonstrated the importance of the paracellular spaces in lithium movement. In this study pre-incubated and loaded rat everted jejunal rings were placed in a buffer containing lithium and PEG-900 for 30 min. The rings were then washed and the efflux of lithium and PEG into fresh buffer measured over time. The movement of lithium was found to be very similar to that of the extracellular marker, suggesting lithium was primarily present in the extra-cellular spaces. Further work by Phillips *et al.* (1988b) used PEG-900 to calculate the extracellular fluid volume associated with the tissue. These calculations confirmed lithium only to be associated with the extracellular fluids, with negligible true tissue uptake actually taking place.

In all the above the studies were performed on lithium naive animals, that is, experiments were performed with acute tissue exposure to lithium. Phillips *et al.* (1990) have recently repeated some of this work on animals chronically exposed to lithium (25 mM) in their drinking water for at least three weeks prior to study. Results from this investigation demonstrated lithium to have accumulated in the tissue itself. This contrasts with the results of acute experiments *in vitro* (Phillips *et al.*, 1988b). These experiments may indicate a change, or adaptation, in the cellular membrane with time. It may be similar to that seen in the gradual inhibition of lithium extrusion from red cells due to changes in lithium–sodium counter transport (Meltzer *et al.*, 1977; Ehrlich and Diamond, 1980).

Gastrointestinal Side Effects of Lithium Ingestion

Gastrointestinal side effects frequently accompany the initiation of lithium treatment (Schou *et al.*, 1970). Over one-third of patients have been found to suffer abdominal discomfort during the first weeks of treatment although most problems do not persist. Common side effects include nausea, diarrhoea, muscular weakness and tremor. These symptoms have generally been found to coincide with the absorption peaks (Amdisen and Sjögren, 1968). In an effort to reduce or eliminate these side effects, many lithium formulations have been developed with 'slowed' or 'sustained' release properties. These formulations have varying degrees of success. Generally, the slow release of a drug will reduce the side effects due to intestinal irritation or related high blood concentration peaks. On the other hand, if the rate of release is substantially delayed (over 6 h), then an increased frequency of diarrhoea is likely (Borg *et al.*, 1974). This results from significant amounts of lithium reaching the colon where it acts as an irritant and induces a purgative effect (Tyrer *et al.*, 1976).

In contrast to lithium-induced diarrhoea, the use of lithium taken orally in the treatment of chronic secretory diarrhoea of life-threatening importance has been reported (Owyang, 1984). Its precise mode of action is not known but it is suggested that it is the effect of lithium on intracellular cyclic adenosine monophosphate which is responsible for its antisecretory effect in the intestine. Increased colonic water absorption has been observed also in rats fed a lithium-supplemented diet (Feldman *et al.*, 1981). This was associated with increased sodium absorption. Both observations support the use of lithium in the treatment of certain types of secretory diarrhoea.

Conclusions

The intestinal absorption of lithium is a passive process occurring via the paracellular pathways in leaky epithelia. Absorption occurs primarily in the small intestine, with the amount of lithium absorption being dependent on the initial luminal lithium concentration. Although metabolic energy is not required for the movement of lithium *per se*, it may be necessary for the maintenance of the intracellular tight junctions through which the lithium passes. In lithium therapy, the characteristics of intestinal absorption may be significantly influenced by the nature of lithium preparation. Very slow release tablets may delay absorption until the drug is in the distal intestine, where it may induce unpleasant side effects.

References

Amdisen, A. and Schou, M. (1967). Biochemistry of depression. *Lancet* **i**, 507.

Amdisen, A. and Sjögren, J. (1968). Lithium absorption from sustained-release tablets (Duretter®). *Acta Pharm. Suecica* **5**, 465–472.

Birch, N.J., Coleman, I.P.L. and Karim, A.R. (1983). The transfer of lithium across everted sacs of rat small intestine. *Br. J. Pharmacol.* **80**, 443P.

Birch, N.J., Coleman, I.P.L., Hilburn, M.E. and Karim, A.R. (1984). Factors affecting lithium absorption in rat small intestine. *Br. J. Pharmacol.* **81**, 158P.

Birch, N.J., Coleman, I.P.L., Karim, A.R. and Mann, C.J. (1985). The uptake of lithium by guinea pig isolated intestinal epithelial cells. *Biochem. Soc. Trans.* **13**, 250.

Borg, K.O., Jeppsson, J. and Sjögren, J. (1974). Influence of the dissolution rate of lithium tablets on side effect. *Acta Pharm. Suecica* **11**, 133–140.

Davie, R.J., Coleman, I.P.L. and Birch, N.J. (1987). Lithium uptake into rodent intestine. *Biochem. Soc. Trans.* **15**, 1170.

Davie, R.J., Coleman, I.P.L. and Partridge, S. (1988). Lithium transport in isolated epithelial preparations. *In* "Lithium: Inorganic Pharmacology and Psychiatric Use" (ed. N.J. Birch), pp. 107–111. IRL Press, Oxford.

Diamond, J.M., Ehrlich, B.E., Morawski, S.G., Santa Ana, C.A. and Fordtran. J.S. (1983). Lithium absorption in tight and leaky segments of intestine. *J. Membrane Biol.* **72**, 153–159.

Ehrlich, B.E. and Diamond, J.M. (1980). Lithium, membranes and manic-depressive illness. *J. Membrane Biol.* **52**, 187–200.

Feldman, G.M., Mann, J.J. and Charney, A.N. (1981). Effect of lithium ingestion on water and electrolyte transport in rat intestine. *Gastroenterology* **81**, 892–897.

Hullin, R.P., Swinscoe, J.C., McDonald, R. and Dransfield, G.A. (1968). Metabolic balance studies on the effect of lithium salts in manic-depressive psychosis. *Br. J. Psychiat.* **114**, 1561–1573.

Kersten, L., Fleck, Ch. and Braunlich, H. (1986). Evidence for an intestinal lithium excretion in rats. *Exp. Pathol.* **29**, 55–63.

Lauterbach, F. (1977). Passive permeabilities of luminal and basolateral membranes in the isolated mucosa of guinea pig small intestine. *N.S. Arch. Pharm.* **297**, 201–212.

Meltzer, H.L., Kassir, S., Dunner, D.L. and Fieve, R.R. (1977). Repression of a lithium pump as a consequence of lithium ingestion by manic depressive-subjects. *Psychopharmacology* **54**, 113–118.

Owyang, C. (1984). Treatment of chronic secretory diarrhoea of unknown origin by lithium carbonate. *Gastroenterology* **87**, 714–718.

Phillips, J.D., Davie, R.J., Kmiot, W.A., Poxon, V.A., Keighley, M.R.B. and Birch, N.J. (1988a). Lithium transport in human ileum. *Br. J. Pharmacol.* **96**, 253P.

Phillips, J.D., Davie, R.J. and Birch, N.J. (1988b). Low intracellular lithium concentrations: a consequence of paracellular transport? *Br. J. Pharmacol.* **95**, 836P.

Phillips, J.D., Davie, R.J. and Birch, N.J. (1990). Tissue uptake of lithium in guinea pig isolated mucosa after chronic lithium ingestion. *Biochem. Soc. Trans.* **18**, 653–654.

Poust, I.R., Mallinger, A.G., Mallinger, J., Himmelhock, J.M. and Hanin, I. (1976).

Pharmacokinetics of lithium in human plasma and erythrocytes. *Psychopharmacol. Commun.* **2**, 91–103.

Ramsey, E.J., Carey, K.V., Peterson, W.L., Jackson, J.J., Murphy, F.K., Reed, N.W., Taylor, K.B., Trier, J.S. and Fordtran, J.S. (1979). Epidemic gastritis with hypochlorhydria. *Gastroenterology* **76**, 932–938.

Schou, M., Baastrup, P.C., Grof, P., Weis, P. and Angst, J. (1977). Pharmacological and clinical problems of lithium prophylaxis. *Br. J. Psychiat.* **116**, 615–619.

Tyrer, S.P., Hullin, R.P., Birch, N.J. and Goodwin, J.C. (1976). Absorption of lithium following slow-release and conventional preparations. *Psycholog. Med.* **6**, 51–58.

Tyrer, S. (1978). The choice of lithium preparation and how to give it. *In* "Lithium in Medical Practice" (ed. F.N. Johnson and S. Johnson), pp. 395–405. MTP Press, Lancaster.

14 Renal Elimination of Lithium

MICHAEL SHALMI[1] and KLAUS THOMSEN[2]

[1] Department of Pharmacology, University of Copenhagen, Denmark

[2] Psychopharmacology Research Unit, University of Aarhus, Denmark

Introduction

When treatment with lithium was first instituted (Cade, 1949), little was known about its therapeutic index and pharmacokinetics. John Cade therefore decided to administer lithium in increasing dosages until side effects were observed and subsequently to reduce the dosage until manic episodes recurred. In this way, the therapeutic dosage became confined to an interposition between one causing toxicity and one being non-therapeutic.

Later it was discovered that lithium also prevented recurrence of manic-depressive illness (Baastrup and Schou, 1967), and extensive studies of the pharmacokinetics of lithium were initiated. It soon became apparent that lithium is eliminated almost exclusively by the kidneys, and urinary recovery ratios close to 100% of a daily dosage were found (Amdisen and Sjögren, 1968). Studies on the renal elimination of lithium in man suggested that lithium is reabsorbed exclusively in the proximal tubules (Thomsen and Schou, 1968). This observation generated the hypothesis that the renal lithium clearance could be used as an estimate of the delivery of sodium and water from the renal proximal tubules (Thomsen et al., 1969). This theory has been confirmed in a variety of experiments, and in 1989 the First International Conference on Lithium in Renal Physiology was held in Utrecht, Holland.

Knowledge of the renal handling of lithium in man originates from two sources: patients given lithium therapeutically or prophylactically, and volunteers given lithium as a small test dose on the day before the experiment. Previously, data from the former category dominated the literature, but our current information on the renal handling of lithium is derived almost exclusively from subjects given a small test dose. In this chapter, data from the two sources will not be dealt with separately but described together.

Today, the studies of the renal lithium excretion have two main assignments:

(1)   to serve as a research tool in studies on renal function; and
(2)   to further increase the therapeutic safety for patients in chronic lithium therapy;

and are thus of relevance for psychiatrists, nephrologists, renal and clinical physiologists, and pharmacologists.

Renal Handling of Lithium

Historical background

Along with the initiation of lithium treatment to prevent recurrences of manic-depressive disorder, methods of increasing the elimination rate of lithium became necessary in the treatment of lithium intoxication which occasionally occurred. During studies on renal lithium elimination in man it was noted that the renal lithium clearance (C_{Li}) was between 20 and 25% of the glomerular filtration rate (GFR) and was affected differently by various diuretics (Thomsen and Schou, 1968). Diuretic drugs acting mainly in the proximal tubules increased C_{Li}, whereas diuretics acting mainly in the loop of Henle or more distally left it unchanged. Together, these results suggested that lithium is reabsorbed in the proximal tubules to the same degree as sodium and water, whereas it is neither reabsorbed nor secreted distally (Thomsen and Schou, 1968). On this basis, Thomsen et al. (1969) proposed the use of lithium as a marker of the outflow from the proximal tubules (V_{prox}).

V_{prox} is regulated via hormonal and nervous systems and is furthermore essential for determining the absolute and fractional reabsorption in the proximal and the distal tubule segments. An estimation of renal water and sodium handling in the various segments of the nephron during different physiological and pathophysiological conditions is desirable from a clinical as well as from a physiological viewpoint.

V_{prox} is traditionally studied in two ways: by conventional clearance techniques in conscious humans and animals or by micropuncture in laparotomized anaesthetized animals. When the latter technique is used, proximal tubular fluid reabsorption is estimated by collection of tubular fluid through thin capillary tubes inserted into the lumen of a tubule, but ethical considerations invalidate its use in humans. Among the clearance techniques, phosphate and urea clearances have been suggested as markers of V_{prox} but both have proved unsuitable. The urine flow or the free water clearance plus chloride clearance during maximal water diuresis have also been proposed. However, these methods clearly overestimate the approximate 75% reabsorption of sodium and water in the proximal tubules due to free water back-diffusion along the distal nephron segment. In addition, these methods require water loading in order to suppress antidiuretic hormone-mediated water reabsorption in the kidney. The traditional methods of estimating V_{prox} were accordingly far from ideal and a clearance marker which remained valid during most conditions, as does C_{Li}, was therefore most welcome.

The study by Steele *et al.* (1975) was the first to support the idea that C_{Li} reflects V_{prox} in humans. However, these authors could not exclude the possibility that a significant amount of lithium was reabsorbed along the distal nephron segment. Ten years after the publication of the hypothesis on which the lithium clearance technique is based, Hayslett and Kashgarian (1979), in a micropuncture study on the renal handling of lithium, confirmed that lithium is reabsorbed exclusively in the proximal tubules to the same extent as water and sodium. During the following five years, two additional studies (Thomsen *et al.*, 1981; Shirley *et al.*, 1983) verified the method in different experimental protocols. These methodologically important studies were needed to stimulate the use of the technique. Today, the C_{Li} method has been generally accepted as a more or less reliable estimate of V_{prox}, with marked technical advantages over other methods. This appears from the increasing number of publications involving the lithium clearance technique: 44 studies in 1986, 73 in 1987, 66 in 1988 and 126 in 1989. However, the large number of studies using C_{Li} should not be taken as evidence of a 100% validity of C_{Li} as a marker of V_{prox}.

Validation of the lithium clearance technique

The lithium clearance technique is based on three simple assumptions regarding the renal handling of lithium:

(1) It is freely filtered and reabsorbed to the same extent as sodium and water in the proximal tubules including pars recta.
(2) It is not secreted.
(3) It is not reabsorbed in the distal nephron segments, i.e. in the loop of Henle, the distal convoluted tubule, the connecting tubule or the collecting duct.

Figure 1 indicates the proposed handling of lithium together with that of sodium and water. If Assumptions 2 and 3 are valid, it is obvious that the amount of lithium leaving the proximal tubules per minute is identical with the recovery of lithium in the urine per minute. If lithium furthermore is reabsorbed in the same proportion as sodium and water in the proximal

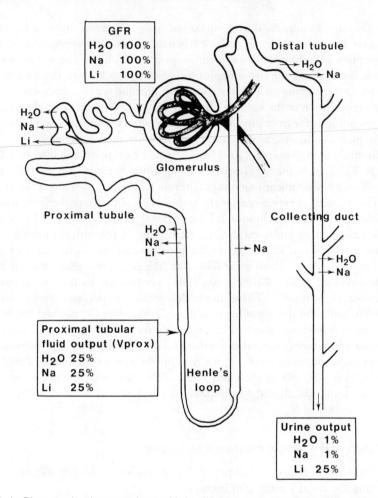

FIG. 1 Diagram showing a nephron with its different segments. For water, sodium and lithium the percentages of the filtered load which is delivered from the proximal tubules to the loop of Henle are indicated. The percentages that are excreted in the urine are also shown. From Thomsen (1987), with permission.

tubules (i.e. isosmotic), the fractional proximal reabsorption of lithium, $1 - C_{Li}/GFR$, is equal to the fractional proximal reabsorption of water, $1 - V_{prox}/GFR$, and of sodium, $1 - C_{Na\,prox}/GFR$; consequently, C_{Li} reflects V_{prox} and the clearance of sodium from the proximal tubules ($C_{Na\,prox}$).

The lithium clearance technique can probably never be completely verified or invalidated, primarily because no ideal method for comparison exists. Furthermore, a direct verification is hampered by the anatomical structure of the kidney. Many of the interesting studies which have been published during the last decade in support of or against the validity of the technique are therefore not unequivocal. However, by bringing different pieces of information together, a still more exact picture of its usefulness can be obtained.

As far as the lithium clearance technique is valid, it enables both clinicians and theoretical researchers to easily estimate V_{prox} from the whole nephron population in conscious animals and humans, without much preceding training and expensive equipment.

Support for the validity of the lithium clearance method has come from three different sources: micropuncture experiments in animals; comparative studies in man and animals with other markers of V_{prox}; and studies with different diuretic drugs.

Micropuncture studies

When experiments are designed to compare micropuncture data with lithium clearance observations, the major problem lies in estimating what is actually reabsorbed in the inaccessible part of the proximal tubules (i.e. the pars recta). This problem can be overcome if it is assumed that the absolute reabsorption of pars recta is about 30% of that in the pars convoluta (Thomsen *et al.*, 1981).

Hayslett and Kashgarian (1979) examined the reabsorption of lithium by measuring its concentration in tubular fluid from different parts of the nephron. They showed that lithium was reabsorbed to the same extent as sodium and water in the proximal tubules, the tubular fluid to plasma lithium ratio, $(TF/P)_{Li}$, being 1.0 at the end of the proximal convoluted tubule (verification of Assumption 1). It was also demonstrated that no net reabsorption or secretion occurred in the distal convoluted tubules and the collecting ducts (partial verification of Assumptions 2 and 3). Until now, this is the only study which validates all three assumptions *en bloc*. It lacks, like all other studies, exact information about tubular lithium reabsorption in pars recta of the proximal tubules and the descending limb and ascending limb of Henle's loop. However, the data fit the supposition that the reabsorption of lithium between end proximal convoluted tubules and early distal tubules is 30% of that in pars convoluta, leaving tubular lithium reabsorption in Henle's loop close to zero.

Thomsen *et al.* (1981) was the second study to validate the lithium clearance technique. C_{Li} was compared with micropuncture and a third method, the occlusion time–transit time (OT-TT), under three different experimental conditions, i.e. aortic constriction, hydropenia, and saline expansion. Under all three circumstances, C_{Li} was linearly correlated with the two other methods, with a high correlation coefficient.

Towards the end of 1989, 10 additional papers or abstracts with a total of 34 experimental groups had examined the relationship between estimates of V_{prox} obtained with C_{Li} and micropuncture or the OT-TT method (Shirley *et al.*, 1983; Dieperink *et al.*, 1986; Haas *et al.*, 1986; Shirley and Walter, 1986, 1989; Kirchner, 1987; Daugaard *et al.*, 1988; Holstein-Rathlou *et al.*, 1988; Walter and Shirley, 1988; Leyssac *et al.*, 1990b). In Fig. 2, the values of fractional reabsorption in the proximal tubules calculated from lithium clearance data are plotted against their respective micropuncture or OT-TT values. When both micropuncture and OT-TT were performed, the average between the two values is presented. Each point illustrates one experimental group. The experimental conditions varying from treatment with nephrotoxic drugs through salt depletion to saline expansion give a good idea of the validity of lithium clearance over a wide variety of conditions. Two experimental conditions led to deviation from the regression line, i.e. a low sodium intake (Kirchner, 1987) and furosemide infusion (Kirchner, 1987; Shirley and Walter, 1989). When these groups were excluded, the correlation coefficient was $r = 0.91$, with $p < 0.001$.

Although at first sight the relationship seems impressive, it gives little guarantee for the reliability of the lithium clearance method. The confidence

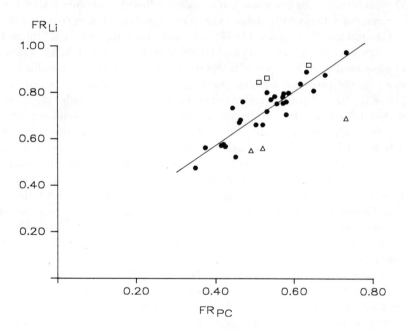

FIG. 2 Plot of the fractional reabsorption of lithium (FR_{Li}) versus the fractional reabsorption of tubular fluid in the superficial proximal convoluted tubules (FR_{PC}) in rats. □, Low sodium diet; △, furosemide. The regression line obtained by the method of least squares after exclusion of rats given a low sodium diet or furosemide is $FR_{Li} = 1.17 FR_{PC} + 0.11$.

interval of 95% was 0.10, which shows that the fractional reabsorption of lithium should increase or decrease by more than this value to be recognized as an outlier. On the other hand, the accuracy may also be better than estimated from the confidence interval and even close to 100% because errors stemming from the micropuncture or the OT-TT method also contribute to the scatter of the data.

Maximal urine flow

During experimental conditions where antidiuretic hormone (ADH) production is either absent or abolished, water reabsorption along the distal nephron segment is minimal. Under these circumstances, changes in V_{prox} will cause a parallel alteration of the urine flow if the osmotic gradient remains unaltered.

Lithium clearance has been compared with the urine flow in rats with hereditary diabetes insipidus (Brattleboro DI rats). In this experimental setting it was discovered that during severe sodium deprivation C_{Li} decreased relatively more than the urine flow and the urine to plasma concentration ratio for lithium fell below 1.0, indicating reabsorption of lithium distal to the proximal tubules (Thomsen and Leyssac, 1986a).

In humans, the knowledge about V_{prox} prior to introduction of the lithium clearance technique almost exclusively relied on measurements of the urine flow during maximal water diuresis (V_{max}) and its related expressions derived from the free water clearance technique. Water loading *per se* does not influence C_{Li} to any significant degree (Thomsen and Schou, 1968; Boer *et al.*, 1988a), and high correlations between C_{Li} and V_{max} ($r = 0.62$ to $r = 0.88$) have been found (Thomsen and Olesen, 1984; Roos *et al.*, 1985; Atherton *et al.*, 1986; Hla-Yee-Yee and Shirley, 1987; Rombola *et al.*, 1987; Boer *et al.*, 1989b).

Lithium clearance has been compared with V_{max} in a number of different clinical conditions and control situations (Thomsen and Olesen, 1984; Roos *et al.*, 1985; Boer *et al.*, 1987a, 1988a,b, 1989a; Gaillard *et al.*, 1987, 1988, 1989; Hla-Yee-Yee and Shirley, 1987; Solomon *et al.*, 1988a; Wetzels *et al.*, 1988; McMurray *et al.*, 1989; Rabelink *et al.*, 1989a,b). Figure 3 (A) illustrates a plot of C_{Li} versus V_{max} obtained from the above-quoted published papers (abstracts are not included in this plot). Each point represents the average of an experimental group and consists of a number of individual measurements. The correlation coefficient for the regression between C_{Li} and V_{max} is $r = 0.94$, with $p < 0.001$. This graph also contains groups of measurements where the subjects had been on a sodium-restricted diet or had been treated with furosemide (Roos *et al.*, 1985; Atherton *et al.*, 1987; Boer *et al.*, 1988b; Wetzels *et al.*, 1988), but no set of data fell outside the 95% confidence limit. This indicates that man and rat respond differently to sodium deprivation and furosemide and suggests that in humans reabsorption of lithium at nephron

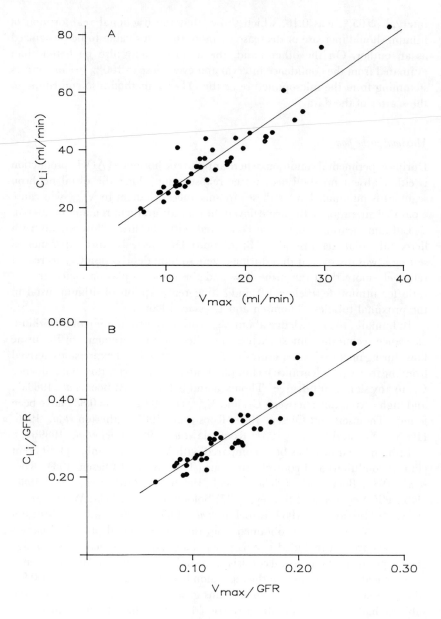

FIG. 3   (A) Plot of the lithium clearance (C_{Li}) versus the urine flow rate during maximal water diuresis (V_{max}) in humans. The regression line obtained by the least-squares method is: $C_{Li} = 1.89V_{max} + 6.4$. (B) Plot of the fractional lithium excretion (C_{Li}/GFR) versus the fractional urine flow rate during maximal water diuresis (V_{max}/GFR) in humans. The regression line is: $C_{Li}/GFR = 1.76V_{max}/GFR + 0.072$.

sites distal to the proximal tubules may not be initiated by sodium deprivation. Division by GFR (Fig. 3(B)) estimated either by inulin clearance or creatinine clearance did not lower the high correlation coefficient significantly ($r = 0.87$, $P < 0.001$). This figure further strengthens the presumption that in man lithium is not reabsorbed to any significant degree along the distal nephron segment. However, as is the case for Fig. 2, the high correlation is not a 100% proof of the validity of the lithium clearance concept, since lithium and water may have been reabsorbed in the same proportion along the distal nephron segment. Some studies suggest that reabsorption of lithium in the loop of Henle can be induced by indomethacin or DDAVP administration, but in these studies a decrease in the outflow from the proximal tubules could also explain the findings, and since this possibility was not excluded they are not conclusive (Boer *et al.*, 1988a; Rabelink *et al.*, 1989a).

In conclusion, comparative studies between C_{Li} and V_{max} have provided important information about the validity of the lithium clearance method. Studies have demonstrated that reabsorption of lithium in the distal nephron segment occurs in sodium-restricted rats, thus invalidating the lithium clearance method, whereas quantitatively significant distal lithium reabsorption has not been explicitly observed in man.

Studies with diuretics

Administration of diuretics is another way to verify the lithium clearance method. Ideally, if segment-specific diuretics for each part of the nephron existed it would be possible to allocate a lithiuretic action to a certain segment. The specific transport mechanisms on which different diuretics act are, however, rarely localized to just one part of the nephron. It is therefore difficult to conclude whether a change in C_{Li} or C_{Li}/GFR in response to administration of a diuretic is due to failure of one of the three lithium clearance assumptions or to an as yet not characterized effect of the drug on the proximal tubules.

Another problem in interpreting a lithiuretic effect, or lack of effect, of diuretics is that the kidneys cannot be studied as an isolated organ with a simple response to administration of a diuretic. A fine balance between the size and composition of the extracellular fluid volume and the renal water and electrolyte excretion rate is maintained via nervous and humoral control mechanisms. When a diuretic is administered, the extracellular volume (ECV) diminishes causing an activation of the control systems. This activation can only be prevented if urinary losses are adequately replaced. If they are not, the clearance results obtained will be a combination of a direct action of the diuretic opposed by compensatory mechanisms triggered by the ECV control systems. This has been clearly demonstrated by Christensen *et al.* (1986). When furosemide was infused constantly to rats without replacement of urinary

M. Shalmi and K. Thomsen

output, C_{Li} and C_{Li}/GFR increased rapidly and thereafter decreased to control levels despite continued furosemide infusion, while it remained elevated in the volume-replaced group (Fig. 4).

Loop diuretics such as furosemide, bumetanide, and piretanide have all been shown to increase C_{Li} in both man and rats (Steele *et al.*, 1975, 1976; Christensen *et al.*, 1986; Atherton *et al.*, 1987, 1988a; Lu *et al.*, 1988; Shalmi *et al.*, 1989, 1990). These observations are in contradiction with the preliminary observations by Thomsen and Schou (1968), but this is simply due to differences in the length of the experimental period. While current studies involving diuretics have a shorter duration, Thomsen and Schou (1968) used a 6-h urine collection period and their initial observation that furosemide did not change C_{Li} is now accepted as an artefact due to the study protocol. In a recent study it was found that the lithiuretic effect of 80 mg furosemide given once a day p.o. for 8 days to healthy volunteers only lasted between 2 and 3 h after the ingestion (Shalmi *et al.*, 1990). The observation that loop diuretics increase C_{Li} suggests that lithium is either reabsorbed in the loop of

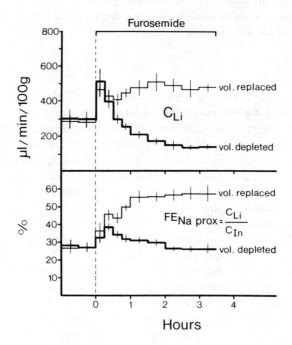

FIG. 4 Effects of constant furosemide infusion (7.5 mg kg^{-1} h^{-1}) on lithium clearance and fractional lithium excretion in rats with and without replacement of volume losses by saline infusion. The data are mean values $\pm$ SE ($n = 6$). In the volume-replaced group a significant increase of C_{Li}/GFR occurred after furosemide infusion for 15 min ($p < 0.05$ versus control group). From Christensen *et al.* (1986), with permission.

Henle (invalidation of Assumption 3) or that the tubular fluid reabsorption in the proximal tubules is also affected. When fractional proximal tubular sodium reabsorption is maximally depressed by massive saline infusion, furosemide can still further increase the sodium clearance while C_{Li} remains unaltered (Atherton *et al.*, 1987; Christensen *et al.*, 1988). This experiment separates the actions of furosemide into a loop of Henle component and a proximal tubule component. Since the former component is present even when the proximal tubular sodium reabsorption is maximally inhibited by the saline infusion, lithium is most probably not reabsorbed to any significant degree in the loop of Henle. However, it has recently been argued that during experimental circumstances where the flow to the ascending limb of Henle's loop is lower than during saline expansion, the transepithelial lumen positive electrical gradient might favour passive paracellular lithium reabsorption (Greger, 1990).

Diuretics from the thiazide group can be divided into those with carbonic anhydrase inhibiting activity and those without. The former (e.g. chloro-thiazide) could theoretically affect proximal tubular sodium reabsorption like acetazolamide, while the latter (e.g. bendroflumethiazide) would be confined to act on the distal convoluted tubule. This has been demonstrated by Boer *et al.* (1989a), who studied the effects of chlorothiazide and bendro-flumethiazide on C_{Li} in healthy sodium-restricted volunteers undergoing water diuresis. They report that while bendroflumethiazide left C_{Li} unchanged, chlorothiazide increased it. This supports the notion that lithium is not reabsorbed in the distal convoluted tubule, even when sodium intake is reduced.

Amiloride and other potassium-sparing diuretics such as spironolactone and canrenone, act primarily in the late parts of the distal tubule and in the collecting ducts. It has been demonstrated that amiloride inhibits the low sodium-induced distal lithium reabsorption in rats and dogs without altering V_{prox} (Thomsen and Leyssac, 1986b; Boer *et al.*, 1987b; Kirchner, 1987). Studies on the effects of amiloride in sodium-restricted humans are conflicting; Atherton *et al.* (1987) observed a small increase in absolute and fractional lithium clearance, while Boer *et al.* (1988b) and Bruun *et al.* (1989) did not. During conditions with a normal sodium intake, no increase in C_{Li} following amiloride administration has been observed in either rats or humans (Thomsen and Leyssac, 1986b; Atherton *et al.*, 1987; Boer *et al.*, 1988b; Rabelink *et al.*, 1989c). It is therefore unlikely that lithium is reabsorbed in the amiloride-sensitive part of the nephron during normal sodium intake.

In summary, studies with diuretics have shown that a definite reabsorption of lithium in the distal nephron segment only occurs during sodium deprivation of animals. In man, no experimental setting has definitely caused an explicit distal reabsorption of lithium.

Conclusions

The studies presented are in accordance with the hypothesis that lithium is reabsorbed exclusively in the proximal tubules, and that C_{Li} is a reliable measure of V_{prox}. Since the lithium clearance method can be used in humans and conscious animals, it offers obvious advantages over other existing methods for determination of the delivery of tubular fluid from the proximal tubules.

During low sodium intake, lithium may be reabsorbed distally by an amiloride-sensitive transport mechanism. This has been observed in rats and dogs, and cannot be totally excluded in humans. Subtle amounts of lithium may also be reabsorbed in the loop of Henle to varying degrees, but so far this has not been unequivocally demonstrated. The possibility of distal reabsorption of lithium either in the loop of Henle or more distally should always be kept in mind and calls for cautious interpretations of changes in C_{Li}, but it should not prevent the use of the method. Other conditions with distal reabsorption of lithium may be recognized as more experience with the method is gained.

Lithium Transport in the Proximal Tubules

The anatomical structure of the nephron (Fig. 1) makes it impossible to assess directly the reabsorption of water and electrolytes in pars recta of the proximal tubules *in situ*. In addition, studies on the tubular handling of lithium are rare, because research in this field has been hampered by technical difficulties in estimating lithium concentration in nanolitre samples. Hayslett and Kashgarian (1979) found in saline-infused rats a $(TF/P)_{Li}$ of 1.0 at the end of pars convoluta. This indicates that lithium is reabsorbed in the same proportion as sodium and water. However, during osmotic Mannitol diuresis, the $(TF/P)_{Li}$ ratio never falls below 1, while the $(TF/P)_{Na}$ ratio is reduced to 0.9, indicating that lithium follows water rather than sodium under these conditions (Leyssac *et al.*, 1990b). Whether experimental conditions other than osmotic diuresis can separate lithium from sodium reabsorption in the proximal tubules remains to be established.

The tubular lithium reabsorption could be either passive and follow the paracellular route or it could be an active transcellular transport. Due to the high conductance of the epithelium in the proximal tubules, passive reabsorption of lithium across the epithelium is likely to play an important role. Urea is only passively reabsorbed in the proximal tubules and has the same passive permeability as lithium. The $(TF/P)_{Urea}$ ratio reaches a value of 1.5 early in the proximal convoluted tubule and remains at this level throughout, whereas the $(TF/P)_{Li}$ is almost equal to 1.0 (Lassiter *et al.*, 1961; Hayslett and Kashgarian, 1979). Passive, paracellular cation transport alone can therefore hardly account for the lithium reabsorption, so an active transport through the cells may also be involved (Leyssac *et al.*, 1990b).

Lithium is easily transported from the lumen into the cell, because the cytosol has a negative potential compared with the lumen. The Na/H antiporter located in the brush border has been shown to possess an affinity to lithium (Aronson and Igarishi, 1986; Garty and Benos, 1988). However, since lithium has little affinity for the basolateral Na/K-ATPase, it can hardly be transported actively across the basolateral membrane from the cytosol to the interstitium by the same mechanism as sodium (Leyssac *et al.*, 199a). Most likely, lithium is transported out of the cell with a secondary active process by the Na/H antiporter but the exact pathway by which lithium passes from lumen of the proximal tubule to the intertubular interstitium is at present not known.

Methodological Considerations in Lithium Clearance Determinations

When lithium clearance is used as a research tool to determine V_{prox}, certain precautions should be taken to minimize the risk of dubious conclusions.

Lithium administration

Lithium itself may affect the renal tubular sodium and water reabsorption (Murphy *et al.*, 1969). Even when plasma lithium concentrations are relatively low, such as after administration of a test dose of lithium, marked natriuretic and diuretic effects have been observed by Peart *et al.* (1987) after administration of 1020 mg lithium carbonate and confirmed by Luik *et al.* (1989) after administration of 600 mg. Strazzullo *et al.* (1988) found that 300 mg lithium carbonate did not affect renal function, indicating a certain dose dependence of the renal effects of lithium. We have shown that C_{Li} in rats is affected by the mode of lithium administration. A 40% increase of C_{Li} was observed when lithium was administered intravenously instead of by dietary intake for two days, even though the plasma lithium concentration was identical in the two groups (Shalmi and Thomsen, 1989).

Jeffrey *et al.* (1988) suggested that lithium (750 mg lithium carbonate administered in the evening before the study) interacts with dopamine at the proximal tubules and thereby invalidates the use of C_{Li} in assessment of tubular actions of dopaminergic agents. The study compared the renal response to gludopa infusion in a group of healthy volunteers given or not given lithium. Its conclusion is, however, doubtful because the baseline sodium excretion rate of the two groups was different at the beginning of the two experiments, and the sodium excretion rate during gludopa infusion did not differ markedly with and without lithium pretreatment. In a subsequent study, Girbes *et al.* (1989) showed that when only 300 mg lithium carbonate was administered, no difference could be detected in the natriuretic response after infusion of a selective agonist of dopaminergic type 1 receptor with and without lithium pretreatment.

The observations that changes in renal function can be induced by lithium administration do not invalidate the C_{Li} technique as claimed by Luik *et al.* (1989), but call for careful considerations when a study protocol involving C_{Li} is planned. The plasma lithium concentration should preferably be kept as low as possible (< 0.2 mmol 1^{-1}), and lithium should preferably be given for two days. If sustained-release tablets are administered instead of ordinary lithium carbonate tablets, a more prolonged plasma lithium concentration profile without a high peak is achieved (Amdisen, 1975). This may reduce the changes in renal water and electrolyte reabsorption following lithium administration. One way to eliminate completely the problem of a natriuretic effect of exogenously administered lithium is to measure the endogenous C_{Li}. This technique has marked advantages over the exogenous C_{Li} in that it allows rapid determination of C_{Li} without interfering effects from the lithium test dose. However, the method of measuring plasma and urinary lithium concentrations in nanomolar concentrations is new and requires flameless atomic absorption spectrometry (Durr *et al.*, 1989).

Effects of posture

Posture influences renal function, which should be taken into account when a C_{Li} experiment is performed. Changes in posture from upright to recumbent increase the fractional excretion of lithium from about 0.2 to about 0.3 (Hla-Yee-Yee and Shirley, 1987; Smith and Shimizu, 1976). In line with this, Kamper *et al.* (1988) showed that sitting gave results that were intermediate between walking and resting in a supine position.

Length of clearance period

The length of the clearance periods must be adjusted according to the purpose of the study. If no acute changes in renal function are induced, a clearance period of 3–6 h is suitable to ensure a sufficiently large urine production to minimize bladder-emptying errors. If, on the other hand, changes in renal function are induced, shorter clearance periods may be necessary in order to detect them. It should be considered, however, that short-lasting periods are influenced by intraindividual spontaneous variations of C_{Li} and therefore not necessarily represent the average kidney function of that particular subject (Thomsen *et al.*, 1988).

Dietary considerations

Lithium clearance in man is affected by the dietary sodium intake (Roos *et al.*, 1985; Atherton *et al.*, 1987; Boer *et al.*, 1988b). Interindividual variations in salt intake may therefore contribute to the scatter of the C_{Li} measurements, and fixed dietary regimens may be recommended in order to reduce the scatter.

Caffeine-containing beverages like coffee, tea, and cocoa should not be ingested on the morning of the experiment, because methyl-xanthine derivatives such as aminophylline have been shown to increase C_{Li} (Thomsen and Schou, 1968).

Intake of proteins increases GFR and C_{Li} (Solomon *et al.*, 1988b). Meals should therefore be avoided during an experiment; if this is not feasible, they should be standardized.

Calculations

By using the lithium clearance technique it is possible to subdivide the nephron into two parts: the proximal tubules (pars convoluta and pars recta) and the distal nephron segment (loop of Henle, the distal convoluted tubule and the collecting duct). If GFR, C_{Li}, sodium clearance (C_{Na}), and urine flow (V) are determined simultaneously, the following mathematical expressions of the absolute rate of reabsorption can be derived:

Reabsorption of water in the proximal tubules $= GFR - C_{Li}$
Reabsorption of water in the distal nephron $= C_{Li} - V$
Reabsorption of sodium in the proximal tubules $= (GFR - C_{Li}) \times P_{Na}$
Reabsorption of sodium in the distal nephron $= (C_{Li} - C_{Na}) \times P_{Na}$

Fractional segmental reabsorption is calculated by dividing the absolute reabsorption rate by GFR, and fractional segmental excretions by subtracting the fractional segmental reabsorption from 1. P_{Na} is the plasma sodium concentration.

Factors Influencing the Serum Lithium Concentration during Maintenance Treatment

Although seldom, episodes of lithium intoxication do occur (Schou *et al.*, 1989). In the catchment area of the psychiatric hospital in Aarhus, a total of 24 cases out of 4900 patient years was observed in the period 1979–1987. Fifteen were caused by deliberate self-poisoning, three were due to patient non-compliance, one was iatrogenic, and five were caused by an altered water and electrolyte intake or excretion. In no case did long-standing prophylactic lithium treatment *per se* induce deterioration of renal function. In a previous report from the same geographical area of Denmark, Hansen and Amdisen (1978) found that 18 out of 23 cases of lithium intoxication during the period 1969–1976 were caused by an altered water and electrolyte balance, while only one was due to self-poisoning. This illustrates that the most frequent reasons for lithium intoxication have changed with time, possibly due to a better understanding of the pharmacokinetics of lithium together with a more close lithium control regimen and a lowering of the therapeutic lithium dosage (Schou and Vestergaard, 1988).

Lithium is almost exclusively eliminated through the kidneys (Amdisen and Sjögren, 1968). Changes in plasma lithium concentration during prophylactic maintenance treatment with a constant dose regimen are therefore due to changes in the renal elimination of lithium. Since the renal C_{Li} is closely and practically solely related to the outflow from the proximal tubules, any situation or circumstance which alters V_{prox} could cause an undesirable increase or decrease in the plasma lithium concentration.

V_{prox} is determined by the GFR and the proximal tubular fluid reabsorption. If GFR is decreased or fluid reabsorption in the proximal tubules is increased, V_{prox} decreases. The V_{prox} is primarily influenced by changes in the volume and constitution of the extracellular fluid. As a model, the ECV can be comprehended as a compartment of electrolytes and water, which is in an equilibrium established by input from the gastrointestinal duct and outputs through the kidneys and sweat glands, with the kidney being in control of the major bulk of output. If this equilibrium is altered, nervous and humoral regulatory mechanisms are triggered in order to reestablish its original status by a decrease or an increase of the renal water and electrolyte excretion.

Lithium

Lithium itself alters renal water and electrolyte excretion (Murphy *et al.*, 1969). Polyuria is a well-known side effect of lithium treatment and necessitates a close control of the daily fluid intake. If urinary losses are not adequately replaced, dehydration occurs with a subsequent lowering of V_{prox} and hence of C_{Li} (Hansen and Amdisen, 1978; Thomsen and Olesen, 1979; Gross *et al.*, 1989).

In rats, lithium intoxication is primarily induced by renal salt loss, which triggers a lowering of C_{Li}. As a result of the reduced C_{Li}, intake of the same amount of lithium will cause an increase in the plasma lithium concentration, which further strengthens the lithium-induced natriuresis with a subsequent lowering of C_{Li}. This vicious circle proceeds until a stage of renal failure is reached (Thomsen, 1976).

While the symptoms of end stage lithium intoxication in rats is a circulatory collapse, lithium intoxication in man is characterized by neurological disorders like tremor, muscular rigidity or fasciculation and apathy gradually increasing in severity towards unconsciousness (Hansen, 1981). In rats, C_{Li} is reduced more than GFR, whereas in humans C_{Li} and GFR are often equally reduced (Thomsen, 1973; Hansen and Amdisen, 1978). The reasons for these species differences are unknown.

Physiological and pathophysiological changes of GFR and V_{prox}

Lithium treatment is usually stopped during pregnancy due to the risk of teratogenicity, but some women nevertheless resume their lithium treatment when the first trimester has been passed (Schou and Weinstein, 1980). During pregnancy, the ECV is expanded and GFR and C_{Li} are increased by about 30–40% (Atherton *et al.*, 1988b).

A reduced input to the ECV compartment can be caused by rigorous slimming or during intake of a sodium-deprived diet. The input is also reduced when patients suffer from physical illness causing vomiting and diarrhoea.

Physical illness with fever represents a condition where the ECV has an increased extrarenal waste due to the loss of sodium with the sweat and is the most common cause of lithium intoxication if the dose regimen is not altered appropriately (Schou *et al.*, 1989).

A decrease of GFR with parallel changes in C_{Li} may also be due to pathophysiological processes such as pyelonephritis or glomerulonephritis (Thomsen *et al.*, 1969). With advancing age, a decrease in GFR due to a decreased amount of functioning nephrons has also been observed (Schou *et al.*, 1986).

Diuretics

Treatment with diuretics was previously the most frequent iatrogenic reason for lithium intoxication. This has led to a general warning from medical authorities against prescription of diuretics to patients in lithium treatment. Nevertheless, certain patients may be in need of both kinds of treatment.

It has been shown that chronic treatment with thiazide diuretics lowers C_{Li} (Petersen *et al.*, 1974). Shirley *et al.* (1983), using micropuncture technique in rats, demonstrated that the thiazide-induced lowering of C_{Li} was due to an increased proximal tubular fluid reabsorption. Treatment with thiazides to patients in lithium therapy has nevertheless been successful, provided there is a close monitoring of plasma lithium concentration and an adequate reduction of the lithium dosage (Solomon, 1978).

The effect of furosemide on C_{Li} differs from that of thiazides. Studies have shown that chronic furosemide treatment does not alter the plasma lithium concentration (Jefferson and Kalin, 1979; Saffer and Coppen, 1983). In a recent investigation in chronically furosemide-treated healthy volunteers it was found that due to a very short period of action, furosemide causes marked diurnal variations of C_{Li}. After administration of 80 mg furosemide p.o., C_{Li} rose above its control level for 2–3 h, while during the rest of the 24 h it remained at a lower level (Shalmi *et al.*, 1990). This biphasic movement of C_{Li} resulted in an unaltered 24-h C_{Li} irrespective of furosemide treatment.

Even though furosemide seems to be a safer diuretic for patients in prophylactic lithium treatment than thiazides, a close control of the plasma lithium concentration is recommended whenever diuretic treatment is needed.

Other drug interactions

Case reports on lithium intoxication caused by treatment with different drugs occasionally occur in the literature. It can be difficult to uncover whether such intoxications are due to direct effects of the drugs on the GFR or the fractional proximal reabsorption resulting in a lowering of V_{prox} or are secondary to a reduced input to the ECV compartment. However, it has been shown that non-steroidal anti-inflammatory drugs such as indomethacin, ibuprofen and phenylbutazone have caused unintentional rises of the plasma lithium concentration.

All nephrotoxic drugs affecting GFR or the proximal tubular fluid reabsorption will also influence C_{Li}. A lowering of GFR and C_{Li} has been observed in patients treated with cyclosporin-A (Dieperink *et al.*, 1987). As for the treatment with diuretics, caution is recommended when nephrotoxic drugs are administered to patients in lithium therapy.

Drug interactions causing a decrease in P_{Li} have also been described. Perry *et al.* (1984) demonstrated that theophylline used in the treatment of asthmatic patients increases C_{Li}. This confirms the observation of Thomsen and Schou (1968) that methyl-xanthine derivatives increase C_{Li}.

Concluding Remarks

Studies on the renal lithium elimination have provided useful information on circumstances where unintended lithium intoxication may arise, and on the delivery of fluid and electrolytes from the proximal tubules under various circumstances.

The lithium intoxications are most often induced by disturbances of the fluid and electrolyte balance. Lithium-treated patients and clinicians should therefore be aware of conditions which may alter the size and constitution of the ECV and/or the renal water and electrolyte excretion. Fortunately, the frequency of these intoxications is still decreasing, presumably due to a better understanding of the renal elimination of lithium in addition to a decreased frequency of lithium-induced polyuria.

The use of the C_{Li} as a research tool is new, and much more experience with the method is needed. For this reason, lithium clearance determinations as a clinical examination are still not a common hospital practice, but Whiting *et al.* (1989) recently suggested measurements of C_{Li} as a useful adjunct to the diagnosis of cyclosporin-A-induced nephrotoxicity. Future research may reveal other clinical conditions where C_{Li} determination could be useful. The technique provides information which cannot be obtained by other methods and is furthermore inexpensive, not unpleasant to the patients and easy to use. So far, only one condition has been found to invalidate its use, i.e. administration of an extremely low salt diet to rats and dogs. With the new technique of flameless atomic absorption photometry, which markedly improves the sensitivity of the lithium determinations, a new promising area in the lithium clearance research field has been introduced.

References

Amdisen, A. (1975). Sustained release preparations of lithium. *In* "Lithium Research and Therapy" (ed. F.N. Johnson), pp. 197–210. Academic Press, London.

Amdisen, A. and Sjögren J. (1968). Lithium absorption from sustained-release tablets (Duretter®). *Acta Pharm. Suec.* **5**, 465–472.

Aronson, P.S. and Igarashi, P. (1986). Molecular properties and physiological roles of the renal Na^+-H^+ exchanger. *In* "Na^+-H^+-Exchange. Intracellular pH and Cell Function. Current Topics in Membranes and Transport" (ed. P.S. Aronson and W.F. Boron), Ch. 4. Academic Press, London.

Atherton, J.C., Hussain, R. and Jones, R. (1986). Assessment of proximal and distal nephron water and sodium reabsorption in man. *Renal Physiol.* **9**, 74–75.

Atherton, J.C., Green, R., Hughes, S., McFall, V., Sharples, J.A., Solomon, L.R. and Wilson, L. (1987). Lithium clearance in man: effects of dietary salt intake, acute changes in extracellular fluid volume, amiloride and frusemide. *Clin. Sci.* **73**, 645–651.

Atherton, J.C., Baker, S., Harrison, P. and Solomon, L.R. (1988a). Lithium clearance: effects of frusemide in water-loaded man. *J. Physiol.* **403**, 14P.

Atherton, J.C., Bielinska, A., Davidson, J.M., Haddon, I., Kay, C. and Samuels, R. (1988b). Sodium and water reabsorption in the proximal and distal nephron in conscious pregnant rats and third trimester women. *J. Physiol.* **396**, 457–470.

Baastrup, P.C. and Schou, M. (1967). Lithium as a prophylactic agent. Its effect against recurrent depressions and manic-depressive psychosis. *Arch. Gen. Psychiat.* **16**, 162–172.

Boer, W.H., Koomans, H.A. and Dorhout Mees, E.J. (1987a). Lithium clearance in mineralocorticoid escape in humans. *Amer. J. Physiol.* **252**, F382-F386.

Boer, W.H., Joles, J.A., Koomans, H.A. and Dorhout Mees, E.J. (1987b). Decreased lithium clearance due to distal tubular lithium reabsorption in sodium-depleted dogs. *Renal Physiol.* **10**, 65–68.

Boer, W.H., Koomans, H.A. and Dorhout Mees, E.J. (1988a). Renal lithium handling during water loading and subsequent d-DAVP-induced anti-diuresis. *Eur. J. Clin. Invest.* **18**, 273–278.

Boer, W.H., Koomans, H.A., Dorhout Mees, E.J., Gaillard, C.A. and Rabelink, A.J. (1988b). Lithium clearance during variations in sodium intake in man: effects of sodium restriction and amiloride. *Eur. J. Clin. Invest.* **18**, 279–283.

Boer, W.H., Koomans, H.A. and Dorhout Mees, E.J. (1989a). Acute effects of thiazides, with and without carbonic anhydrase inhibiting activity, on lithium and free water clearance in man. *Clin. Sci.* **76**, 539–545.

Boer, W.H., Koomans, H.A., Beutler, J.J., Gaillard, C.A., Rabelink, A.J. and Dorhout Mees, E.J. (1989b). Small intra- and large inter-individual variability in lithium clearance in humans. *Kidney Int.* **35**, 1183–1188.

Bruun, N.E., Skøtt, P., Lønborg-Jensen, H. and Giese, J. (1989). Unchanged lithium clearance during acute amiloride treatment in sodium-depleted man. *Scan. J. Clin. Lab. Invest.* **49**, 259–263.

Cade, J.F.J. (1949). Lithium salts in the treatment of psychotic excitement. *Med. J. Austr.* **36**, 349–352.

Christensen, S., Steiness, E. and Christensen, H. (1986). Tubular sites of furosemide natriuresis in volume-replaced and volume-depleted conscious rats. *J. Pharmacol. Exp. Ther.* **239**, 211–218.

Christensen, S., Shalmi, M. and Petersen, J.S. (1988). Lithium clearance as an indicator of proximal tubular sodium handling during furosemide diuresis. *J. Pharmacol. Exp. Ther.* **246**, 753–757.

Daugaard, G., Holstein-Rathlou, N.-H. and Leyssac, P.P. (1988). Effect of cisplatin on proximal convoluted and straight segments of the rat kidney. *J. Pharmacol. Exp. Ther.* **244**, 1081–1085.

Dieperink, H., Leyssac, P.P., Starklint, H. and Kemp, E. (1986). Nephrotoxicity of cyclosporin-A. A lithium clearance and micropuncture study in rats. *Eur. J. Clin. Invest.* **16**, 69–77.

Dieperink, H., Leyssac, P.P., Kemp, E., Starklint, H., Frandsen, N.E., Tvede, N., Møller, J., Buchler Frederiksen, P. and Rossing, N. (1987). The nephrotoxicity of cyclosporin A in humans. Effects of glomerular filtration and tubular reabsorption rates. *Eur. J. Clin. Invest.* **17**, 493–496.

Durr, J.A., Miller, N.L. and Alfrey, A.C. (1989). The electrothermal atomic absorption method for trace lithium levels in serum and urine, a potential tool for research in physiology. Congr. "Lithium in Renal Physiology", Utrecht, Abstr. 20.

Gaillard, C.A., Koomans, H.A., Rabelink, A.J. and Dorhout Mees, E.J. (1987). Effects of indomethacin on renal response to atrial natriuretic peptide. *Amer. J. Physiol.* **253**, F868–F873.

Gaillard, C.A., Koomans, H.A., Rabelink, T.J., Braam, B., Boer, P. and Dorhout Mees, E.J. (1988). Enhanced natriuretic effect of atrial natriuretic factor during mineralocorticoid escape in humans. *Hypertension* **12**, 450–456.

Gaillard, C.A., Koomans, H.A., Rabelink, A.J., Boer, P. and Dorhout Mees, E.J. (1989). Renal response to infusion versus repeated bolus injections of atrial natriuretic factor in man. *Eur. J. Clin. Pharmacol.* **36**, 195–197.

Garty, H. and Benos, D.J. (1988). Characteristics and regulatory mechanisms of the amiloride-blockable Na^+ channel. *Physiol. Rev.* **68**, 309–373.

Girbes, A.R.J., Smit, A.J., Meijer, S. and Reitsma, W.D. (1989). Lithium does not influence renal effects of fenoldopam. Congr. "Lithium in Renal Physiology", Utrecht, Abstr. 22.

Greger, R. (1990). Possibility of lithium transport in and beyond the thick ascending limb of Henle's loop. Proc. Congr. "Lithium in Renal Physiology", Utrecht, *Kidney Int.* **37** suppl. 28, 10–16.

Gross, P., Lang, R., Ketteler, M., Hausmann, C., Rascher, W., Ritz, E. and Favre, H. (1989). Natriuretic factors and lithium clearance in patients with the syndrome of inappropriate antidiuretic hormone (SIADH). *Eur. J. Clin. Invest.* **19**, 11–19.

Haas, J.A., Granger, J.P. and Knox, F.G. (1986). Effect of renal perfusion pressure on sodium reabsorption from proximal tubules of superficial and deep nephrons. *Amer. J. Physiol.* **250**, F425–F429.

Hansen, H.E. (1981). Renal toxicity of lithium. *Drugs* **22**, 461–476.

Hansen, H.E. and Amdisen, A. (1978). Lithium intoxication. (Report of 23 cases and review of 100 cases from the literature). *Q. J. Med.* **47**, 123–144.

Hayslett, J.P. and Kashgarian, M. (1979). A micropuncture study of the renal handling of lithium. *Pflügers Arch.* **380**, 159–163.

Hla-Yee-Yee and Shirley, D.G. (1987). The influence of posture on lithium clearance in man. *J. Physiol.* **364**, 167P.

Holstein-Rathlou, N.H., Kanters, J.K. and Leyssac P.P. (1988). Exaggerated natriuresis and lithium clearance in spontaneously hypertensive rats. *J. Hypertension* **6**, 889–895.

Jefferson, J.W. and Kalin, N.H. (1979). Serum lithium levels and long-term diuretic use. *J. Amer. Med. Assoc.* **241**, 1134–1136.

Jeffrey, R.F., MacDonald, T.M., Brown, J., Rae, P.W.H. and Lee, M.R. (1988). The effect of lithium on the renal response to the dopamine prodrug gludopa in normal man. *Br. J. Clin. Pharmacol.* **25**, 725–732.

Kamper, A.L., Strandgaard, S., Holstein-Rathlou, N.-H., Munck, O. and Leyssac, P.P. (1988). The influence of body posture on lithium clearance. *Scand. J. Clin. Lab. Invest.* **48**, 509–512.

Kirchner, K.A. (1987). Lithium as a marker for proximal tubular delivery during low salt intake and diuretic infusion. *Amer. J. Physiol.* **253**, F188–F196.

Lassiter, W.E., Gottschalk, C.W. and Mylle, M. (1961). Micropuncture study of net transtubular movement of water and urea in nondiuretic mammalian kidney. *Amer. J. Physiol.* **200**, 1139–1146.

Leyssac, P.P., Frederiksen, O. and Holstein-Rathlou, N.-H. (1990a). Renal tubular transport of lithium. *In* "Lithium and the Kidney" (ed. S. Christensen), pp. 19–33. Karger, Basel.

Leyssac, P.P., Holstein-Rathlou, N.-H., Skøtt, P. and Alfrey, A.C. (1990b). A micropuncture study of proximal tubular transport of lithium during osmotic diuresis. *Amer. J. Physiol.* **258**, F1090–F1095.

Lu, W., Endoh, M., Katayama, K., Kakemi, M. and Koizumi, T. (1988). Pharmacokinetic and pharmacodynamic studies of piretanide in rabbits. II: Effects on the proximal tubules and the loop of Henle. *J. Pharmacobio-Dyn.* **11**, 95–105.

Luik, A.J., Straub, J.P., Martens, H.J.M. and Donker, A.J.M. (1989). Lithium carbonate orally increases renal sodium excretion. Congr. "Lithium in Renal Physiology", Utrecht, Abstr. 29.

McMurray, J., Seidelin, P.H. and Struthers, A.D. (1989). Evidence for a proximal and distal nephron action of atrial natriuretic factor in man. *Nephron* **51**, 39–43.

Murphy, D.L., Goodwin, F.K. and Bunney, W.E. (1969). Aldosterone and sodium response to lithium administration in man. *Lancet* **ii**, 458–461.

Peart, W.S., Reid, P. and Sutters, M. (1967). Urinary sodium, potassium and water excretion in man after a single dose of 13.8 mmol of lithium carbonate. *J. Physiol. (Lond.)* **390**, 204P.

Petersen, V., Hvidt, S., Thomsen, K. and Schou, M. (1974). Effect of prolonged thiazide treatment on renal lithium clearance. *Br. Med. J.* **3**, 143–145.

Perry, P.J., Calloway, R.A., Cook, B.L. and Smith, R.E. (1984). Theophylline precipitated alterations of lithium clearance. *Acta Psychiat. Scand.* **69**, 528–537.

Rabelink, A.J., Koomans, H.A., Boer, W.H., Dorhout Mees, E.J. and van Rijn, H.J.M. (1989a). Indomethacin increases renal lithium reabsorption in man. *Nephrol. Dial. Transplant.* **4**, 27–31.

Rabelink, T.J., Koomans, H.A., Boer, W.H., van Rijn, H.J. and Dorhout Mees, E.J. (1989b). Lithium clearance in water immersion-induced natriuresis in humans. *J. Appl. Physiol.* **66**, 1744–1748.

Rabelink, A.J., Koomans, H.A., Boer, W.H. and Dorhout Mees, E.J. (1989c). No effect of distal diuretics on lithium clearance in humans. Congr. "Lithium in Renal Physiology", Utrecht, Abstr. 38.

Rombola, G., Colussi, G., De Ferrari, M.E., Surian, M., Malberti, F. and Minetti, L. (1987). Clinical evaluation of segmental tubular reabsorption of sodium and fluid in man: lithium vs free water clearances. *Nephrol. Dial. Transplant.* **2**, 212–218.

Roos, J.C., Koomans, H.A., Dorhout Mees, E.J. and Delawi, I.M.K. (1985). Renal sodium handling in normal humans subjected to low, normal, and extremely high sodium supplies. *Amer. J. Physiol.* **249**, F941–F947.

Saffer, D. and Coppen, A. (1983). Frusemide: a safe diuretic during lithium therapy? *J. Affect. Dis.* **5**, 289–292.

Schou, M. and Vestergaard, P. (1988). Prospective studies on a lithium cohort. 2. Renal function. Water and electrolyte metabolism. *Acta Psychiat. Scand.* **78**, 427–433.

Schou, M. and Weinstein, M.R. (1980). Problems of lithium maintenance treatment during pregnancy, delivery and lactation. *Agressologie* **21** A, 7–9.

Schou, M., Thomsen, K. and Vestergaard, P. (1986). The renal lithium clearance and its correlations with other biological variables: observations in a large group of physically healthy persons. *Clin. Nephrol.* **25**, 207–211.

Schou, M., Hansen, H.E., Thomsen, K. and Vestergaard, P. (1989). Lithium treatment in Aarhus. 2. Risk of renal failure and of intoxication. *Pharmacopsychiat.* **22**, 101–103.

Shalmi, M. and Thomsen, K. (1989). Alterations of lithium clearance in rats by different modes of lithium administration. *Renal Physiol. Biochem.* **12**, 273–280.

Shalmi, M., Petersen, J.S. and Christensen, S. (1989). Effects of intravenous bumetanide administration on renal haemodynamics and proximal and distal tubular sodium reabsorption in conscious rats. *Pharmacol. Toxicol.* **65**, 313–317.

Shalmi, M., Rasmusen, H., Amtorp, O. and Christensen, S. (1990). Effect of chronic oral furosemide administration on the 24-hour cycle in lithium clearance and electrolyte excretion in humans. *Eur. J. Clin. Pharmacol.* **38**, 275–280.

Shirley, D.G. and Walter, S.J. (1986). Proximal tubular function in potassium-depleted rats. *J. Physiol. (Lond.)* **381**, 27P.

Shirley, D.G. and Walter, S.J. (1989). The effect of frusemide on lithium clearance: a micropuncture study. Congr. "Lithium in Renal Physiology", Utrecht, Abstr. 41.

Shirley, D.G., Walter, S.J. and Thomsen, K. (1983). A comparison of micropuncture and lithium clearance methods in the assessment of renal tubular function in rats with diabetes insipidus. *Pflügers Arch.* **399**, 266–270.

Smith, D.F. and Shimizu, M. (1976). Effect of posture on renal lithium clearance. *Clin. Sci. Molec. Med.* **51**, 103–105.

Solomon, K. (1978). Combined use of lithium and diuretics. *South. Med. J.* **71**, 1098–1100.

Solomon, L.R., Atherton, J.C., Bobinski, H., Hillier, V. and Green, R. (1988a). Effect of low dose infusion of atrial natriuretic peptide on renal function in man. *Clin. Sci.* **75**, 403–410.

Solomon, L.R., Atherton, J.C., Bobinski, H., Cottam, S.L., Gray, C., Green, R. and Watts, H.J. (1988b). Effect of a meal containing protein on lithium clearance and plasma immunoreactive atrial natriuretic peptide in man. *Clin. Sci.* **75**, 151–157.

Steele, T.H., Manuel, M.A., Newton, M. and Boner, G. (1975). Renal lithium reabsorption in man: physiologic and pharmacologic determinants. *Amer. J. Med. Sci.* **269**, 349–363.

Steele, T.H., Dudgeon, K.L. and Larmore, C.K. (1976). Pharmacological characterization of lithium reabsorption in the rat. *J. Pharmacol. Exp. Ther.* **196**, 188–193.

Strazzullo, P., Iacoviello, L., Iacone, R. and Giorgione, N. (1988). Use of fractional lithium clearance in clinical and epidemiological investigation: a methodological assessment. *Clin. Sci.* **74**, 651–657.

Thomsen, K. (1973). The effect of sodium chloride on kidney function in rats with lithium intoxication. *Acta Pharmacol. Toxicol.* **33**, 92–102.

Thomsen, K. (1976). Renal elimination of lithium in rats with lithium intoxication. *J. Pharmacol. Exp. Ther.* **199**, 483–489.

Thomsen, K. (1987). Excretion. *In* "Mania and Depression. Modern Lithium Therapy" (ed. F.N. Johnson), pp. 75–78. IRL Press, Oxford.

Thomsen, K. and Leyssac, P.P. (1986a). Effect of dietary sodium content on renal handling of lithium. Experiments in conscious diabetes insipidus rats. *Pflügers Arch.* **407**, 55–58.

Thomsen, K. and Leyssac, P.P. (1986b). Acute effects of various diuretics on lithium clearance. Studies in rats on medium and low sodium diet. *Renal Physiol. (Basel)* **9**, 1–8.

Thomsen, K. and Olesen, O.V. (1979). The effect of water deprivation on lithium clearance and lithium excretion fraction in lithium-polyuric rats. *J. Pharmacol. Exp. Ther.* **209**, 327–329.

Thomsen, K. and Olesen, O.V. (1984). Renal lithium clearance as a measure of the delivery of water and sodium from the proximal tubule in humans. *Amer. J. Med. Sci.* **288**, 158–161.

Thomsen, K. and Schou, M. (1968). Renal lithium excretion in man. *Amer. J. Physiol.* **215**, 823–827.

Thomsen, K., Schou, M., Steiness, I. and Hansen, H.E. (1969). Lithium as an indicator of proximal sodium reabsorption. *Pflügers Arch.* **308**, 180–184.

Thomsen, K., Holstein-Rathlou, N.-H. and Leyssac, P.P. (1981). Comparison of three measures of proximal tubular reabsorption: lithium clearance, occlusion time, and micropuncture. *Amer. J. Physiol.* **241**, F348–F355.

Thomsen, K., Schou, M. and Vestergaard, P. (1988). Distinction between proximal and distal regulations of sodium and potassium excretion in humans. *Clin. Nephrol.* **29**, 12–18.

Walter, S.J. and Shirley, D.G. (1988). Changes in proximal tubular function following unilateral nephrectomy. *Acta Physiol. Pharmacol. (Bulg.)* **14** (Suppl.), Abst. 162.

Wetzels, J.F.M., Wiltink, P.G., Hoitsma, A.J., Huysmans, F.Th.M. and Koene, R.A.P. (1988). Diuretic and natriuretic effects of nifedipine in healthy persons. *Br. J. Clin. Pharmacol.* **25**, 547–553.

Whiting, P.H., Towler, H.M.A., Cliffe, A.M. and Forrester, J.V. (1989). Lithium clearance in cyclosporin (CsA) treated uveitis patients. Congr. "Lithium in Renal Physiology", Utrecht, Abstr. 48.

15 Lithium and Essential Fatty Acid Metabolism in Psychiatric Disorders, Herpes Virus Infections and Seborrhoeic Dermatitis

DAVID F. HORROBIN

Efamol Research Institute, P.O. Box 818, Kentville, Nova Scotia, Canada, B4N 4H8

Introduction

In spite of over 30 years intensive research there is no agreed mechanism whereby lithium exerts its therapeutic effects in manic-depression. Actions on sodium/potassium transport, on second-messenger functions (especially those related to cyclic-AMP) and on phosphatidyl-inositol metabolism have all provided more or less plausible approaches to understanding. This chapter raises the possibility of an alternative mechanism, relating to the interaction between lithium and the metabolism of essential fatty acids (EFAs) and their derivative eicosanoids. Since the eicosanoids are important in the regulation of cyclic nucleotide metabolism, and since the EFAs are essential components of phosphatidyl-inositol, the EFA concept of lithium action may be related to those two other hypotheses. The effects of lithium on second-messenger systems and phosphoinositides are discussed in the chapter by Professor W. Sherman (Chapter 8, this volume).

The fatty acids have four main functions in the body: they can be oxidized to supply energy; they can be incorporated into adipose tissue which has both storage and packing and lubricating functions; they form major parts of the structure of all membranes; and they act as precursors for a variety of short-lived substances with second-messenger type actions, such as prostaglandins, leukotrienes, hydroxy-acids and other eicosanoids. The fatty acids are found in the body in four main chemical forms, free fatty acids, triglycerides, cholesterol esters and phospholipids. The triglycerides are the main storage form, while the phospholipids and cholesterol esters are used for both fatty acid transport and membrane structure. It is usually necessary to

LITHIUM AND THE CELL
ISBN 0-12-099300-7

convert the fatty acids into the free form before they can be metabolized to prostaglandins (PGs) and other eicosanoids.

In the saturated fatty acids, as the name implies, all the carbon atoms are saturated and there are no double bonds linking adjacent carbon atoms. In the unsaturated fatty acids, there are double bonds linking one or more pairs of carbon atoms. Fatty acids with one double bond are known as mono-unsaturates and those with two or more double bonds are known as polyunsaturates. Most unsaturated fatty acids important in humans have between one and six double bonds. The fatty acids are vital for normal membrane structure where an appropriate blend of saturates, monounsaturates and polyunsaturates is required. Saturated fatty acids tend to make membranes rigid and viscous, whereas polyunsaturated fatty acids have the opposite effects.

The main saturated and monounsaturated fatty acids can be made within the body from other fatty acids or from non-lipid sources. However, two groups of fatty acids cannot be synthesised within the body. These are the polyunsaturated EFAs derived from linoleic acid (18:2n-6) and from α-linolenic acid (ALA) (18:3n-3). All EFAs are polyunsaturated but not all polyunsaturated fatty acids (PUFAs) are EFAs. The EFAs must have double bonds in precisely specified positions and must have *all* those double bonds in the *cis* as opposed to the *trans* configuration. The abbreviated nomenclature is simple. The first number (18 in the case of linoleic acid) indicates the number of carbon atoms in the molecule. The second number (2 in the case of linoleic acid) indicates the number of double bonds in the molecule. The third number (6 in the case of linoleic acid) indicates the carbon atom where the first double bond is situated, starting from the methyl end of the molecule.

The main dietary EFAs are linoleic acid and ALA, although all the other EFAs are found in some foods to some extent. Linoleic acid and ALA are metabolized by the alternating desaturation (introducing a further double bond) and elongation (adding two carbon atoms) reactions shown in Fig. 1. The linoleic acid (n-6) series of EFAs is much more important than the n-3 series (Horrobin, 1989b), although the n-3 series does have several essential roles in the body, notably in the retina and brain. The remainder of this chapter will concentrate mainly on the n-6 series.

Dietary deprivation of n-6 EFAs eventually leads to severe disruption of all body systems (Rivers and Frankel, 1981). This is because EFAs have at least four important roles within the body:

(1)   They are required for the normal structure of all cell membranes and in their absence membranes lack normal fluidity and flexibility. As a result membrane functions are altered as are the behaviours of membrane-bound proteins such as receptors, ion channels, second-messenger systems and ATPases.

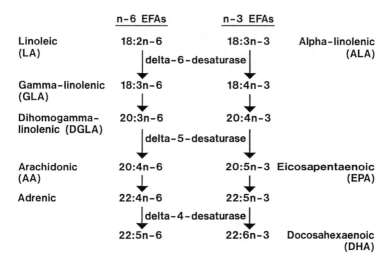

FIG. 1 An outline of the pathway of essential fatty acid metabolism. DGLA, AA and EPA are all important precursors of eicosanoids. It is believed that both n-3 and n-6 EFAs are metabolized by the same series of alternating desaturation and elongation steps. The desaturation steps remove two hydrogen atoms and so add a double bond, while the elongation steps add two carbon atoms.

(2) They are required as precursors of a variety of short-lived molecules which exert many regulatory actions within and between cells. Particularly important among these regulatory molecules are the prostaglandins of the 1 series derived from dihomogammalinolenic acid (DGLA, 20:3n-6) and of the 2 series derived from arachidonic acid (AA, 20:4n-6).

(3) They are required to regulate the normal water permeability (or impermeability) of the skin and other membranes.

(4) They are required for the transport of cholesterol around the body.

It is clear that the EFAs and their metabolites play important roles in every tissue within the body. It is suggested that they may be one of the targets for the therapeutic effects of lithium, particularly in manic-depressive psychosis, in herpes, and in seborrhoeic dermatitis.

Lithium Effects on EFAs and Prostaglandins

To date the effects of lithium on the EFA-prostaglandin system have been studied in only one system, the perfused superior mesenteric vascular bed of the rat. Since many aspects of the metabolism of both EFAs and PGs are known to be both species-specific and organ-specific, investigations in other tissues are urgently required and the results must be interpreted with caution. However, it may be pointed out that most other proposals concerning the mechanism of action of lithium are based on a similar narrow platform of experimental evidence. It is generally true of all the proposed mechanisms that an expansion of the experimental base to other systems and other species would be desirable.

The effects of lithium were first noted in the course of experiments on the effects of prolactin and vasopressin on vascular reactivity. The prolactin effects appear to be related to the synthesis of 1 series PGs, notably PGE_1, derived from DGLA (Manku *et al.*, 1979a), whereas the vasopressin effects appear related to the synthesis of 2 series PGs and thromboxane from AA (Karmazyn *et al.*, 1978). Low concentrations of lithium of 2 mmol l^{-1} or less, which are relevant to the psychiatric actions of the element, selectively inhibited the effects of prolactin but not those of vasopressin. Higher concentrations of lithium inhibited the actions of both hormones. This was taken to mean that low lithium levels had a selective inhibitory effect on the formation of PGE_1 whereas higher concentrations non-selectively inhibited the formation of both 1 and 2 series eicosanoids from both DGLA and AA (Horrobin *et al.*, 1977; Horrobin, 1979b).

These indirect conclusions were later investigated by the direct measurement of the effects of various concentrations of lithium on the release of fatty acids, PGs and thromboxane from the mesenteric vascular bed (Horrobin *et al.*, 1988; Horrobin, 1990a,b). The experiments were conducted by removing the superior mesenteric vascular bed and perfusing it in isolation using Krebs-bicarbonate buffer bubbled with 95% O_2 and 5% CO_2. Lithium in various concentrations could be added to the perfusing buffer. The effluent emerging from the vascular bed in the presence of various lithium levels was collected. The concentrations of PGE_1 derived from DGLA, and of PGE_2, 6-keto-PGF_1, and of thromboxane B_2 derived from AA, were measured by specific radioimmunoassays. 6-keto-PGF_1 (6kPGF) is the stable metabolite of prostacyclin or PGI_2 while thromboxane B_2 (TxB_2) is the stable metabolite of thromboxane A_2(TxA_2), a potent vasoconstrictor and pro-aggregatory agent. The fatty acids in the effluent were also extracted, transmethylated and the amounts measured by gas chromatography using appropriate standards. Almost all the fatty acids present in the effluent were in the free fatty acid form, presumably having been released from the phospholipids, triglycerides and cholesterol esters present in the tissue.

The main conclusions of these and other unpublished studies are as follows:

(1)   At concentrations possibly relevant in psychiatry (0–2 mmol l^{-1}) lithium had variable and statistically non-significant effects on release of all fatty acids, both essential and non-essential. There is a consistent

tendency for release of fatty acids to be slightly increased but only rarely does this reach statistical significance.

(2) At concentrations probably irrelevant to psychiatry, but certainly relevant to the actions of topically applied lithium, of $50 \, \text{mmol} \, l^{-1}$ and above, the release of all fatty acids, saturates, monounsaturates and EFAs, is inhibited substantially and highly significantly.

(3) At concentrations in the $0.4 - 1.6 \, \text{mmol} \, l^{-1}$ range there is a consistent significant inhibition of the release of TxB_2 which is present at $0.4 \, \text{mmol} \, l^{-1}$ and does not increase with increasing lithium concentrations above that. Over this concentration range there is also a selective inhibition of PGE_1 synthesis which is detectable at 0.4 and near complete at $1.6 \, \text{mmol} \, l^{-1}$. Over this concentration range levels of 6KPGF and PGE_2 are not significantly changed by lithium. Since lithium within this range does not consistently alter release of DGLA or AA, the inhibitory effects on PGE_1 and TxB_2 cannot be related to reduced availability of precursors. The first step in conversion of AA to its derivatives, the formation of endoperoxide, is common to TxB_2; PGE_2 and 6kPGF. The lack of effects of lithium on PGE_2 and 6kPGF presumably means that the inhibited step is the conversion of endoperoxide to TxA_2.

(4) At concentrations of $50 \, \text{mmol} \, l^{-1}$ and above, the levels of all four eicosanoids measured are profoundly reduced by lithium. This is probably largely due to the inhibitory effect of these lithium concentrations on the release of all fatty acids, including the eicosanoid precursors DGLA and AA.

The remaining sections of this chapter will attempt to relate these biochemical observations to the clinical effects of lithium in manic-depression, in alcoholism, in herpes and in seborrhoeic dermatitis.

Lithium in Psychiatric Disorders

The main use of lithium in psychiatry is in the long-term *prevention* of recurrent episodes of both mania and depression. Traditionally this action has been achieved by the maintenance of plasma lithium concentrations in the $0.8-1.2 \, \text{mmol} \, l^{-1}$ range. More recently the target range for plasma lithium has dropped to 0.6 (or even 0.4) $- 1.0 \, \text{mmol} \, l^{-1}$. Lithium also has an established place in the *treatment* of existing mania but its role in the *treatment* of existing depression is controversial and there is doubt as to whether this effect exists. Although not widely used for this purpose, there is good evidence from controlled trials that lithium may be used to prevent relapse in alcoholics, especially when the alcoholism is associated with depression (Kline *et al.*, 1974; Merry *et al.*, 1976).

The EFAs and PGs are good candidates for roles as contributors to the biochemical basis of psychiatric disorders, including manic-depression and alcoholism (Horrobin, 1977, 1979a, 1980a, 1982, 1985b, 1987, 1989a, 1990a,c; Horrobin *et al.*, 1978; Horrobin and Manku, 1980). The EFAs make up around 20% of the dry matter of the brain. The EFAs determine the fluidity and flexibility of neuronal membranes and are therefore likely to modulate neuronal function. PGE_1 at low concentrations enhances nerve conduction and at high concentrations inhibits it (Horrobin *et al.*, 1977). In various systems PGE_1 antagonizes the action of dopamine (Horrobin, 1985b, 1990c). PGE_1 itself, quite apart from the EFAs from which it is derived, modulates membrane fluidity at extraordinary low concentrations (Rasmussen *et al.*, 1975). PGE_1 has a wide range of effects in many systems and is a potent stimulator of cyclic-AMP formation (Kirtland, 1988; Horrobin, 1988). None of these effects provides a specific reason for linking EFAs or PGs to mania or to depression, but they do show that the effects they exert on membranes, on neuronal function and on neurotransmitter release and action are not incompatible with the idea that they may play roles in psychiatric disorders.

The formation of 1 and 2 series PGs from the main dietary precursor, linoleic acid, deserves further consideration. The series of reactions shown in Fig. 1 is subject to influence by many factors. In general, the desaturating steps are very slow and rate-limiting, particularly in humans (Brenner, 1982; Horrobin, 1983, 1989b) whereas the elongation steps are very rapid. Thus in the conversion of linoleic acid to AA, the first and third steps (conversion of LA to γ-linolenic acid (GLA) and of DGLA to AA) are extremely slow, while conversion of GLA to DGLA is rapid. Of particular interest is the fact that slow conversion of linoleic acid to GLA can be further inhibited by alcohol in both animals and humans (Nervi *et al.*, 1980; Glen *et al.*, 1984, 1987, 1990; Johnson *et al.*, 1985).

Also of interest is the fact that the very slow conversion of DGLA to AA means that the formation of 1 series and of 2 series PGs can be independently regulated. PGE_1 stimulates the formation of cyclic AMP which in turn inhibits the phospholipase systems which mobilize AA from membrane phospholipids. It was therefore proposed that measures to elevate PGE_1, such as provision of its precursors GLA or DGLA, would simultaneously elevate the formation of 1 series PGs and inhibit the formation of 2 series PGs, the effects of phospholipase being more important that any enhanced conversion of DGLA to AA (Horrobin, 1980b). Conversely, it was suggested that mild EFA deficiency, far from being associated with reduced formation of 2 series PGs from AA, would actually be associated with excessive 2 series PG production. There are very large stores of AA within the body but much smaller stores of DGLA: a moderate EFA deficiency state which led to a fall in DGLA and PGE_1 levels would be expected to lead to release of stored AA and excessive formation of 2 series PGs.

These theoretical predictions have been borne out in practice. Experimental EFA deficiency initially leads to increased 2 series PG formation (Huang *et al.*, 1984). Patients with cystic fibrosis, who tend to be EFA deficient, have excessively high rates of 2 series PG production which can be brought towards normal by the provision of linoleic acid or GLA (Chase and Dupont, 1978; Dodge, 1990). Provision of GLA in several situations in both humans and animals has been shown to inhibit rather than stimulate formation of 2 series eicosanoids (Belch *et al.*, 1985; Puolakka *et al.*, 1985; Abou El-Ela *et al.*, 1987; Horrobin, 1989b; Bunce and Abou El-Ela, 1990).

EFA and PG levels in mania, depression and alcoholism

The first work in this field was published in 1975 (Abdulla and Hamadah, 1975). Platelet preparations from patients with mania, depression and schizophrenia, and also from normal individuals, were incubated with DGLA and with ADP to stimulate conversion of DGLA to PGE_1. At half maximal levels of ADP, the platelet preparations from patients with depression and with schizophrenia produced significantly less PGE_1 than normal, while those from patients with mania produced significantly more PGE_1 than normal. At maximal ADP concentrations the differences between normal individuals and those with mania and depression disappeared, while platelets from schizophrenia still produced much less PGE_1.

Unfortunately measurements of PGs in biological fluids are difficult to perform and are at present relatively insensitive and unreliable so that no determinations of PGE_1 concentrations in plasma or other body fluids have yet been made in manic-depression or in alcoholism. However, there are two other sets of observations which are relevant to PGE_1. Alcohol, at concentrations within the range which produces intoxication, specifically stimulates the formation of PGE_1 from platelets without enhancing the production of 2 series PGs: indeed the production of TxB_2 is modestly reduced (Manku *et al.*, 1979b; Pennington *et al.*, 1982; Segarnick *et al.*, 1985). Second, PGE_1 infusions are now extensively used in the management of peripheral vascular disease. Although this has not formally been written up, several clinicians have reported to me that patients may feel mildly euphoric during PGE_1 infusions. This elevation of mood does not appear to be secondary to the peripheral vasodilation because prostacyclin (PGI_2) which produces similar vascular effects does not appear to be associated with similar mood elevation and may indeed produce dysphoria.

Thus several observations point to the interesting possibility that PGE_1

may be a mood regulator. The elevated PGE_1 production of platelets from manic patients, the reduced PGE_1 formation by platelets from those who are depressive, the temporary elevation in PGE_1 formation produced by alcohol, and the mild euphoria which is sometimes associated with PGE_1 infusions, all suggest that PGE_1 may be able to produce a euphoric state, while deficiency of PGE_1 may produce depression. These observations are clearly tentative and require much more intensive exploration.

There are a number of observations on AA derivatives in various body fluids in patients with depression. Both PGE_2 and TxB_2 levels in plasma are significantly elevated in depressed individuals (Lieb *et al.*, 1983). Similarly cerebrospiral fluid concentrations of a stable PGE_2 metabolite are above normal in depression (Calabrese *et al.*, 1986). Different forms of stress in humans have been shown to elevate blood concentrations of both PGE_2 and thromboxane B_2 (Mest *et al.*, 1982).

To my knowledge no-one has measured blood levels of eicosanoids in patients with alcoholism. However, plasma EFA levels have been measured by several groups (Alling *et al.*, 1979; Glen *et al.*, 1984, 1987, 1990; Johnson *et al.*, 1985; Horrobin, 1987). Consistently concentrations of all EFAs are below normal, in part indicating inadequate EFA intake because of poor diet and inadequate lipid absorption because of impaired pancreatic and hepatic function. However, the levels of the metabolites of linoleic acid are further reduced than are those of linoleic acid itself, indicating impaired conversion of linoleic acid to GLA as described earlier.

Lithium and EFA metabolism in mania and depression

The key clinical observations which must be explained by any theory of lithium action in manic-depression are that lithium can be used to treat mania but not depression, whereas it is effective in preventing both mania and depression. Both these effects can be explained by a PGE_1 theory of manic-depression (Horrobin and Manku, 1980).

DGLA is the immediate precursor of PGE_1. Stores of DGLA in the body are very limited. The rate of conversion of dietary linoleic acid through to DGLA is very slow because of the severely rate-limiting 6-desaturation step. Evidence outlined above indicates that PGE_1 may regulate mood, with elevated levels producing a euphoric or happy state, and reduced levels producing a depressed mood. Because PGE_1 controls the mobilization of AA, the euphoric state will be associated with reduced formation of AA metabolites and the depressed state with increased production of AA metabolites such as PGE_2 and TxB_2.

Although the physiological systems controlling PGE_1 formation are largely unknown, it is safe to assume that they must exist, so potent are the biological actions of PGE_1 (Horrobin, 1988; Kirtland, 1988). Suppose that in mania the primary event is a loss of control over the conversion of DGLA to PGE_1. Then excess levels of PGE_1 will be produced, and, *as long as DGLA stores are adequate*, this PGE_1 will produce an elevated mood state. However if the rate of DGLA loss to PGE_1 formation exceeds the rate of DGLA formation from linoleic acid, the time will come when the DGLA pool being used for PGE_1 formation will be depleted and PGE_1 production will crash from being above normal to being below normal. If PGE_1 is important in regulating mood, then a depressed state will ensue. Thus on this hypothesis mania will be associated with excess PGE_1 formation and depression with reduced PGE_1 production consequent upon depletion of the relevant DGLA pool.

Lithium inhibits the conversion of DGLA to PGE_1. It is therefore clear why lithium can treat mania: it will reduce the elevated PGE_1 levels, since that is an effect of lithium which occurs at the clinically relevant concentrations. It is also clear why it will not improve depression and indeed that it could make depression worse: if PGE_1 levels are already depressed, lithium can only lower them further. However, the continued presence of lithium in an individual who has achieved normality will prevent the swings in PGE_1 production associated with both mania and depression. Lithium will prevent the rise in PGE_1 formation which the hypothesis states is associated with the manic state. Because there is no excessive conversion of DGLA to PGE_1, then DGLA will be conserved and the relevant pool will never reach a critically depleted state. Thus no crash in PGE_1 formation and no state of depression will ensue.

It must be emphasized that this remains a hypothesis and much more testing is required. However, to my knowledge it is the only hypothesis which can account for the major clinical observations on the effects of lithium. The hypothesis makes interesting predictions about the nature of cycling and the duration of the cycle. The length of the manic phase will depend on the rate of conversion of DGLA to PGE_1, on the size of the DGLA pool to start with and on the rate at which DGLA is formed. The duration of the depressed state will depend on whether or not the rate of DGLA formation ever rises to a level which can compensate for the rate of PGE_1 formation. Either the manic phase will be prolonged or the depressed phase shortened by increasing the rate of DGLA formation, for example, by by-passing the rate-limiting step and providing GLA or DGLA directly. The reverse effects will occur if the rate of DGLA formation is reduced, as may occur in alcoholics.

Lithium and alcoholism

Much less is known about the effects of lithium in alcoholism but it does seem that there is a modest therapeutic effect, especially in depressed alcoholics

(Kline *et al.*, 1974; Merry *et al.*, 1976). Alcohol has a dual effect on PGE_1 formation: (1) enhancement of conversion of DGLA to PGE_1; and (2) inhibition of DGLA formation as a result of general EFA depletion and specific blockade of conversion of linoleic acid to GLA (see earlier section). Thus in the presence of adequate DGLA stores, alcohol acutely is likely to stimulate PGE_1 production and elevate mood. Conversely, when the stimulating effect of alcohol has worn off, the depletion of DGLA as a result of increased PGE_1 formation is likely then to lead to reduced PGE_1 production until the DGLA can be replenished, or until its conversion to PGE_1 can again be stimulated, perhaps by alcohol. This provides a plausible explanation for the well-known phenomena of euphoria during early stages of alcoholic intoxication followed by depression in the late stages or after intoxication has worn off. This depression, associated with PGE_1 depletion, is then likely to make the sufferer seek further alcohol (Horrobin and Manku, 1980; Horrobin, 1982). In the long term, inhibition of DGLA formation by alcohol-induced blockade of conversion of linoleic acid to DGLA is likely to lead to chronic severe depletion of DGLA and a near permanent state of depression.

Thus in some individuals who become alcoholics, drinking may occur in order to induce euphoria or to relieve a depressed mood. However, the elevation of mood can only be temporary and will be followed by depression, which requires alcohol for its relief. As DGLA is progressively depleted, the euphoria will become more and more difficult to achieve, and the depression will last longer and longer.

Lithium could act in this cycle much as it is postulated to do in manic-depression. Lithium will make the elevation in mood in response to alcohol less marked, because it will limit the rise in PGE_1. Such a reduction of the effect of alcohol or mood in the presence of lithium has already been observed (Judd *et al.*, 1979). This may make the effect of alcohol less attractive. At the same time lithium will prevent excessive depletion of DGLA and so prevent depression occurring after alcohol consumption. This mechanism could explain the modest effect of lithium in treating alcoholics.

None of the foregoing is intended in any way to downplay or exclude other mechanisms of manic-depression and alcoholism which have been proposed. The aim instead is to offer a new perspective which may supplement those already available. In particular it offers a new therapeutic modality which could be used in support of lithium therapy, namely the provision of GLA or DGLA to prevent DGLA depletion and to aid in its replenishment.

Topical Effects in Dermatological Disorders

The use of topical lithium in dermatological disorders is a relatively new aspect of the therapeutic use of lithium. The concentration of lithium in the topical preparations used is of the order of 600 mmol l^{-1}. It is uncertain what concentration is achieved in the underlying skin, but it is likely to be in the vicinity of $50-200 \text{ mmol l}^{-1}$. At those concentrations in rat vascular tissue, as described earlier, lithium inhibits the release of all fatty acids, both essential and non-essential. Probably largely because of the blockade of release of the EFA PG precursors, lithium at these concentrations substantially reduces the secretion of all eicosanoids so far measured, in addition to its apparently selective effects on PGE_1 and TxB_2.

Topical lithium has now been shown to be effective in two dermatological disorders, seborrhoeic dermatitis and herpes virus infections. The lithium has been used in the form of lithium succinate, a salt which seems compatible with the skin surface and which causes little or no irritancy or sensitization.

Oral and genital herpes

Oral (cold sores) and genital herpes are common infections caused by herpes simplex virus I and II. Lieb (1979) first noted that manic-depressive patients who also had recurrent herpes infections and who were treated with lithium sometimes experienced remission of their chronic herpes. At the same time, and quite independently, Skinner *et al.* (1980) observed inhibition by lithium ions of herpes simplex virus replication *in vitro* at concentrations which did not destroy the host cells. Horrobin (1985a) formulated a topical lithium preparation based on the work of Sherwin (1972). When tested in genital herpes in a randomized, double-blind, placebo-controlled trial, the topical lithium was effective both in relieving pain and inflammation and shortening the clinical course of the disease and also in reducing viral shedding (Skinner, 1983; Skinner *et al.*, 1990).

The effects of high levels of lithium in blocking release of free fatty acids may contribute to the clinical effect in herpes, Prostaglandin synthesis seems to be stimulated by viral infections and inhibitors of PG formation have been found to block the multiplication of herpes viruses (Harbour *et al.*, 1978; Newton, 1979; Baker *et al.*, 1982). Secondly, the clinical symptoms associated with viral infections are often related to a local inflammatory reaction to the viral-infected cells. Since derivatives of fatty acids, including PGs, leukotrienes and hydroxy-acids, are important contributors to such inflammatory responses, lithium by blocking fatty acid release is likely to have a local anti-inflammatory effect. This will be separate from and in addition to any direct antiviral action. Thus it is possible that lithium may find more general use as a topical anti-inflammatory and anti-pruritic agent (Lieb, 1981).

Seborrhoeic dermatitis

Seborrhoeic dermatitis (SD) is a common inflammatory condition affecting the sebum-producing areas of the skin. In its more severe forms it presents as an inflamed area of the face, especially the eyebrows and nose, with thick yellowish scales which flake off to leave a bright red area underneath. Any sebum-producing skin can be affected although, contrary to common belief, the inflamed areas do not produce excessive amounts of sebum. In its milder forms SD presents as dandruff.

It is now generally thought that the condition is caused by excessive proliferation of a common skin commensal fungus, *Pityrosporum ovale* (Ford *et al.*, 1984; Horrobin, 1990b). In individuals with SD for as yet unknown reasons, *Pityrosporum* proliferates excessively, provoking the typical skin reaction. SD responds well to treatment with various antifungal agents although these may have important side effects.

During studies on the lithium succinate ointment, it was accidentally noted that SD appeared to respond well. Subsequent double-blind, placebo-controlled trials have shown that topical lithium is extraordinary effective in SD (Boyle *et al.*, 1986; Gould *et al.*, 1988; Horrobin, 1990b,d). It clears the skin and at the same time produces a dramatic fall in the skin population of *Pityrosporum* in spite of having little direct antifungal action demonstratable *in vitro* (Boyle *et al.*, 1986; Horrobin, 1990b). It therefore appears to be working by a new type of action.

The possible mechanism of lithium action in SD emerged as the resultant of the work on lithium and fatty acid release, and of some old studies on the nutrient requirements of *Pityrosporum* (Shifrine and Marr, 1968; Wilde and Stewart, 1968). It appears that *Pityrosporum* thrives on the sebum producing areas of skin because it has an absolute requirement for free fatty acids with more than 10 carbon atoms. When deprived of such free fatty acids, even if provided with fatty acids in the forms of glycerides and phospholipids, it grows poorly. Since lithium at the concentrations used topically blocks release of free fatty acids in the rat vascular model, it seems probable that lithium acts in SD by depriving the fungus of its required nutrients. This would explain why the skin population of *Pityrosporum* can fall drastically even though lithium has only a weak, if any, direct antifungal action. The general anti-inflammatory actions of lithium will contribute to the clinical therapeutic effect.

Other skin conditions

Lithium has a bad reputation in relation to the skin because of reports that oral lithium may exacerbate skin conditions, notably psoriasis and acne. However, both lithium treatment and skin diseases are common and associations of the two will frequently occur by chance. A recent review has shown that given the likelihood of such occurrences, reports of the exacerbation of skin diseases, by lithium are surprisingly few (Horrobin, 1990d). The association with psoriasis, while possibly real in a few patients, has probably been emphasized because it fits in with a now unlikely theory that psoriasis is caused by low cyclic-AMP levels in the skin. Since lithium reduces cyclic-AMP levels (Wolff *et al.*, 1970; Ebstein *et al.*, 1976), possibly partly by reducing synthesis of PGE_1 which stimulates cyclic-AMP formation, lithium exacerbation of psoriasis would have provided support for the theory.

Conclusions

The interactions between lithium and fatty acid metabolism offer new insights into the mechanisms of action of both oral and topical lithium. It must be emphasized that to date this effect of lithium has been studied in only one model system in one species. Until it has been shown to be more generally applicable, the results should be interpreted with caution.

References

Abdulla, Y.H. and Hamadah, K. (1975). Effect of ADP on PGE formation in blood platelets from patients with depression, mania and schizophrenia. *Br. J. Psychiat.* **127**, 591–595.

Abou El-Ela, S.H., Prasse, K.W., Carroll, R. and Bunce, O.R. (1987). Effects of dietary primrose oil on mammary tumorigenesis induced by 7,12-dimethylbenz(a)-anthracene. *Lipids* **22**, 1041–1044.

Alling, C., Aspenstrom, G., Dencker, S.J. and Svennerholm, L. (1979). Essential fatty acids in chronic alcoholism. *Acta Med. Scand. (Suppl.)* **631**, 1–38.

Baker, D., Thomas, J., Epstein, J., Posilico, D. and Stone, M.L. (1982). The effect of prostaglandins on the multiplication and cell to cell spread of herpes simplex virus type 2 *in vitro. Amer. J. Obstet. Gyn.* **144**, 346–349.

Belch, J.J.F., Shaw, B., O'Dowd, A., Saniabadi, A., Leiberman, P., Sturrock, R.D. and Forbes, C.D. (1985). Evening primrose oil (Efamol) in the treatment of Raynaud's phenomenon: a double blind study. *Thromb. Haemost.* **54**, 490–494.

Boyle, J., Burton, J.L. and Faergemann, J. (1986). Use of topical lithium succinate for seborrhoeic dermatitis. *Br. Med. J.* **292**, 28.

Brenner, R.R. (1982). Nutritional and hormonal factors influencing desaturation of essential fatty acids. *Progr. Lipid Res.* **20**, 41–48.

Bunce, O.R. and Abou El-Ela, S.H. (1990). Eicosanoid synthesis and ornithine decarboxylase activity in mammary tumors of rats fed varying levels and types of n-3 and /or n-6 fatty acids. *Prostaglandins Leukotr. EFAs*, in press.

Calabrese, J.R., Skwerer, R.G., Barna, B. *et al.* (1986). Depression immunocompetence and prostaglandins of the E series. *Psychiat. Res.* **17**, 41–47.

Chase, H.P. and Dupont, J. (1978). Abnormal levels of prostaglandins and fatty acids in blood of children with cystic fibrosis. *Lancet* **ii**, 236–238.

Dodge, J.A. (1990). Essential fatty acids in cystic fibrosis. *In* "Omega-6 Essential Fatty Acids: Pathophysiology and Roles in Clinical Medicine" (ed. D.F. Horrobin), pp. 427–435. Allan R. Liss, New York.

Ebstein, R., Belmaker, R. and Grunhaus, L. (1976). Lithium inhibition of adrenaline-stimulated adenylate cyclase in humans. *Nature (London)* **259**, 411–413.

Ford, G.P., Farr, P.M., Ive, F.A. and Shuster, S. (1984). The response of seborrhoeic dermatitis to ketoconazole. *Br. J. Dermatol.* **111**, 603–607.

Glen, I., Glen, E., MacDonnell, L. and MacKenzie, J. (1984). Possible pharmacological approaches to the prevention and treatment of alcohol related CNS impairment: results of a double blind trial of essential fatty acids. *In* "Pharmacological Treatments for Alcoholism" (ed. G. Edwards and J. Littleton), pp. 331–350. Croom Helm, London.

Glen, A.I.M., Skinner, F., Glen, E.M.T. and MacDonnell, L. (1987). The role of essential fatty acids in alcohol dependence and tissue damage. *Alcoholism Clin. Exp. Res.* **11**, 37–41.

Glen, A.I.M., Glen, E.M.T., MacDonell, L.E.F. and Skinner, F.K. (1990). Essential fatty acids and alcoholism. *In* "Omega-6 Fatty Acids: Pathophysiology and Roles in Clinical Medicine" (ed. D.F. Horrobin), pp. 321–332. Alan R. Liss, New York.

Gould, D.J., Davies, M.G., Kersey, P.J.W. *et al.* (1988). Topical lithium succinate, a safe and effective treatment for seborrhoeic dermatitis in adults. British Association of Dermatologists, Annual Meeting, 6–9 July.

Harbour, D.A., Blyth, W.A. and Hill, T.J. (1978). Prostaglandins enhance spread of herpes simplex virus in cell cultures. *J. Gen. Virol.* **41**, 87–95.

Horrobin, D.F. (1977). Schizophrenia as a prostaglandin deficiency disease. *Lancet* **i**, 936–937.

Horrobin, D.F. (1979a). Schizophrenia: reconciliation of the dopamine, prostaglandin and opioid concepts and the role of the pineal. *Lancet* **i**, 529–531.

Horrobin, D.F. (1979b). Lithium as a regulator of prostaglandin synthesis. *In* "Lithium" (ed. T. Cooper, N.S. Kline and M. Schou), pp. 854–880. Excerpta Medica, Amsterdam.

Horrobin, D.F. (1980a). A biochemical basis for alcoholism and alcohol-induced damage, including the fetal alcohol syndrome and cirrhosis: interference with essential fatty acid and prostaglandin metabolism. *Med. Hypotheses* **6**, 785–800.

Horrobin, D.F. (1980b). The regulation of prostaglandin biosynthesis: Negative feedback mechanisms and the selective control of formation of 1 and 2 series prostaglandins: relevance to inflammation and immunity. *Med. Hypotheses* **6**, 687–709.

Horrobin, D.F. (1982). A possible biochemical basis for alcoholism and for schizoid reactions during alcohol withdrawal. *In* "Biological Aspects of Schizophrenia and Addiction" (ed. G. Hemmings), pp. 163–171. John Wiley & Sons Ltd, London.

Horrobin, D.F. (1983). The regulation of prostaglandin biosynthesis by the manipulation of essential fatty acid metabolism. *Rev. Pure Appl. Pharmacol. Sci.* **4**, 339–342.

Horrobin, D.F. (1985a). Lithium in the control of herpes virus infections. *In* "Lithium: Current Applications in Science, Medicine and Technology" (ed. R.O. Bach), pp. 397–406. John Wiley & Sons Inc., New York.

Horrobin, D.F. (1985b). Essential fatty acids and prostaglandins in schizophrenia and alcoholism. *In* "Biological Psychiatry" (ed. C. Shagass). Elsevier, Amsterdam.

Horrobin, D.F. (1987). Essential fatty acids, prostaglandins, and alcoholism: An overview. *Alcoholism: Clin. Exp. Res.* **11**, 2–9.

Horrobin, D.F. (1988). Prostaglandin E_1: Physiological significance and clinical use. *Wein. Klin. Wochenschr.* **100**, 472–477.

Horrobin, D.F. (1989a). Essential fatty acids, psychiatric disorders, and neuropathies. *In* "Omega-6 Essential Fatty Acids: Pathophysiology and Roles in Clinical

Medicine" (ed. D.F. Horrobin), pp. 305–320. Alan R. Liss, New York.

Horrobin, D.F. (1989b). Clinical biochemistry of essential fatty acids. *In* "Omega-6 Essential Fatty Acids: Pathophysiology and Roles in Clinical Medicine" (ed. D.F. Horrobin), pp. 21–54. Alan R. Liss, New York.

Horrobin, D.F. (1990a). Effects of lithium on essential fatty acid and prostaglandin metabolism. *In* "Lithium and Cell Physiology" (ed. R.O. Bach and V.S. Gallichio), pp. 137–149. Springer-Verlag, New York.

Horrobin, D.F. (1990b). Lithium, fatty acids and seborrhoeic dermatitis: A mechanism of lithium action and a new treatment for seborrhoeic dermatitis. *Lithium* **1**, 149–155.

Horrobin, D.F. (1990c). Essential fatty acids, prostaglandins and schizophrenia. Proceedings of the World Congress of Psychiatry, Athens, October 1989, in press.

Horrobin, D.F. (1990d). Lithium and dermatological disorders. *In* "Lithium and Cell Physiology" (ed. R.O. Bach and V.S. Gallichio), pp. 158–168. Springer-Verlag, New York.

Horrobin, D.F. and Manku, M.S. (1980). Possible role of prostaglandin E_1 in the affective disorders and in alcoholism. *Br. Med. J.* **280**, 1363–1366.

Horrobin, D.F., Durand, L.G. and Manku, M.S. (1977a). Prostaglandin E_1 modifies nerve conduction and interferes with local anaesthetic action. *Prostaglandins* **14**, 103–108.

Horrobin, D.F., Mtabaji, J.P., Manku, M.S. and Karmazyn, M. (1977b). Lithium as a regulator of hormone-stimulated prostaglandin synthesis. *In* "Lithium in Medical Practice" (ed. F.N. Johnson and S. Johnson), pp. 243–246. MTP Press, Lancaster.

Horrobin, D.F., Ally, A.J., Karmali, R.A. *et al.* (1978). Prostaglandins and schizophrenia: further discussion of the evidence. *Psychol. Med.* **8**, 43–8.

Horrobin, D.F., Jenkins, D.K., Mitchell, J. and Manku, M.S. (1988). Lithium effects on essential fatty acid and prostaglandin metabolism. *In* "Lithium: Inorganic Pharmacology and Psychiatric Use" (ed. N.J. Nirch), pp. 173–176. IRL Press, Oxford.

Huang, Y.S., Mitchell, J., Jenkins, K., Manku, M.S. and Horrobin, D.F. (1984). Effect of dietary depletion and repletion of linoleic acid on renal fatty acid composition and urinary prostaglandin excretion. *Prostaglandins Leukotr. Med.* **15**, 223–228.

Johnson, S.B., Gordon, E., McClain, C., Low, G. and Holman, R.T. (1985). Abnormal polyunsaturated fatty acid patterns of serum lipids in alcoholism and cirrhosis: arachidonic acid deficiency in cirrhosis. *Proc. Natl. Acad. Sci. USA* **82**, 1815–1818.

Judd, L.L., Hubbard, B., Janowsky, D.S., Huety, L.Y., Abrams, A.A, Riney, W.B. and Pendery, M.M. (1979). Ethanol-lithium interactions in alcoholics. *In* "Alcoholism and Affective Disorders" (ed. D.W. Goodwin and C.K. Erickson), pp. 109–136. SP Inc., New York.

Karmazyn, M., Manku, M.S. and Horrobin, D.F. (1978). Changes of vascular reactivity induced by low vasopressin concentrations: Interactions with cortisol and lithium and possible involvement of prostaglandins. *Endocrinology* **102**, 1230–1236.

Kirtland, S.J. (1988). Prostaglandin E_1: a review. *Prostaglandins Leukotr. EFAs* **32**, 165–174.

Kline, N.S., Wren, J.C., Cooper, T.B., Varga, E. and Canal, O. (1974). Evaluation of lithium therapy in chronic and periodic alcoholism. *Amer. J. Med. Sci.* **268**, 15–22.

Lieb, J. (1979). Remission of recurrent herpes infection during therapy with lithium. *New Engl. J. Med.* **301**, 942.

Lieb, J. (1981). Immunopotentiation and inhibition of herpes virus activation during therapy with lithium carbonate. *Med. Hypotheses* **7**, 885–890.

Lieb, J., Karmali, R.A. and Horrobin, D.F. (1983). Elevated levels of prostaglandin E$_2$ and thromboxane B$_2$ in depression. *Prostaglandins Leukotr. Med.* **10**, 361–367.

Manku, M.S., Horrobin, D.F., Karmazyn, M. *et al.* (1979a). Prolactin and zinc effects on rat vascular reactivity: Possible relationship to dihomogammalinolenic acid and to prostaglandin synthesis. *Endocrinology* **194**, 774–779.

Manku, M.S., Oka, M. and Horrobin, D.F. (1979b). Differential regulation of the formation of prostaglandins and related substances from arachidonic acid and from dihomogammalinolenic acid. I. Effects of ethanol. *Prostaglandins Med.* **3**, 119–128.

Merry, J., Reynolds, C.M., Bailey, J. and Coppen, A. (1976). Prophylactic treatment of alcoholism by lithium carbonate. *Lancet* **ii**, 481–482.

Mest, H.-J., Zehl, U., Sziegoleit, W. *et al.* (1982). Influence of mental stress on plasma level of prostaglandins, thromboxane B$_2$ and on circulating platelet aggregates in man. *Prostaglandins Leukotr. Med.* **8**, 553–563.

Nervi, A.M., Peluffo, R.O. and Brenner, R.R. (1980). Effect of ethanol administration of fatty acid desaturation. *Lipids* **15**, 263–268.

Newton, A. (1979). Inhibitors of prostaglandin synthesis as inhibitors of herpes simplex virus replication. *Adv. Ophthalmol.* **38**, 58–63.

Pennington, S.N., Woody, D.G. and Rumbley, R.A. (1982). Ethanol-induced changes in the oxidative metabolism of arachidonic acid. *Prostaglandins Leukotr. Med.* **9**, 151–157.

Puolakka, J., Makarainen, L., Viinikka, L. and Ylikorkala, O. (1985). Biochemical and clinical effects of treating the premenstrual syndrome with prostaglandin synthesis precursors. *J. Reprod. Med.* **30**, 149–153.

Rasmussen, H., Lake, W. and Allen, J.E. (1975). The effect of catecholamines and prostaglandins upon human and rat erythrocytes. *Biochim. Biophys. Acta* **411**, 63–73.

Reitz, R.C. (1979). The effects of ethanol ingestion on lipid metabolism. *Progr. Lipid Res.* **18**, 87–115.

Rivers, J.P.W. and Frankel, T.L. (1981). Essential fatty acid deficiency. *Br. Med. Bull.* **37**, 59–64.

Segarnick, D.J., Ryer, H., Rotrosen, H. and Rotrosen, J. (1985). Precursor and pool-dependent differential effects of ethanol on human platelet prostanoid synthesis. *Biochem. Pharmacol.* **34**, 1343–1346.

Sherwin, L. (1972). Compositions containing lithium succinate. U.S. Patent 3,639,625 (1972).

Shifrine, M. and Marr, A.G. (1968). The requirement of fatty acids by *Pityrosporum ovale*. *J. Gen. Microbiol.* **32**, 268–270.

Skinner, G.R.B. (1983). Lithium ointment for genital herpes. *Lancet* **ii**, 288.

Skinner, G.R.B., Buchanan, A., Hartley, C.E. *et al.* (1980). The preparation, efficacy and safety of antigenoid vaccine NFU$_1$ (S.L.) MRC toward prevention of Herpes simplex virus infections in human subjects. *Med. Microbiol. Immunol.* **169**, 39–51.

Skinner, G.R.B., Randall, S.L., Billstrom, M.A. and Buchan, A. (1990). Antiviral actions of essential fatty acids *in vitro*. *In* "Omega-6 Essential Fatty Acids: Pathophysiology and Roles in Clinical Medicine" (ed. D.F. Horrobin), pp. 261–273. Alan R. Liss, New York.

Wilde, P.F. and Stewart, P.S. (1968). A study of the fatty acid metabolism of the yeast *Pityrosporum ovale*. *Biochem. J.* **108**, 225–231.

Wolff, J., Berens, S.C. and Jones, A.H. (1970). Inhibition of thyrotropin-stimulated adenyl cyclase activity of beef thyroid membranes by low concentrations of lithium ion. *Biochem. Biophys. Res. Commun.* **39**, 77–82.

16 Lithium, Lymphocytes and Labyrinths: Insights into Biological Regulation and Diversity

DAVID A. HART

Faculty of Medicine, University of Calgary Health Sciences Centre, Calgary, Alberta, Canada T2N 4N1

Introduction

Over the past 20 years our understanding of immune regulation and the complexity of the elements of this system has increased dramatically (Cantor *et al.*, 1984; Sell, 1987; Vitetta *et al.*, 1989). The internal regulation of the antigen-specific elements of this system, with its myriad of T-lymphocyte and B-lymphocyte subsets, is almost overwhelming even to immunologists. The ontogeny of sets of lymphocytes at different anatomic sites (thymus, bone marrow, Peyer's patches, etc.) and their integration into a functional immune system which then interfaces with a number of effector cells is truly a marvellous organizational feat. The level of complexity which exists in the immune system may be analogous to that required to initiate and maintain neuronal and neuroendocrine networks. The analogy can be extended further if we consider, admittedly in a simplistic fashion, that the two systems are designed to perceive and respond to stimuli in a similar, organized fashion (Blalock, 1984; Hart, 1986a) and both systems exhibit memory. Both systems also appear to be dynamic, with many oscillating activities which presumably contribute to their flexibility in responding to varied and diverse stimuli.

In recent years, not too surprisingly, evidence has started to accumulate which indicates that the immune system and the neuro/neuroendocrine systems are intimately interactive and there appears to be a bidirectional flow of information between the systems. In fact, the two systems have the potential to use many similar mediators to transmit the information (Ader, 1981; Blalock and Smith, 1985; Weigent and Blalock, 1987; Blalock and Bost, 1988). Thus, there is the potential for informational exchange between the two systems and the neuro/neuroendocrine system can serve as an external regulator of the immune system (Jankovic *et al.*, 1987; Pierpaoli and Spector, 1988; Blalock and Bost, 1988). The dominance of specific regulatory influences over others and the 'set-point' for the homeostasis between these two systems is very likely derived from a number of variables, including species-specific considerations, as well as other external regulatory systems and individual genetic variations.

Salts of lithium have been used for a number of years in the treatment of a subset of patients with affective disorders (Cade, 1949; Johnson and Johnson, 1977; Cooper et al., 1979; Johnson, 1984; Maj et al., 1989). It is particularly effective in the treatment of bipolar manic-depression. Interestingly, not all patients with this diagnosis respond to lithium, and some of those that do respond favourably with regard to their affective disorder experience side effects which preclude continued use of this chemically simple drug. One set of side effects involves the immune system (reviewed in Johnson, 1979; Hart, 1986a, 1990a; Lieb, 1987). Based on the points raised earlier, it is not possible to conclude that lithium is impacting on the immune system directly, on the neuroimmune interface, or both, since they are interrelated. However, these observations, determined from patients receiving lithium for affective disorders, have provided the impetus for many investigators to explore in more detail the pharmacological activity of lithium on immune system elements, immuno-regulation and the neuroimmune interface. This chapter will focus on integrating the information derived from studies carried out both *in vivo* and *in vitro* with a number of species, and then attempt to synthesize a set of hypotheses which may be of use in future research directions.

Before embarking on a discussion of lithium, lymphocytes and the immune system, it should be pointed out (and is probably evident from other chapters in this volume) that lithium can influence a number of cell types and cellular processes where proliferation and differentiation occur and therefore the reader should not construe the following discussion to imply that the observations, particularly *in vitro* findings, are necessarily unique to the immune system or lymphocytes. Lithium (McMahon, 1974; Pandey and Davis, 1980; Hart, 1986a) and other simple ions (McMahon, 1974) can influence growth and development, responsiveness of cells to hormones and growth factors, and modify intracellular second-messenger systems used in most eukaryotic cells (i.e. cAMP, ITP, ion pumps, etc.). Conversely, the failure to detect an influence of lithium on an immune process may also provide further insights into cell-specific dependence on the intracellular messengers available.

Biochemistry of Lymphocyte Activation

Lymphocytes can be activated in an antigen-specific fashion via mechanisms involving the T-lymphocyte receptor and cell surface immunoglobulin on B-lymphocytes, or they can be activated in a non-antigen-specific manner by certain agents or reagents termed mitogens. Mitogens can show some specificity for subsets of lymphocytes and offer the advantage to investigators of stimulating sufficient numbers of cells to detect biochemical changes. The obvious disadvantage of using mitogens is that the biochemical mechanisms involved in mitogen stimulation may only partially reflect those utilized during antigen-driven responses. In spite of this limitation, there are many similarities between the two systems once the response has been initiated. Both systems usually depend on the presence of accessory cells such as macrophages or a macrophage-like cell and also depend on the generation of cytokines (growth and differentiation factors). While the immune system relies on antigen-specific events to initiate responses, once initiated the immune response is dependent on and regulated by a cytokine network (Balkwill and Burke, 1989). This cytokine network is not unlike neuroendocrine networks and in fact, some of these cytokines are believed to be components of the larger neuroimmune network (Blalock, 1984; Blalock and Smith; 1985; Weigent and Blalock, 1987; Jankovic, 1989). Some of these cytokines express a myriad of biological activities and their expression is not limited to cells of the immune system, while others may be more specific for cells of this system. The second level of regulation that also involves the cytokine network is the regulation of cytokine receptor expression. Similar to most other systems involving hormones, transmitters, or growth factors, the cytokine network utilizes receptor-mediated events to generate intracellular messengers in order to convey information. The second-messenger systems implicated in lymphocyte activation and the cytokine network involve those that are also well studied in other systems. They include cyclic-AMP, arachidonic acid metabolites such as prostaglandins, inositol triphosphate metabolism, protein kinase C metabolism, tyrosine kinase and phosphatase activity, ion pumps such as the Na/K ATPase, and activation of certain oncogenes (reviewed in Pandey and Davis, 1980; Hart, 1986a; Moller, 1987; Cambier *et al.*, 1988; Alexander and Cantrell, 1989). It is thus apparent that the immune system uses common intracellular pathways and exerts system-specific regulation via the cytokine–cytokine receptor network and antigen-specific receptors.

Lymphocyte Regulation and Immunoregulation

The immune system consists of lymphocytes, which can be divided into a number of subsets, accessory cells such as antigen-presenting macrophages, and effector cells, all of which coexist as a highly dynamic, highly mobile and interactive system. As discussed above, regulation at a restricted level is believed to involve antigens and cytokines. At the larger systemic level, internal regulation very likely involves suppressor cell networks (Moller, 1975) as well as elements of a humoral idiotype network (Jerne, 1974; Green and Nisonoff,

1984). This latter 'network' evolves during development of the immune system and is influenced by the genetics of the host. Under antigenic stimulation, oscillations in the idiotype–anti-idiotype interactions can occur (reviewed in Cerney and Kelsoe, 1984), conceptually not unlike other biological oscillators except that it does not necessarily maintain constant amplitude. Such a network can be manipulated by exogenous anti-idiotypic antibodies (Hart *et al.*, 1972; Pawlak *et al.*, 1973) and it may play a role in autoimmune states. Further immunoregulation is accomplished by segregating certain functions to specific anatomical sites such as the spleen, lymph nodes or the mucosal immune system. Not only do such sites serve specific immune functions with regard to immune reactivities, but these immunological organs also serve as sites of neuroimmune interactions since they are innervated (Calvo, 1968; Bulloch and Moore, 1981; Williams and Felton, 1981; Singh, 1984; Felton *et al.*, 1984), and denervation can modify immune reactivity (Kasahara *et al.*, 1977; Giron *et al.*, 1980; Hall *et al.*, 1982).

Much of our information regarding immune regulation has come from the study of animal models, particularly the mouse. Studies of human systems has been more restricted due to tissue availability and ethical considerations. However, considerable information regarding the human system has come from studies of immunodeficiency states, responses to vaccines, autoimmune diseases, surgical procedures (i.e. splenectomy, thymectomy), cancer patients, transplant patients and more recently AIDS patients. While there are many similarities detected between species, it is readily apparent that each species has some unique features regarding immunoregulation and that much more work is required before we can begin to understand all the variables involved in human immunoregulation, not the least of which are the genetic variables. These species differences and the other considerations discussed in this introduction are relevant to the discussion to be presented in the following sections. Information regarding the pharmacological activity of lithium on lymphocytes and on immune regulation will be presented and then I will attempt to integrate the information and propose some speculations regarding future research directions.

Lithium and Human Lymphocyte Activity and Regulation

Observations from lithium treatment groups

As stated earlier, lithium administration for affective disorders, particularly bipolar illness, is a very effective treatment modality. However, as with most treatment modalities, this 'drug' is not universally successful in the treatment of mood disorders and a number of patients have experienced adverse physiological responses during treatment (discussed in Lydiard and Gelenberg, 1982). Lithium treatment can lead to unwanted influences on the central nervous system, renal effects, cardiovascular effects, gastrointestinal effects, endocrine effects and haematological/immunological effects (discussed in Lydiard and Gelenberg, 1982; Hart, 1986a, 1990a; Lieb, 1987). Some patients respond to the treatment and experience/complain of no side effects, others respond and complain of side effects, and with still others they stop taking the drug because of the side effects and it is not known if their affective disorder was responsive to the treatment (Cooper *et al.*, 1979; Lydiard and Gelenberg, 1982). This heterogeneous response to lithium with regard to both the affective disorder and the side effects, emphasizes the difficulty in interpreting reports in the literature and generalizing from them with regard to lithium effects on immunoregulation and lymphocyte function in patients. It is obvious that genetic heterogeneity has a strong influence on the responses to lithium. Part of the reason for such variation may also be related to the fact that efficacious doses of lithium border toxicity levels.

From the literature, it does not appear that lithium treatment has a consistent positive effect on lymphocyte-based host defence systems. To my knowledge, there are no reports in the literature which indicate that lithium-treated patients have fewer episodes of viral or bacterial infections than do the general population. However, Lieb (1981, 1987) has reported on a group of patients whose herpes infections and respiratory infections remitted during lithium treatment for affective disorders. Whether this was due to an effect on the immune system or a direct effect of lithium on viral replication and activity (Ziaie and Kelalides, 1989), is not clear. There is also a report in the literature which indicated that a patient with recurrent infections improved after initiation of lithium treatment for affective disorder (Perez *et al.*, 1980). However, this proved to be an effect on the patient's neutrophil function rather than on lymphocyte function. Her neutrophils failed to exhibit normal chemotactic responses, due to elevated intracellular levels of cAMP, and this defect was apparently corrected by the lithium.

In contrast to positive effects on immune-mediated host defence, there are numerous reports in the literature regarding autoimmune disease and lithium treatments. These include reports of exacerbations of clinically diagnosed autoimmune disease as well as reports of the 'induction' of autoimmune disease in patients previously without clinically apparent symptoms (reviewed in Johnson, 1979; Hart, 1986a, 1990a; Lieb, 1987). Whether the latter group of patients had underlying subclinical immune dysfunction prior to lithium treatment is impossible to ascertain. Some immunodeficiency syndromes and immune 'dysfunctions' can occur with reasonably high frequency but do not lead to overt clinically defined diseases. Lithium treatment in such individuals could shift regulation to a point where clinical manifestations become apparent.

Lithium has been reported to be associated with the exacerbation of a number of autoimmune diseases such as psoriasis (Carter, 1972; Evans and Martin, 1979), diabetes mellitus (Craig *et al.*, 1977; Johnson, 1977), myasthenia gravis (Neil *et al.*, 1976), systemic lupus erythematosus (Presley *et al.*, 1976), as well as a number of other endocrine disorders (reviewed in Lydiard and Gelenberg, 1982; Calabrese *et al.*, 1985; Salata and Klein, 1987; Lieb, 1987). Whether all of these changes are due to effects of lithium on lymphocyte function and activity is not well defined. For instance, lithium may potentiate neutrophil degranulation in psoriatic patients (Bloomfield and Young, 1983), due to *in vivo* activation during the disease process, and thus exacerbate inflammatory aspects of the disease. However, in some of the other disorders alterations in lymphocyte activity is implied since changes in antibody levels have been detected. Lithium treatment has also been reported to induce the appearance of antinuclear, antithyroid and antigastric antibodies in individuals with no previous autoimmune disease (Presley *et al.*, 1976). Interestingly, in such individuals the level of autoantibodies declined once the lithium treatment was stopped, so the transient induction of autoantibodies was reversible and did not lead to sustained autoimmune disease. Thus, lithium treatment impacted on immunoregulation in certain individuals to alter elements that control the expression of antiself reactivities. However, as discussed below, this interpretation may not be generally applicable, and other considerations are also involved.

Lazarus *et al.* (1986) have investigated this phenomenon in more detail. These authors designed a prospective study of thyroid immune status of manic-depressive patients before and after initiation of lithium therapy. Sixteen of 37 patients (43%) had autoantibodies prior to initiation of lithium therapy. Thus there was a high incidence of thyroid and/or immune dysfunction in these patients even before lithium therapy. Following lithium treatment none of the patients negative for autoantibody developed antithyroid titres. Interestingly, those patients with detectable autoantibody titres prior to lithium treatment exhibited a very heterogeneous immunological response to lithium therapy. Some experienced a rise in autoantibody titre, while others experienced a decline in titre or the titre remained unchanged. Interestingly, Wahlin *et al.* (1984) have reported a very heterogeneous response to lithium in both patients and normals with regard to T-lymphocyte subsets. Prelithium ratios of Helper/Suppressor T-cells in patients were not different from controls. The ratios following lithium treatment were altered in both patients and normals, but the distribution of the changes observed was different when the patients were compared with the controls. Perhaps such heterogeneity could be correlated with the heterogeneity observed by Lazarus *et al.* for the antithyroid antibody response. Whatever the interpretation, based on these findings it would appear that manic-depressive patients have a higher than normal incidence of underlying thyroid/immune dysfunction. As Lazarus *et al.* (1986) point out, such patients should be immunologically evaluated prior to initiation of lithium therapy. Additional studies are obviously needed to clarify whether the patient population is unique in having such underlying disease or if lithium can actually cause antithyroid immune dysfunction.

A second point that is of interest with regard to 'development' of autoimmune side effects during lithium treatment is that of frequency and immunogenetics. For many autoimmune diseases there is a higher frequency in females than in males (Inman, 1978; Ahmed and Talal, 1989). Sex hormones have been implicated in this gender-dependence and women tend to be more immunologically reactive than men (Inman, 1978; Ahmed and Talal, 1989). One might, therefore, predict that the frequency of autoimmune disease occurring in patients receiving lithium would exhibit some gender-dependence if lithium were exerting a general impact on lymphocyte-mediated immunoregulatory circuits. In the study by Presley *et al.* (1976), lithium treatment of a group of 50 patients (19 men, 31 women) led to the appearance of antinuclear antibodies (ANA) in 7 women (23%) and 3 men (16%). One female with SLE prior to lithium therapy experienced an exacerbation of her disease. Interestingly, in this study there was also a higher incidence of antiparietal cell antibodies in the lithium group, but no mention of any gender-dependence was made in the report. In a second study, reported by Whalley *et al.* (1981), a group of 54 patients receiving lithium for affective disorders was investigated. All patients had been on lithium for at least 6 months and were not taking other drugs. Twenty per cent of the patients expressed detectable levels of antinuclear antibodies. Of the 11 positive patients, 9 were females and 2 were males. The incidence in the female population was 9/35 (26%) and that in the male population was 2/19 (11%). Thus in both studies there was a tendency for a higher incidence in the female population, but further studies are needed before we can conclude that any gender-dependent variables are operative.

Another variable that is apparently operative in autoimmune disease is that of immunogenetics. Certain autoimmune diseases exhibit immunogenetic associations with specific major histocompatibility complex loci (Cruse and Lewis, 1988; Farid, 1988). However, Lazarus *et al.* (1986) found that the HLA type of patients with antithyroid antibodies was no different from that of the control population. There was no increase in the frequency of antigens reported to predispose to thyroid disease (HLA: B8, DR3, DR4). Likewise, in the study of Whalley *et al.* (1981), the investigators could not detect any association between the induction of antinuclear antibodies and particular HLA types. Therefore, while such associations are not absolute and may be complex, at present there is no evidence presently available to link autoimmune side effects of lithium with any immunogenetic phenotypes. However, additional studies with larger numbers of patients are probably warranted.

As has been discussed, there are numerous reports in the literature regarding lithium and certain autoimmune phenomena. It is also important to point out that there is a deficiency of reports of lithium effects on other autoimmune diseases where autoantibodies have been detected and lymphocyte-mediated derangements have been implicated. For instance, exacerbation of inflammatory bowel disease, ulcerative colitis, or Crohn's disease has not been reported (reviewed in Lydiard and Gelenberg, 1982). Furthermore, lithium has been used to treat affective disorders in multiple sclerosis patients (Weizscher and Tegeler, 1984); to my knowledge there are no reports in the literature regarding lithium treatment and exacerbations of the disease. Since this disease is of the relapsing–remitting type, it may be difficult to identify lithium as the triggering event in the induction of a relapse. However, patients with MS appear to have dysfunctions in their T-lymphocyte regulatory circuits (Chofflon *et al.*, 1989; Salonen *et al.*, 1989), and one might have expected to observe an impact of lithium on disease progression if the ion was capable of influencing immunoregulation via a general mechanism. In that regard, recent preliminary observations of a patient with MS and being treated with lithium for affective disorder have been made (L. Metz, S. Johnson, P. Seland, J. Lamarre and D. Hart, unpubl. obsv.). While on lithium, the MS progressed rapidly, and when the lithium was discontinued disease progression stabilized. Therefore, in this patient lithium may have had an influence on the autoimmune process, but such a conclusion remains only speculative at this point.

Based on this discussion of lithium's apparent impact on lymphocyte function and immunoregulation, one would have to conclude that no generalized effect by lithium can be ascertained. Given this conclusion, then we should ask what is the meaning of the available data on the spectrum and frequency of autoimmune diseases and autoantibodies expressed under lithium's influence. Particularly intriguing is the development of antithyroid, anti-islet, antigastric and antinuclear antibodies. This pattern is not dissimilar to a pattern of autoimmune reactivity that can occur in patients with Addison's disease (Jackson and Eisenbarth, 1989). Patients with idiopathic loss of adrenal function (Addison's disease) can develop multiglandular endocrine disease, usually with more than one of the autoantibodies mentioned above. A number of authors (reviewed in Skutsch, 1985) have described possible involvement of adrenal function in affective disorders. Skutsch (1985) has hypothesized that manic-depression is a multiple hormone disorder not unlike that which may occur in Addison's disease. However, it is obvious that most patients being treated with lithium do not develop multiglandular autoimmune disease, so rather than having a primary role, the appearance of this spectrum of autoantibodies in some patients may define a subset of bipolar affective disorders. Likewise, if adrenal dysfunction were playing a central role in bipolar illness, then one might expect patients with Addison's disease to be at risk for such syndromes. While many of these patients do exhibit some psychiatric disorders (Cleghorn, 1951; Martin *et al.*, 1977; Lever and Stansfeld, 1983; Leigh and Kramer, 1984; Varadaraj and Cooper, 1986, discussed in Tobin *et al.*, 1989), it is not specific for bipolar affective disorders.

Treatment of normal individuals or patients with lymphocyte-mediated immune deficiencies with lithium has yielded results that support the conclusion that this ion is not a general modulator of immunoregulation. Greco (1980) has reported that lithium had little or no influence on general immune reactivity when given to normal controls. Dosch *et al.* (1980) have reported on the treatment of individuals exhibiting immunodeficiency due to excess suppressor T-cell activity. Following treatment with lithium, there was an increase in the number of circulating surface Ig-positive B-lymphocytes, a decrease in the number of suppressor T-lymphocytes, but no increase in the serum levels of IgM, IgA or IgE. Therefore, lithium appears to correct some of the defects in such deficiency states but had no obvious effect on the essential variable, antibody secretion. Interestingly, Kovithavongs *et al.* (1983) have reported on a transplant patient who was treated with lithium and then promptly rejected the transplanted kidney. Since suppressor cells are believed to be important in preventing rejection, the hypothesis was raised that lithium inhibition of suppressor cell function contributed to the rejection of the transplant.

Therefore, the perception that lithium can modulate lymphocyte activity and immunoregulation *in vivo* needs to be evaluated by additional prospective studies in order to define the mechanisms involved and to better identify affective disorder patients at risk. Without some well controlled studies, and more thorough monitoring of patients on lithium, the immunopharmacology of this ion will continue to be in a state of flux and biased by anecdotal reports.

Observations with human peripheral blood lymphocytes *in vitro*

Interest in studying the effect of lithium on human lymphocytes *in vitro* has been maintained by the perception that lithium has immunomodulating effects *in vivo* as well as the known involvement of lithium-sensitive intracellular pathways in lymphocyte activation. This work has used cells derived from peripheral blood almost exclusively and this compartment contains lymphocytes that are in the recirculating pool of cells. In addition, the cells of the macrophage–monocyte lineage present in peripheral blood are also of the recirculating pool, which may be quite distinct from cells of the same lineage which present antigen in tissues such as spleen and lymph nodes.

In 1978, Shenkman *et al.* (1978) reported that addition of lithium to cultures of human peripheral blood lymphocytes enhanced the proliferative response of these cells to the mitogen PHA. Concentrations of lithium from 1.25 to 5 mM were found to be effective. The response to a second lectin, ConA, was also enhanced by lithium (Wadler *et al.*, 1979; Shenkman *et al.*, 1980). In contrast to the ConA and PHA results, it was found that lithium was less effective in enhancing the two-way mixed lymphocyte reaction (Shenkman *et al.*, 1978). Only the highest concentration of lithium tested, 5 mM, exhibited an enhancing effect. In other non-proliferation assays of lymphocytes, the same authors also presented evidence that lithium could enhance the percentage of sheep red-blood cell rosette-forming cells (T-lymphocytes) detected in peripheral blood mononuclear cell populations. Further experiments indicated that the suppressor T-lymphocyte population of cells was the primary cell type influenced by this ion (Wadler *et al.*, 1979). From experiments performed with compounds which elevate cAMP levels, theophylline and prostaglandin PGE_2, it was concluded that the effect of lithium on lymphocytes was mediated by its ability to inhibit adenyl cyclase (Shenkman *et al.*, 1978). Interestingly, lymphocytes were not the only cell population in the peripheral blood preparations influenced by lithium (Shenkman *et al.*, 1978). Cells of the monocyte/macrophage lineage were also influenced by exposure to lithium resulting in increased phagocytic activity. Whether this activation by exposure to lithium also leads to increased secretion of monokines and/or elevated antigen-presenting activity should be more thoroughly investigated.

Experiments similar to those described above have also been reported by Gelfand *et al.* (1979). However, this group has pursued the influence of lithium on suppressor T-lymphocyte activity in more detail. With regard to mitogen stimulation of peripheral blood lymphocytes, Gelfand *et al.* (1979) have reported that lithium could enhance [^{3}H]thymidine incorporation by approximately 1.7-fold. Lithium could also reverse the theophylline-induced suppression of the PHA response, again supporting the conclusion that this ion was acting at the level of adenyl cyclase. The same concentrations of lithium similarly could reverse the inhibition of IgM secretion induced by agents which elevate intracellular cAMP levels. Additional reports from the same group have indicated that lithium can inhibit suppressor T-cells *in vitro* (Dosch *et al.*, 1980). Utilizing cells from immunodeficiency patients with excess suppressor activity, these authors found that lithium was capable of functionally inactivating the excess activity *in vitro*. Interestingly, when the same patients were treated with lithium *in vivo*, the immunodeficiency was not functionally corrected but some immune parameters were altered (Dosch *et al.*, 1980).

Other investigators have also investigated the pharmacological effects of lithium on human lymphocytes *in vitro*. Licastro *et al.* (1983) reported that lithium enhanced suboptimal mitogen stimulation of lymphocytes from older individuals and individuals with Down's syndrome to a much greater degree than cells from normal young adults. Since older individuals and those with

Down's syndrome have reported alterations in immune regulation (suppressor cell activity), lithium would appear to 'correct' such influences. However, the enhancement by lithium was not specific for this ion, so the interpretation of this report is still somewhat open. Other studies have confirmed the conclusion that lithium is effective in enhancing lymphocyte stimulation by suboptimal concentrations of stimuli (Fernandez and Fox, 1980; Bray *et al.*, 1981). Fernandez and Fox (1980) have reported this observation and concluded that the effect was again at the suppressor cell level. Bray *et al.* (1981) reported that lithium enhanced stimulation by PHA and PWM and it was also effective in cultures that had been made monocyte-deficient. Weetman *et al.* (1982) have reported that lithium treatment of PWM stimulated human peripheral blood lymphocytes leads to increased secretion of IgG, IgM and IgA. These authors also reported that there were also elevated levels of spontaneous plaque-forming cells (indicative of Ig secretion) in the peripheral blood of patients receiving lithium for affective disorders. Some patients had normal levels of PFC while others were elevated. Kucharz *et al.* (1988) have reported that lithium enhances the release of interleukin-2 (IL-2, T-cell growth factor) from cultured lymphocytes. This ion also enhanced the responsiveness of a T-cell line to IL-2 and the change was not due to increased IL-2 receptor expression. In contrast, the response of human peripheral blood lymphocytes to recombinant IL-2 was recently found not to be enhanced by lithium (D. Matheson and D.A. Hart, unpubl. obsv.). Such differences may reflect differences in the indicator systems used. Kucharz *et al.* (1988) also have reported that lithium enhances IL-1 secretion from human monocytes. We have found that lithium will not replace monocytes and that lithium does not enhance lymphocyte stimulation under monocyte-deficient conditions (D. Matheson and D.A. Hart, unpubl. obsv.). Therefore, it does not appear that the mechanism of action of lithium is simply to enhance growth factor expression.

In the human system, it is obvious that the response to lithium is very heterogeneous and no firm conclusions can be drawn based on the data available. In fact, such heterogeneity raises several issues which remain to be resolved. For instance, if lithium is modifying common intracellular second-messenger systems, why does the ion not influence other systems when it is effective in specific patients with affective disorders? In addition, if indeed it is effective in modulating the immune system directly, why do patients with some autoimmune diseases experience exacerbations of the disease and others are apparently not influenced? Further, why is there a dichotomy between the concentrations of the ion which are effective *in vivo* and those required to induce detectable changes *in vitro*? Is it possible that lithium *in vivo* is not influencing the immune system directly, but is rather influencing a system that interfaces with the immune system, such as the neuroimmune interface?

Since lithium is so effective in restoring regulation of systems involved in mood, can we attribute its mechanism of action to just one site (Berridge *et al.*, 1989) or does it owe its effectiveness to multiple sites of action (Hart, 1988a)? The keys to this puzzle must await further understanding of the human system as well animal studies, some of which will be discussed below.

Influence of Lithium on Lymphocytes and the Immune System of Animal Model Systems

Mice

As mentioned above, mice have been extensively utilized by immunologists over the past 20 years to analyse the immune system and the availability of inbred strains of mice has been critical to our understanding of many areas. In spite of the intensive experimentation on mice, the literature regarding the effect of lithium on lymphocytes and immune regulation in this species is not extensive. Ishizaka and Moller (1982) reported that lithium chloride (10 mM) could enhance both the proliferative response and the antibody secretion response of mouse splenocytes to bacterial lipopolysaccharide (LPS). The enhancement was most dramatic with cells from the normally LPS-resistant C3H/HeJ strain but cells from other strains were partially influenced. In experiments performed in this laboratory (D.A. Hart, unpubl. obsv.), it was not possible to demonstrate a reproducible effect of 1–10 mM lithium chloride on mitogen stimulation of splenocytes or lymph node cells from a number of strains of mice. The mitogens LPS, ConA and PHA were used for these studies and both optimal and suboptimal concentrations of the mitogens were tested. Fuggetta *et al.* (1988) have reported that lithium lactate could enhance natural killer cell activity of mouse splenocytes when the animals were exposed to the compound *in vivo*, but that lithium was relatively ineffective when the cells were exposed *in vitro*. Jankovic's group has reported (Jankovic *et al.*, 1978a; Popeskovic *et al.*, 1979) that treatment of mice with lithium salts leads to depressed responsiveness to antigens. Based on these reports, we have undertaken a study of the effect of lithium on autoimmune disease development in murine models of systemic lupus erythematosus (Krause *et al.*, in prep.). Such mice are genetically predisposed spontaneously to develop a spectrum of autoantibodies, usually anti-DNA, antinuclear and others, which leads to glomerulonephritis, renal failure, vascular involvement and death (Theofilopoulos and Dixon, 1985; Kyogoku and Wigzell, 1987; Hart *et al.*, 1990). Groups of 20 female animals were given daily injections of either lithium-6 chloride or lithium-7 chloride from prior to development of disease (10 weeks of age) until death. Treatment of the animals with either isotope of lithium protected the animals from disease progression. After 7 months of injection, when all of the control animals were dead, 70% of the Li-7 group and 50% of the Li-6 group were still alive. Based on results from sequential bleedings of these animals, treatment of the animals with lithium only slightly influenced the levels of some antinuclear antibody specificities and in fact, caused a 2.5-fold elevation in the plasma levels of other autoantibody specificities. Thus, in contrast to the findings in humans discussed previously, in this murine model of SLE, lithium did not cause exacerbation of the disease and only partially impacted on systems responsible for regulating autoantibody production. Interestingly, no detectable differences between the two isotopes of lithium were detected at the level of autoantibody production, splenomegaly, kidney function (kidney failure) or lethality. We are currently examining the kidneys of treated and untreated animals morphologically to determine if other changes have occurred. In addition, we are examining the effect of lithium on animals after initiation of the disease process has started, as well as other murine models of SLE and inflammatory processes.

From the studies discussed above on autoimmune mice, it was apparent that one aspect of the disease process, namely lymphocyte-mediated auto-antibody production, was variably influenced by lithium, but the disease process was interfered with by events altered by lithium. One potential area of influence is at the level of inflammatory effector cells such as macrophages and polymorphonuclear leukocytes (PMN). Recent experiments have indicated that treatment of mice with a variety of inflammatory stimuli leads to similar changes in host defence systems (Hart, 1986b, 1987a, 1988b,c,d; Hart *et al.*, 1990). These include some bacterial stimuli (killed *Coryne bacterium parvum*), but not others (such as BCG or purified LPS), tumours, and autoimmune diseases. These observations appear to be associated with changes in macro-phages and PMN, and only indirectly due to lymphocyte activity. The response to *C. parvum* is independent of lymphochtes, and is therefore likely to be a pure inflammatory cell response. This response cannot be inhibited by high dose steroids, tuftsin or indomethacin (Hart, 1987b; Hart *et al.*, 1989). However, the induction of splenomegaly, hepatomegaly, or elevated levels of plasma proteinase activity by *C. parvum* could be suppressed up to 80% by injection of either lithium-7 or lithium-6, but not by injection of equivalent concentrations of either rubidium or potassium (Hart, in prep.). The suppression was dose dependent and the genetics of the mouse strain used in the experiments appeared to influence the degree of suppression observed. Since macrophages are critical to immune responses, as well as functioning as inflammatory cells, the finding that lithium can modify the activity of mouse macrophages could also be relevant to the previously mentioned work from Jankovic's group (Jankovic *et al.*, 1978a; Popeskovic *et al.*, 1979). Since mitogen stimulation, antibody responses and DTH reactions are all macrophage-dependent, if lithium altered macrophage function, all of these responses could be influenced. These findings would not, however, explain our previously discussed results in the autoimmune mice where lithium treatment led to elevated levels of antibody. In the autoimmune mice, we do not know what drives the autoimmune process and whether it uses the same mechanisms as are operative in traditional immune responses to exogenous antigens.

While the results obtained thus far with this species do not allow us to make firm conclusions with regard to the site and mechanism of action of lithium on the immune system, they do indicate that the ion can impact on both lymphocyte and macrophage function in vivo and that further experimentation is warranted. The availability of genetically defined strains of mice may also be very useful in furthering our understanding of the heterogeneity observed in the immune status of patients receiving lithium.

Rats and guinea pigs

Like the previous results with mice, administration of lithium to rats usually leads to the suppression of lymphocyte-mediated responses such as antibody formation and delayed type hypersensitivity reactions (Jankovic et al., 1978a,b, 1982). In one of these studies it was found that treatment of Wistar male rats with lithium inhibited the development of experimental autoimmune thyroiditis (Jankovic et al., 1982). Rats were immunized with exogenous thyroid antigen and lithium prevented the induction of the disease process. Whether the point of interference was at the lymphocyte level or at the antigen-processing macrophage level was not identified. Hassman et al. (1985) have also studied the effect of lithium on the development of experimental thyroiditis in female August rats. These authors found that lithium enhanced the level of antithyroid antibodies detected if it was given during development of the disease and depressed autoantibody levels if given during the resolution of disease. Interestingly, lithium treatment of normal August rats did not induce antithyroid antibodies. Therefore, lithium could exert both positive and negative influences in this strain of rat. In the same study, Hassman et al. (1985) also found that splenocytes from immunized rats treated with lithium responded significantly better in vitro to mitogens than did splenocytes from rats that were only immunized. We have attempted to determine whether lithium could influence the stimulation of splenocytes and lymph node cells, from a number of strains of rats, by mitogens in vitro, and have been unable to demonstrate any reproducible, specific effects with this ion (D.A. Hart, unpubl. obsv.). Thus, there is a dichotomy between the effects in vitro of Li on naive cells and the effects in vivo observed by Hassman et al. (1985).

Likewise, we have been unable to demonstrate any reproducible, specific effects with lithium on mitogen stimulation of guinea pig cells (spleen and

LNC) or antigen stimulated cells (LNC) (Hart, 1990b, and unpubl. obsv.). We have experience with cells from this species (Hart *et al.*, 1973; Hart, 1977; Stein-Streilein and Hart, 1979) and do not think the failure to find an effect was due to technical considerations. The failure to detect any influence on lymphocyte activity at low concentrations of lithium in this species is especially interesting since Cade (1949) found that treatment of guinea pigs with lithium did influence their behaviour.

Therefore, in these two species lithium does not appear to be able to exert a positive effect on lymphocyte activities in most situations *in vitro*, even though it can influence behaviour and some immune parameters *in vivo*. However, these two species have not been investigated very thoroughly, and perhaps should be the subject of more research before conclusions are drawn.

Hamsters

In contrast to many of the other animal models discussed, the immune system of the hamster is probably the least studied but this species may offer some unique insights into biological variation. The Syrian hamsters utilized by most investigators were derived from a restricted genetic stock and the inbred strains are therefore also restricted in both number and diversity (discussed in Streilein *et al.*, 1981). There are a number of other hamsters available for research, but most of them have not been used in the study of lithium, lymphocytes and the immune system.

For the past 14 years, we have been actively investigating the immune system of the Syrian hamster and characterizing lymphocyte activation in this species. This analysis has included the effect of lithium on reactivity *in vitro* of hamster lymphocytes and accessory cells. Early reports characterized the effect of lithium on the lymphoid cell response to mitogens such as PHA (Hart, 1979a), ConA (Hart, 1979b), LPS (Hart, 1982a,b), dextran sulphate (Hart, 1982a), trypsin (Hart, 1982a), zinc ions (Hart, 1979c) and cytochalasin D (Hart, 1987c). Other investigations focused on both soluble (Hart, 1979d, 1988e) and particulate (Hart and Stein-Streilein, 1981) antigens. Still other reports have dealt with analysing the mechanisms by which lithium exerts an influence on these responses (Hart, 1981a,b, 1982b). These studies have been discussed in detail elsewhere (Hart, 1986a, 1990b,c). The most critical observations regarding lymphocyte stimulation with cells from this species are that lithium could dramatically enhance activation by both antigens and most mitogens, particularly when suboptimal concentrations of stimulant were used and when conditions were not optimized. The effect of lithium was specific for this ion, most effective in the early phases of stimulation, and appeared to influence the cells via mechanisms involving cyclic nucleotides and membrane ion pumps (Hart, 1981a,b, 1982b). Cells derived from thymus, spleen and lymph nodes could all be influenced by lithium. Peripheral blood cells were not tested. The optimal concentration of lithium which was effective in the assays *in vitro* was somewhat dependent on the stimulus. For mitogen stimulation the optimal concentration was 10 mM, while the optimal concentrations for soluble and particular antigens were 5 mM and 2.5 mM, respectively. It should be noted that the levels of enhancement observed when the cultures were supplemented with lithium were several times higher than the degree of enhancement observed with human lymphocytes or with other animal species (discussed previously).

In contrast to the other species discussed earlier, there is not much information available on the effect of lithium on immune responses, lymphocyte activation and inflammatory processes *in vivo* with the hamster system. Such information would be very useful for better interpretation of the *in vitro* findings.

Summary of the Immunopharmacological Properties of Lithium

From the above discussion, it is apparent that lithium can exert effects on lymphocytes and the immune system both *in vivo* and *in vitro*. It is also apparent that a dichotomy exists between what is observed *in vivo* in species such as the human, mouse and rat, and what is observed *in vitro* with lymphoid cells from such species. *In vitro*, lithium exerts a modest to insignificant effect on cells at concentrations equal to or in excess of those which influence immune functions *in vivo*. In the one species where lithium can reproducibly enhance lymphocyte activity *in vitro*, namely the hamster, relatively little is known about the immunopharmacology *in vivo*. While this latter data will be important to obtain, one can conclude that the effect of lithium on hamster lymphocytes *in vitro* is unique when compared with the reported observations with cells from the other species tested (human, mouse, rat, guinea pig). Interestingly, cells derived from hamsters optimally stimulated with antigen (Hart, 1988e) were the least responsive to lithium's influence, while 'resting' or normal cells were the most responsive. Thus, in the maximally stimulated state, cells from hamsters responded to lithium in a fashion not unlike 'resting' cells from the other species tested. Therefore, the 'resting' or maintenance state of the cells from hamsters is apparently unique. Why this should be the case is not clear at the present time.

The failure to detect a reproducible immunopharmacological effect of lithium *in vitro* with lymphoid cells from species other than hamsters could have a number of explanations, but the mere fact that lithium did not exhibit an effect is also significant. If this ion is modifying intracellular second-messenger systems common to a number of cell types, why was lithium not effective? This issue remains an enigma which will have to be addressed by future investigations.

Possible Explanations for the Dichotomy between Results *In Vivo* and *In Vitro* and for Species Differences

One of the most obvious explanations of the differences noted between the results obtained with lithium *in vivo* and *in vitro*, particularly those of the human, rat and mouse systems, is that the effects of lithium *in vivo* on lymphocytes are indirect. We know that therapeutic concentrations of lithium can modulate mood and behaviour in patients with certain affective disorders and that such treatment can influence the immune system of such patients. Similar concentrations of lithium have been reported to alter immune regulation in control populations as well, so a direct link between mood and immune regulation need not be invoked although that link has been made by others (discussed in Lieb, 1987; Maestroni *et al.*, 1986, 1987). In contrast, it may be that therapeutic concentrations of lithium exert a pharmacological effect on the immune system by acting at an interface between the immune system and an external regulator such as the neuroendocrine system or the 'hard-wired' innervation system found in the spleen, LNC, thymus, or bone marrow.

A dichotomy between the effects *in vitro* and *in vivo* of a very relevant immunomodulator is not without precedent in the literature. Maestroni *et al.* (1986), have reported that melatonin, a component of the neuroendocrine system, has immunopharmacological effects *in vivo*, but is inactive *in vitro*. Further investigation of this phenomenon revealed that melatonin influences the opoid system and that the effect *in vivo* was mediated by components of the latter system (Maestroni *et al.*, 1987; Pierpaoli and Maestroni, 1987). Whether the opoids were exerting their influence at the level of lymphocytes and/or monocytes is not clear, since both cell types can be influenced (Maestroni *et al.*, 1987; Prieto *et al.*, 1989). In the studies of Maestroni *et al.* (1987), mice were used for the study and the melatonin was only effective when these nocturnal animals were injected in the afternoon (16.00 h). The reasons why this type of study may be relevant are that: (1) melatonin has been implicated in circadian rhythms (Seggie *et al.*, 1985, 1987); (2) dysregulation of circadian rhythms has been implicated and/or hypothesized to be involved in affective disorders (Kripke *et al.*, 1978; Siever and Davis, 1985; Souetre *et al.*, 1989), possibly at the level of signal transduction (Lachman and Papolos, 1989); and (3) lithium can inhibit melatonin synthesis in some experimental systems (Kemali *et al.*, 1989). While it is tempting to try and explain the effects of lithium via mechanisms associated with regulatory aspects of the neuroendocrine system, it remains difficult to visualize how such changes could result in: (1) the specific alterations in immunoregulation that are observed in patients taking lithium; (2) those that were observed in the SLE-prone mice discussed earlier; and (3) the fact that some autoimmune diseases are apparently not influenced by lithium. If lithium does not impact on the neuroendocrine–immune interface to yield a generalized modification of the 'set-point' of this interface, then lithium could also modify the immune system indirectly by acting at the level of the innervation of central immune organs such as the spleen, lymph nodes, bone marrow and thymus. Chemical sympathectomy (denervation) can alter the development and expression of immune reactivity (Kasahara *et al.*, 1977; Hall *et al.*, 1982). In mice, chemical sympathectomy led to a depression of immune responses (Kasahara *et al.*, 1977). Since therapeutic concentrations of lithium appear to allow the expression of certain antibody specificities rather than suppress their expression, one would have to hypothesize that if lithium were functioning in this manner, the ion would have to modify the contribution of the neural component in these tissues in order to accomplish this effect. Again, it is difficult to conceive of the basis for the specificity of the lithium effect on a restricted spectrum of antibodies if this mechanism were operative. However, it is not outside the realm of possibility that part of the role of the innervation is to control specific immune reactivities, such as autoimmune reactivities. Disturbances in the neural component, leading to hyperactivity, could theoretically contribute to the maintenance (Levine *et al.*, 1985; Iversen, 1985), and even development, of diseases such as SLE, rheumatoid arthritis, and others if such a system were operative. Lithium could cause the observed exacerbations of such disease by enhancing the underlying dysregulation, not of the immune system directly, but by modifying the neural regulatory system. Thus some autoimmune diseases or other immune dysfunctions could theoretically be in fact, neurological or neuroendocrine diseases.

While such speculations have been invoked to explain the dichotomy between the observations *in vivo* and *in vitro* in humans and mice, such hypotheses do not address the question of the differences observed between cells of different species *in vitro*. As mentioned above, lymphoid cells from a number of species have been cultured *in vitro* in the presence or absence of lithium and the only species where lithium has a very positive, reproducible effect is the hamster. It is possible that this finding is not unique, and that as more species are tested others will be found that exhibit the same characteristics. However, of the species tested thus far, the hamster is the only species that hibernates. Laboratory Syrian hamsters can be induced to exhibit a hibernating state by altering the photoperiod and the room temperature. However, hamsters are 'optional' hibernators and are not strictly seasonal hibernators (Minor *et al.*, 1978). Hibernation in hamsters is accompanied by a depression in splenic immune responsiveness (Sidkey and Auerbach, 1968) without a marked involution. Hamsters do not apparently elaborate a hibernation induction trigger (HIT) (Minor *et al.*, 1978) which could be responsible for the observed alterations in the immune system. It should be pointed out that at present it is not known whether the hyporesponsiveness noted in hibernating hamsters is due to an effect on lymphocytes or on antigen-presenting accessory cells such as macrophages. A survey of the literature has revealed only one study on the effect of lithium on hibernation in the hamster (Zvolsky *et al.*, 1981) and this investigation did not analyse immune parameters. The regulation of the events controlling the induction and maintenance of hibernation are not well understood, but again the pineal gland and the opiate system have been implicated (Oeltgen *et al.*, 1987; Hastings, 1989). It is obvious that even in the hibernating state the animal would have to maintain some host defence systems and it would also have to have a functioning immune system operative once arousal was initiated. Perhaps hamsters have evolved an adaptation system in their lymphoid cells and/or macrophages to accommodate the hiberation state and such adaptations involve lithium-sensitive steps. If this hypothesis is correct, it would be very interesting to determine the effect of lithium on immune responsiveness *in vivo* using both control and hibernating animals.

Alternatively, the depressed responsiveness observed in hibernating animals could reflect a modification of the neuroregulatory elements at the interface with the immune system. In such a scenario, the 'set-point' of the neuro-regulatory components could be varied according to the metabolic state (normal, hibernating) and the immune system would be required to accommodate this variable set-point. Such accommodation may have required the evolution of unique lithium-sensitive damping/activation mechanisms to adapt to such wide swings in the neuroregulatory elements. Some seasonal changes in innervation patterns have been observed in animals which hiberate (Korneeva and Serdiukova, 1987), but these issues have not been addressed in lymphoid tissues as far as I am aware.

Therefore, studies on the effect of lithium on lymphocytes and the immune system have generated several interesting observations and raised many intriguing questions for immunologists as well as for scientists directed toward understanding brain function and the pharmacological basis of lithium action. At the present time, I believe it is fair to say that we do not understand the immunopharmacology of lithium and that explanations of the mechanism of action are more speculative than factual.

Future Research Directions

From the previous discussion and other chapters in this volume, it is readily apparent that lithium is a very simple 'drug', but its mechanism of action on biological systems is a labyrinth (*labyrinth*: any confusingly intricate state of things or events; an intricate combination of passages in which it is difficult to find one's way or to reach the exit). While still somewhat confusing and not simply interpreted, research performed thus far on lithium, particularly in the area of the immunopharmacology of lithium, has raised some interesting hypotheses which serve to point us toward future research directions.

With regard to the human system, it is apparent that additional, more detailed, prospective studies are needed to characterize the immune status of affective disorder patients prior to initiation of lithium treatment. Such studies should not only provide new insights into the immunopharmacology of lithium, but also provide information on the possible links between immune dysfunction and affective disorders. Without studies of this kind, the literature will continue to expand with anecdotal reports of lithium and immune status, which will be difficult to interpret.

It is also apparent that further animal studies are needed and warranted to further our understanding of how lithium can impact on immunoregulation. The availability of inbred strains of mice, rats and hamsters should allow investigators to identify genetic influences on responsiveness to lithium in particular circumstances. For instance, certain strains of mice develop autoimmune diseases, but the disease processes in the different strains can be different (Hart *et al.*, 1989a). Other strains of mice differ in where and how much melatonin they synthesize (Ebihara *et al.*, 1986) and how they respond to lithium (Possidents and Exner, 1986). As discussed earlier, we have found that lithium can alter the course of autoimmune disease in such mice as well as their response to inflammatory stimuli, and that a genetic component is involved. Perhaps additional and more detailed studies of the impact of lithium on immunoregulation in such animals can be related to the studies proposed above.

The effect of lithium on immunoregulation, both *in vivo* and *in vitro*, should be more thoroughly investigated in the hamster system. In particular, the ability of lithium to modify immune responsiveness in both hibernating and non-hibernating hamsters should be investigated. Likewise, the effect of lithium on lymphocyte activation should be investigated *in vitro* using cells from hibernating and non-hibernating animals. Such studies may shed light on whether the lymphocytes, the macrophages, or the neuroimmune interface plays a primary role in the downregulation of immune reactivity observed in hibernating hamsters.

Finally, characterization of the immunopharmacological activity of lithium on cells from additional species is needed. Thus far there is evidence for genetic heterogeneity in humans, and some evidence of biological diversity using mice, rats, guinea pigs and hamsters. However, additional data using cells from animals that hiberate and others that do not is needed to determine if patterns of biological diversity emerge.

The types of studies outlined above should not only allow us to formulate more firm conclusions regarding the immunopharmacology of lithium, but they may also yield new approaches to the study and management of autoimmune and inflammatory diseases in patient populations.

Acknowledgements

The author thanks Judy Crawford for her excellent secretarial assistance in the preparation of the manuscript. Many of the recent investigations discussed were supported by the Canadian Arthritis Society and earlier studies were supported by the Alberta Heritage Foundation for Medical Research, the Alberta Cancer Board and the American Cancer Society. The author is an AHFMR Scholar.

References

Ader, R. (1981). "Psychoneuroimmunology". Academic Press, New York.

Ahmed, S. and Talal, N. (1989). Sex hormones and autoimmune rheumatic disorders. *Scand. J. Rheumatol.* **18**, 69–76.

Alexander, D. and Cantrell, D. (1989). Kinases and phosphatases in T-cell activation. *Immunol. Today* **10**, 200–205.

Balkwill, F. and Burke, F. (1989). The cytokine network. *Immunol. Today* **10**, 299–305.

Berridge, M., Downes, C. and Hanley, M. (1989). Neural and developmental actions of lithium: A unifying hypothesis. *Cell* **59**, 411–419.

Blalock, J.E. (1984). The immune system as a sensory organ. *J. Immunol.* **132**, 1068–1072.

Blalock, J.E. and Bost, K. (eds) (1988). "Neuroimmunoendocrinology". Karger, Basel.

Blalock, J.E. and Smith, E. (1985). The immune system: our mobile brain? *Immunol. Today* **6**, 115–117.

Bloomfield, F. and Young, M. (1983). Enhanced release of inflammatory mediators from lithium-stimulated neutrophils in psoriasis. *Br. J. Dermatol.* **109**, 9–13.

Bray, J., Turner, A.R. and Dusel, F. (1981). Lithium and the mitogenic response of human lymphocytes. *Clin. Immunol. Immunopathol.* **19**, 284–288.

Bulloch, K. and Moore, R.Y. (1981). Innervation of the thymus gland by brain stem and spinal cord in mouse and rat. *Amer. J. Anat.* **162**, 157–166.

Cade, J.F. (1949). Lithium salts in the treatment of psychotic excitement. *Med. J. Austr.* **36**, 349–353.

Calabrese, J., Gulledge, A., Hahn, K., Skewerer, R., Kotz, M., Schumacher, O., Gupta, M., Krupp, N. and Gold, P.W. (1985). Autoimmune thyroiditis in manic-depressive patients treated with lithium. *Amer. J. Psychiat.* **142**, 1318–1321.

Galvo, W. (1968). The innervation of the bone marrow in laboratory animals. *Amer. J. Anat.* **123**, 315–328.

Cambier, J., Chen, Z., Pasternak, J., Ransom, J., Sandoval, V. and Pickles, H. (1988). Ligand-induced desensitization of B-cell membrane immunoglobulin-mediated Ca^{2+} mobilization and protein kinase C translocation. *Proc. Natl. Acad. Sci. USA* **85**, 6493–6497.

Cantor, H., Chess, L. and Sercarz, E. (1984). "Regulation of the Immune System". Alan Liss, New York.

Carter, T.N. (1972). The relationship of lithium carbonate to psoriasis. *Psychosomatics* **13**, 325–327.

Cerny, J. and Kelsoe, G. (1984). Priority of the anti-idiotypic response after antigen administration: Artefact or intriguing network mechanism? *Immunol. Today* **5**, 61–63.

Chofflon, M., Weiner, H., Morimoto, C. and Hofler, D. (1989). Decrease of suppressor inducer (CD4 + 2H4 +) T cells in multiple sclerosis cerebrospinal fluid. *Ann. Neurol.* **25**, 494–499.

Cleghorn, R.A. (1951). Adrenal cortical insufficiency: psychological and neurological observations. *Can. Med. Assoc. J.* **65**, 449–454.

Cooper, T.B., Gershon, S., Kline, N. and Schou, M. (1979). "Lithium: Controversies and Unresolved Issues". Excerpta Medica, Amsterdam.

Craig, J., Abou-Saleh, M. and Smith, B. (1977). Diabetes mellitus in patients on lithium. *Lancet* ii, 1028.

Cruse, J. and Lewis, R. (eds) (1988). "Genetic Basis of Autoimmune Disease". Karger, Basel.

Dosch, H.-M., Matheson, D., Shuurman, R. and Gelfand, E. (1980). Anti-suppressor cell effects of lithium *in vitro* and *in vivo*. *In* "Lithium Effects on Granulopoiesis and Immune Function" (ed. A. Rossof and W. Robinson), pp. 463–469. Plenum Press, New York.

Ebihara, S., Marks, T., Hudson, D. and Menaker, M. (1986). Genetic control of melatonin synthesis in the pineal gland of the mouse. *Science* **231**, 491–493.

Evans, D. and Martin, W. (1979). Lithium carbonate and psoriasis. *Amer. J. Psychiat.* **136**, 1326–1327.

Farid, N. (1988). "Immunogenetics of Endocrine Disorders". Alan Liss, New York.

Felton, D., Livnet, S., Felton, S., Carlson, S., Bellinger, D. and Yeh, P. (1984). Sympathetic innervation of lymph nodes of mice. *Brain Res. Bull.* **13**, 693–699.

Fernandez, L. and Fox, R. (1980). Perturbation of the human immune system by lithium. *Clin. Exp. Immunol.* **41**, 527–532.

Fuggetta, M., Alvino, E., Romani, L., Grohman, U., Potenza, C. and Giuliani, A. (1988). Increase of natural killer cell activity of mouse lymphocytes following *in vitro* and *in vivo* treatment with lithium. *Immunopharm. Immunotoxicol.* **10**, 79–91.

Gelfand, E., Dosch, H.-M., Hastings, D. and Shore, A. (1979). Lithium: A modulator of cyclic AMP-dependent events in lymphocytes? *Science* **203**, 365–367.

Giron, L., Cratcher, K. and Davis, J. (1980). Lymph nodes — A possible site for sympathetic neuronal regulation of immune responses. *Ann. Neurol.* **8**, 520–525.

Greco, F.A. (1980). Lithium and immune function in man. *In* "Lithium Effects on Granulopoiesis and Immune Functions" (ed. A. Rossof and W. Robinson), pp. 463–469. Plenum Press, New York.

Green, M.I. and Nisonoff, A. (1984). "The Biology of Idiotypes". Plenum Press, New York.

Hall, N., McClure, J., Hu, S.-K., Tare, N., Seals, C. and Goldstein, A.L. (1982). Effects of 6-hydroxydopamine upon primary and secondary thymus dependent immune responses. *Immunopharmacology* **5**, 39–48.

Hart, D.A. (1977). The effect of soybean trypsin inhibitor on concanavalin A and phytohemagglutinin stimulation of hamster, guinea pig, rat and mouse lymphoid cells. *Cell. Immunol.* **32**, 146–159.

Hart, D.A. (1979a). Potentiation of phytohemagglutinin stimulation of lymphoid cells by lithium. *Exp. Cell Res.* **119**, 47–53.

Hart, D.A. (1979b). Modulation of concanavalin A stimulation of hamster lymphoid cells by lithium chloride. *Cell. Immunol.* **43**, 113–122.

Hart, D.A. (1979c). Augmentation of zinc ion stimulation of lymphoid cells by calcium and lithium. *Exp. Cell Res.* **121**, 419–425.

Hart, D.A. (1979d). Modulation of lymphocyte activation by LiCl. *In* "The Molecular Basis of Immune Cell Function (ed. J.G. Kaplan), pp. 408–411. North-Holland, Amsterdam.

Hart, D.A. (1981a). Ability of monovalent cations to replace potassium during stimulation of hamster lymphoid cells. *Cell Immunol.* **57**, 209–218.

Hart, D.A. (1981b). Evidence that lithium ions can modulate lectin stimulation by multiple mechanisms. *Cell Immunol.* **58**, 372–384.

Hart, D.A. (1982a). Differential potentiation of lipopolysaccharide stimulation of lymphoid cells by lithium. *Cell Immunol.* **71**, 159–168.

Hart, D.A. (1982b). Studies on the mechanisms of LiCl enhancement of lipopolysaccharide stimulation of lymphoid cells. *Cell Immunol.* **71**, 169–182.

Hart, D.A. (1986a). Lithium as an *in vitro* modulator of immune cell function: Clues to its *in vivo* biological activities. *IRCS Med. Sci.* **14**, 756–762.

Hart, D.A. (1986b). Evidence that the elevated levels of proteinase activity in the plasma of melanoma-bearing mice may be of host origin. *Haemostasis* **16**, 34–42.

Hart, D.A. (1987a). *C. parvum*, but not *BCG*, induces elevations in plasma proteinase activity similar to those observed in tumor-bearing mice. *Haemostasis* **17**, 79–88.

Hart, D.A. (1987b). Steroids and tuftsin fail to prevent the induction of altered plasma proteinase homeostasis in mice bearing the B16 melanoma or treated with *C. parvum*. *Int. J. Immunopharmacol.* **9**, 669–674.

Hart, D.A. (1987c). Influence of lithium and calcium on cytochalasin D stimulation of lymphocytes. *Med. Sci. Res.* **15**, 667–669.

Hart, D.A. (1988a). Immunopharmacologic aspects of lithium: one aspect of a general role as a modulator of homeostasis. *In* "Lithium: Inorganic Pharmacology and Psychiatric Use" (ed. N. Birch), pp. 99–102. IRL Press, Oxford.

Hart, D.A. (1988b). Differences between beige and Bg/ + mice in the disruption of plasma proteinase regulation in the tumor-bearing state or following *C. parvum* treatment: evidence for the involvement of polymorphonuclear leukocyte proteinases. *Haemostasis* **18**, 154–162.

Hart, D.A. (1988c). Contribution of host factors in the induction of disrupted plasma proteinase homeostasis by treatment of mice with *C. parvum*. *Med. Sci. Res.* **16**, 225–227.

Hart, D.A. (1988d). Further studies on the regulation of plasma proteinase homeostasis during inflammatory responses. *Med. Sci. Res.* **16**, 687–688.

Hart, D.A. (1988e). Lithium potentiates antigen-dependent stimulation of lymphocytes only under suboptimal conditions. *Int. J. Immunopharmacol.* **10**, 153–160.

Hart, D.A. (1990a). Immunoregulation in patients receiving lithium for affective disorders. *In* "Lithium and the Blood" (ed. V. Gallicchio). Karger, Basel, in press.

Hart, D.A. (1990b). Modulation of immune system elements by lithium. *In* "Lithium and Cell Physiology" (ed. R.O. Bach and V. Gallicchio), pp. 58–81. Springer-Verlag, New York.

Hart, D.A. (1990c). Lithium, lymphocyte stimulation and the neuroimmune interface. *In* "Lithium and the Blood" (ed. V. Gallicchio). Karger, Basel, in press.

Hart, D.A. and Stein-Streilein, J.S. (1981). Hamster lymphoid responses *in vitro In* "Hamster Immune Responses in Infectious and Oncological Diseases" (ed. J.W. Streilein, D.A. Hart, J.S. Stein-Streilein, W. Duncan, and R.E. Billingham), pp. 7–22. North-Holland, Amsterdam.

Hart, D.A., Wang, A.-L., Pawlak, L. and Nisonoff, A. (1972). Suppression of idiotypic specificities in adult mice by administration of anti-idiotypic antibody. *J. Exp. Med.* **135**, 1293–1300.

Hart, D.A., Jones, J. and Nisonoff, A. (1973). Mitogenic factor from inbred guinea pigs. *Cell. Immunol.* **9**, 173–185.

Hart, D.A., Garlepp, M. and Fritzler, M. (1989). Effect of endogenous and exogenous steroids on plasma proteinase levels in the MRL-lpr/lpr model. *Med. Sci. Res.* **17**, 543–545.

Hart, D.A., Garlepp, M. and Fritzler, M. (1989a). Plasma proteinase regulation during disease progression in murine models of SLE. *J. Clin. Lab. Immunol.* **30**, 27–34.

Hassman, R., Lazarus, J., Dieguez, C., Weetman, A., Hall, R. and McGregor, A. (1985). The influence of lithium chloride on experimental autoimmune thyroid disease. *Clin. Exp. Immunol.* **61**, 49–57.

Hastings, M. (ed.) (1989). Phylogeny and function of the pineal. *Experientia* **45**, 903–1008.

Inman, R. (1978). Immunologic sex differences and the female predominance in systemic lupus erythematosus. *Arth. Rheumat.* **21**, 849–852.

Ishizaka, S. and Moller, G. (1982). Lithium chloride induces partial responsiveness to LPS in non-responder B cells. *Nature* **299**, 363–365.

Iversen, L. (1985). The possible role of neuropeptides in the pathophysiology of rheumatoid arthritis. *J. Rheumatol.* **12**, 399–400.

Jackson, R. and Eisenbarth, G. (1989). Autoimmune endocrine disorders. In "Textbook of Internal Medicine" (editor-in-chief, W. Kelly), pp. 2229–2232. Lippincott, Philadelphia.

Jankovic, B.D. (1989). Neuroimmunomodulation: facts and dilemmas. *Immunol. Lett.* **21**, 101–118.

Jankovic, B.D., Lenert, P. and Mitrovic, K. (1982). Suppression of experimental allergic thyroiditis in rats treated with lithium chloride. *Immunobiology* **161**, 488–493.

Jankovic, B.D., Popeskovic, L. and Isakovic, K. (1978a). Suppressed immune responses in mouse and rat during treatment with lithium chloride. *Fed. Proc.* **37**, 1651.

Jankovic, B.D., Popeskovic, L. and Isakovic, K. (1978b). Cation-induced immunosuppression: The effect of lithium on arthus reactivity, delayed hypersensitivity and antibody production in the rat. *Adv. Exp. Biol. Med.* **114**, 339–344.

Jankovic, B., Markovic, B. and Spector, N. (eds) (1987). Neuroimmune interactions. *Ann. N.Y. Acad. Sci.* **496**.

Jerne, N. (1974). Towards a network theory of the immune system. *Ann. Immunol. (Inst. Pasteur)* **125C**, 373.

Johnson, B. (1977). Diabetes mellitus in patients on lithium. *Lancet* **ii**, 935–936.

Johnson, F.N. (1979). Immunologic aspects of lithium therapy. *IRCS Med. Sci.* **7**, 375–376.

Johnson, F.N. (1984). "The History of Lithium Therapy". London, Macmillan.

Johnson, F.N. and Johnson, S. (1977). "Lithium in Medical Practice". University Park Press, Baltimore, MD.

Kasahara, K., Tanaka, S., Ito, T. and Hamashima, Y. (1977). Suppression of the primary immune response by chemical sympathectomy. *Res. Commun. Chem. Path. Pharmacol.* **16**, 687–694.

Kemali, M., Monteleone, P., Maj, M., Milici, N. and Kemali, D. (1989). Lithium decreases retinal melatonin levels in the frog. *Neurosci. Lett.* **96**, 235–239.

Korneeva, T. and Serdiukova, E. (1987). Seasonal changes in the adrenergic fibers in the wall of the microvessels of the cheek pouch of the hamster, *Mesocricetus auratus* (translated title). *Zh. Evol. Biokhim. Fiziol. (Russia)* **23**, 642–646.

Kovithavangs, T., Marchuck, L., Schlaut, J., Pazderka, F. and Dossetor, J. (1983). Perturbation of immune function by lithium: a possible explanation for an irreversible renal transplant rejection. *Transplant. Proc.* **15**, 1832–1835.

Krause, G., Zhao, P., Martin, L., Fritzler, M. and Hart, D.A. Lithium modulation of disease progression in murine models of SLE, in prep.

Kripke, D., Mullaney, D., Atkinson, M. and Wolf, S. (1978). Circadian rhythm disorders in manic depressives. *Biol. Psychiat.* **13**, 335–351.

Kucharz, E., Sierakowski, S., Staite, N. and Goodwin, J. (1988). Mechanism of lithium-induced augmentation of T-cell proliferation. *Int. J. Immunopharmacol.* **10**, 253–259.

Kyogoku, M. and Wigzell, H. (eds) (1987). "New Horizons in Animal Models of Autoimmune Disease". Academic Press, Tokyo.

Lachman, H. and Papolos, D. (1989). Abnormal signal transduction: a hypothetical model for bipolar affective disorder. *Life Sci.* **45**, 1413–1426.

Lazarus, J., McGregor, A., Ludgate, M., Dark, C., Creagh, F. and Kingswood, C. (1986). Effect of lithium carbonate therapy on thyroid immune status in manic-depressive patients: a prospective study. *J. Affect. Dis.* **11**, 155–160.

Leigh, H. and Kramer, S. (1984). The psychiatric manifestations of endocrine disease. *Adv. Int. Med.* **29**, 413–442.

Lever, E.G. and Stansfeld, S.A. (1983). Addison's disease, psychosis and the syndrome of inappropriate secretion of antidiuretic hormone. *Br. J. Psychiat.* **143**, 406–410.

Levine, L., Collier, D., Basbaum, A., Moskowitz, M. and Helms, C. (1985). Hypothesis: The nervous system may contribute to the pathophysiology of rheumatoid arthritis. *J. Rheumatology* **12**, 406–411.

Licastro, F., Chiricolo, M., Tabacchi, P., Barboni, F., Zannotti, M. and Franceschi, C. (1983). Enhancing effect of lithium and potassium ions on lectin-induced lymphocyte proliferation in aging and Down's Syndrome subjects. *Cell. Immunol.* **75**, 111–121.

Lieb, J. (1981). Immunopotentiation and inhibition of herpes virus activation during therapy with lithium carbonate. *Med. Hypotheses* **7**, 885–890.

Lieb, J. (1987). Lithium and immune function. *Med. Hypotheses* **23**, 73–93.

Lydiard, R. and Gelenberg, A. (1982). Hazards and adverse effects of lithium. *Ann. Rev. Med.* **33**, 327–344.

Maestroni, G., Conti, A. and Pierpaoli, W. (1986). Role of the pineal gland in immunity. *J. Neuroimmunol.* **13**, 19–30.

Maestroni, G., Conti, A. and Pierpaoli, W. (1987). The role of the pineal gland in immunity: II. Melatonin enhances the antibody response via an opiatergic mechanism. *Clin. Exp. Immunol.* **68**, 384–391.

Maj, M., Pirozzi, R. and Kemali, D. (1989). Long-term outcome of lithium prophylaxis in patients initially classified as complete responders. *Psychopharmacology* **98**, 535–538.

Martin, J., Reichlin, S. and Brown, G. (1977). "Clinical Neuroendocrinology". F.A. Davis, Philadelphia.

McMahon, D. (1974). Chemical messengers in development: a hypothesis. *Science* **185**, 1012–1020.

Minor, J., Bishop, D. and Badger, C. (1978). The golden hamster and the blood-borne hibernation trigger. *Cryobiology* **15**, 557–562.

Moller, G. (ed.) (1975). Suppressor T-lymphocytes. *Transplant Rev.* **26**, 3–205.

Moller, G. (ed.) (1987). Activation antigens and signal transduction in lymphocyte activation. *Immunol. Rev.* **95**, 5–194.

Neil, J.F., Himmelhoch, J. and Licata, S. (1976). Emergence of myasthenia gravis during treatment with lithium carbonate. *Arch. Gen. Psychiat.* **33**, 1090–1092.

Oeltgen, P., Welborn, J., Nuchols, P., Spurrier, W., Bruce, D. and Su, T.-P. (1987). Opioids and hiberation: effects of kappa opioid U69593 on induction of hibernation in summer active ground squirrels by 'hibernation induction trigger'. *Life Sci.* **41**, 2115–2120.

Pandey, G. and Davis, J. (1980). Biology of the lithium ion. *In* "Lithium Effects on Granulopoiesis and Immune Function" (ed. A. Rossof and W. Robinson), pp. 15–59. Plenum Press, New York.

Pawlak, L., Hart, D.A. and Nisonoff, A. (1973). Requirements for prolonged suppression of an idiotypic specificity in adult mice. *J. Exp. Med.* **137**, 1442–1458.

Perez, H., Kaplan, H., Goldstein, I., Shenkman, L. and Borkowsky, W. (1980). Reversal of an abnormality of polymorphonuclear leukocyte chemotaxis with lithium. *Clin. Immunol. Immunopathol.* **16**, 308–315.

Pierpaoli, W. and Maestroni, G. (1987). Melatonin: a principal neuroimmuno-regulatory and anti-stress hormone: Its anti-aging effects. *Immunol. Lett.* **16**, 355–362.

Pierpaoli, W. and Spector, N. (eds) (1988). Neuroimmunomodulation: intervention in aging and cancer. *Ann. N.Y. Acad. Sci.* **521**.

Popeskovic, J., Jankovic, B.D. and Isakovic, K. (1979). Effect of lithium on immunological reactivity in CBA mice. *Period. Biol.* **81**, 179.

Possidents, B. and Exner, R. (1986). Gene-dependent effect of lithium on circadian rhythms in mice *(Mus musculus)*. *Chronobiol. Int.* **3**, 17–21.

Presley, A., Kahn, A. and Williamson, N. (1976). Antinuclear antibodies in patients on lithium carbonate. *Br. Med. J.* **2**, 280–281.

Prieto, J., Subira, M., Castilla, A. and Serrano, M. (1989). Naloxone-reversible monocyte dysfunction in patients with chronic fatigue syndrome. *Scand. J. Immunol.* **30**, 13–20.

Salata, R. and Klein, I. (1987). Effects of lithium on the endocrine system: a review. *J. Lab. Clin. Immunol.* **110**, 130–136.

Salonen, R., Olonen, J., Jageroos, H., Syrjala, H., Nurmi, T. and Reunanen, M. (1989). Lymphocyte subsets in the cerebrospinal fluid in active multiple sclerosis. *Ann. Neurol.* **25**, 500–502.

Seggie, J., Werstiuk, E. and Joshi, M. (1985). Lithium and twenty-four hour rhythms of serum corticosterone, prolactin and growth hormone in pigmented eye rats. *Progr. Neuro-Pharmacol. Biol. Psychiat.* **9**, 755–758.

Seggie, J., Werstiuk, E. and Grota, L. (1987). Lithium and circadian patterns of melatonin in the retina, hypothalamus, pineal and serum. *Progr. Neuro-Pharmacol. Biol. Psychiat.* **11**, 325–334.

Sell, S. (1987). "Immunology, Immunopathology and Immunity". Elsevier, New York.

Shenkman, L., Borkowsky, W., Holzman, R. and Shopsin, B. (1978). Enhancement of lymphocyte and macrophage function *in vitro* by lithium chloride. *Clin. Immunol. Immunopathol.* **10**, 187–192.

Shenkman, L., Borkowsky, W. and Shopsin, B. (1980). Lithium as an immunologic adjuvant. *Med. Hypotheses* **6**, 1–6.

Sidkey, Y. and Auerbach, R. (1968). Effect of hibernation on the hamster spleen immune response reaction *in vitro*. *Proc. Soc. Exp. Biol. Med.* **129**, 122–127.

Siever, L. and Davis, L. (1985). Overview: toward a dysregulation hypothesis of depression. *Amer. J. Psychiat.* **142**, 1017–1031.

Singh, U. (1984). Sympathetic innervation of fetal mouse thymus. *Eur. J. Immunol.* **14**, 757–759.

Skutsch, G. (1985). Manic-depression: a multiple hormone disorder? *Biol. Psychiat.* **20**, 662–668.

Souetre, E., Salvati, E., Belugou, J.L., Prinquey, D., Candito, M., Krebs, B., Ardisson, J.-L. and Darcourt, G. (1989). Circadian rhythms in depression and recovery: Evidence for blunted amplitude as the main chronobiological abnormality. *Psychiat. Res.* **28**, 263–278.

Streilein, J.W., Hart, D.A., Stein-Streilein, J.S., Duncan, W. and Billingham, R. (eds) (1981). "Hamster Immune Responses in Infectious and Oncological Disease". Plenum Press, New York.

Stein-Streilein, J.S. and Hart, D.A. (1979). *In vitro* development of a primary antibody response with dissociated cells from hamsters, guinea pigs and mice: evidence that the cells responsible reside primarily with lymph node cells. *Cell. Immunol.* **45**, 241–248.

Theofilopoulos, A. and Dixon, F. (1985). Murine models of systemic lupus erythematosus. *Adv. Immunol.* **37**, 269–390.

Tobin, M., Aldridge, S., Morris, A., Belchetz, P. and Gillmore, I. (1989). Gastrointestinal manifestations of Addison's disease. *Amer. J. Gastroenterol.* **84**, 1302–1305.

Varadaraj, R. and Cooper, A.J. (1986). Addison's disease presenting with psychiatric symptoms. *Amer. J. Psychiat.* **143**, 552–554.

Vitetta, E.S., Fernandez-Dotran, R., Myers, C. and Sanders, V. (1989). Cellular interactions in the humoral immune response. *Adv. Immunol.* **45**, 1–105.

Wadler, S., Shenkman, L. and Borkowsky, W. (1979). Effects of lithium on suppressor-enriched and suppressor-depleted mononuclear cell preparations. *Clin. Res.* **27**, 339.

Wahlin, A., Von Knorring, L. and Roos, G. (1984). Altered distribution of T lymphocyte subsets in lithium-treated patients. *Neuropsychobiology* **11**, 243–246.

Weetman, A.P., McGregor, A., Lazarus, J., Rees-Smith, B. and Hall, R. (1982). Enhancement of immunoglobulin synthesis by human lymphocytes with lithium. *Clin. Immunol. Immunopathol.* **22**, 400–407.

Weigent, D. and Blalock, J.E. (1987). Interactions between the neuroendocrine and immune systems: Common hormones and receptors. *Immunological Rev.* **100**, 79–108.

Weizscher, M. and Tegeler, J. (1984). Erfolgreiche lithiumbehandlung einer maniformen einer symptomatik bei multipler skelrose. *Nervenarzt* **55**, 214–216.

Whalley, K., Roberts, D., Wentzel, J. and Watson, K. (1981). Antinuclear antibodies and histocompatibility antigens in patients on long-term lithium therapy. *J. Affect. Dis.* **3**, 123–130.

Williams, J. and Felton, D. (1981). Sympathetic innervation of murine thymus and spleen: a comparative histofluorescence study. *Anat. Rec.* **199**, 531–542.

Ziaie, Z. and Kefalides, N. (1989). Lithium chloride restores host protein synthesis in herpes simplex virus-infected endothelial cells. *Biochem. Biophys. Res. Commun.* **160**, 1073–1078.

Zvolsky, P., Jansky, L., Vyskocilova, J. and Grof, P. (1981). Effects of psychotropic drugs on hamster hibernation—pilot study. *Progr. Neuro-Pharmacol. Biol. Psychiat.* **5**, 599–602.

17 Bioavailability of Lithium Formulations

JONATHAN D. PHILLIPS

Biomedical Research Laboratory, School of Health Sciences, Wolverhampton Polytechnic, Wolverhampton WV1 1DJ, UK

Introduction

It is estimated that up to 1 in 1000 of the population in the United Kingdom may be taking lithium for the prophylaxis of recurrent affective disorder (Phillips and Birch, 1990), yet there is still doubt as to the most appropriate formulation for clinical use (Phillips *et al.*, 1990). In view of the extensive use of lithium in psychiatry it is important to resolve these doubts.

Clinical Monitoring of Lithium Therapy

Lithium carbonate is usually administered in tablet form at a total dose of up to 30 mmol per day. Treatment is monitored using regular estimations of serum lithium, taken twelve hours after the previous dose (Vestergaard *et al.*, 1982; Amdisen and Nielsen-Kudsk, 1986). These serum lithium concentrations should lie in the range 0.4 to 0.8 mmol l^{-1}, and higher levels may be associated with toxic side effects, which can include tremor, dizziness, drowsiness and diarrhoea (Coppen *et al.*, 1983; Schou, 1986). More serious toxic effects include coarse hand tremor, slurred speech and vomiting, and these may be followed by coma and death (Bone *et al.*, 1980; Vestergaard *et al.*, 1980). However, serious intoxication is rare if therapy is well controlled, although transient, non-toxic side effects such as dry mouth and nausea may be experienced by patients when serum lithium concentrations are within the therapeutic range (Birch, 1982). These generally occur within four hours of the dose when serum lithium concentrations are at their highest (Johnson, 1980). Since these higher concentrations are often associated with transient side effects, the use of 'controlled release'' or 'sustained release' formulations designed to reduce or delay the lithium peak has been seen as potentially beneficial (Shelley and Silverstone, 1987). However, it has been proposed that lithium could be administered less frequently than the usual daily dose (s), without loss of efficacy. This could minimize some of the more serious toxic side effects of the drug, particularly those which are claimed to occur in the kidney (Plenge and Mellerup, 1986). The longer time interval between doses would result in relatively high 'peak' serum lithium concentrations followed by prolonged low 'troughs', and it has been suggested that a comparatively large flux in serum lithium concentrations may be relevant to the therapeutic activity of lithium (Plenge and Mellerup, 1988).

Routine clinical measurements of lithium are performed using atomic absorption spectrometry (AAS) or flame emission spectroscopy (Blijenberg and Leijinse, 1968). In general, for lithium determination, AAS is regarded as the more reliable of the two techniques. More recent developments include inductively coupled plasma emission spectrometry (ICP), electrothermal atomization atomic absorption spectrometry (ETAAS) and spectrofluorimetric methods. Spectrofluorimetry and ETAAS offer greater sensitivity than the traditional methods and are useful research tools (Barnes, 1978; Wheeling and Christian, 1984). Lithium ion-selective electrodes (ISE) have been developed and are capable of providing rapid serum lithium estimations (Xie and Christian, 1986). A preliminary study has demonstrated the potential for ISE analysers to be sited close to the point of contact with the patient, in a lithium clinic, in order to provide instant monitoring of lithium treatment (Phillips *et al.*, 1989). See also pp. 161–3 of this volume.

Lithium Formulations

Lithium salts are administered by the oral route, both for reasons of economy and ease of dosage regulation. Lithium carbonate tablets are the preferred dosage form in most countries, including the United Kingdom (British National Formulary, 1988), though liquid lithium preparations have recently been subjected to clinical evaluation (Collier *et al.*, 1990). Most patients in the United Kingdom are prescribed one of three preparations of lithium: Camcolit (Norgine Ltd), Camcolit 400 (Norgine Ltd) or Priadel (Delandale Laboratories Ltd). Tabletting generally consists of combination of the lithium salt with suitable excipients followed by compression. The nature of this process may affect the dissolution rate of the tablets and thus the rate at which the drug is presented to the intestinal mucosal surface for absorption (Smith, 1985). Oral drug formulation design may be influenced by perceived advantages of, for example, reducing or increasing the rate of absorption of a drug, or of directing the drug towards specific sites of absorption in the intestine. These objectives may be achieved, in principle, by the design of formulations which modify the rate of release of a drug into solution in the gut lumen. This is often assessed *in vitro* by the use of dissolution tests, but it should be noted that the dissolution rate of a given formulation may not equate with the desired absorption rate *in vivo*. In our laboratory, we have investigated the dissolution and pharmacokinetic properties of the major lithium formulations in the United Kingdom.

Dissolution Properties of Lithium Formulations

Bioavailability of pharmaceutical preparations may be estimated from pharmacokinetic studies which involve the performance of timed serial plasma estimations after administration of the drug to patients or normal volunteers. The results of such studies reflect the rates of both dissolution and absorption of orally administered drugs. If, however, the dissolution rate is slower than that for absorption, then intestinal absorption becomes dissolution rate-limited (Wadke and Jacobson, 1980). In such cases, comparison of dissolution rates *in vitro* has potential value in the assessment of different formulations of the same drug. These principles apply to the study of lithium formulations for which modified release properties are claimed: the design of suitable formulations of lithium tablets involves testing for their dissolution properties.

Standard methods for investigation of the dissolution properties *in vitro* of slow lithium carbonate tablets using a disintegration apparatus have not proved entirely satisfactory, and it has been suggested that there is a need for the development of a more suitable technique (British Pharmacopoea, 1980). An alternative method for estimation of the rate of release of lithium preparations has been described (Birch *et al.*, 1974; Tyrer *et al.*, 1976), and was used as the basis for a series of experiments carried out in our laboratory. The original technique, based on a small flow-through dissolution chamber, was modified to allow greater variability in experimental conditions, which were adjusted to reflect physiological changes in the digestive tract (Phillips and Birch, 1988). Using this technique we showed that the standard lithium formulation (Camcolit 250 mg) dissolved more readily than either of the controlled release formulations (Camcolit 400, Priadel). Furthermore, Camcolit 400 was rather less soluble in simulated intestinal fluid than the other two formulations. However, there are reasons to be cautious about the results of dissolution tests *in vitro*. This method does present some technical problems, notably a tendency for filters to block with certain formulations. For this reason, the technique may not have widespread commercial application in situations where large-scale rapid testing is necessary.

There are other potential problems with dissolution tests. Dissolution properties of drugs are usually measured after exposure to simulated gastric

and intestinal fluids (Trigger and Davies, 1987a). Simulated intestinal fluid
is a phosphate-containing buffer (pH 7.5) which some researchers have found
to retard lithium dissolution rate due to the formation of insoluble trilithium
phosphate (Wall *et al.*, 1986). However, other workers (Trigger and Davies,
1987a) failed to find any differences in lithium carbonate dissolution rate
when media composed of citrate and phosphate buffers were compared, and
suggested the phosphate medium to be satisfactory if the solvent volume for
dissolution tests is sufficiently large. This is not the case with all methods, and
the validity of using simulated intestinal fluid containing phosphate buffer
for lithium tablet dissolution experiments when utilizing the flow-through
apparatus remains unresolved. Results obtained using the phosphate buffer
medium should be interpreted with caution unless confirmed by other data,
ideally from pharmacokinetic experiments. In this respect, the finding of
differences in dissolution properties between standard and controlled release
formulations (Trigger and Davies, 1987a,b; Phillips and Birch, 1988) may
not be reflected in their pharmacokinetics (Phillips, 1989).

Lithium Pharmacokinetics

Lithium, at pharmacological doses, is required to be delivered efficiently to
body fluids, at a rate which avoids toxic concentrations in the blood yet has
sufficient bioavailability. For some years there has been discussion as to the
relative merits of the lithium formulations available in the United Kingdom.
Both Camcolit 400 and Priadel are marketed as 'controlled release'
formulations of lithium carbonate, yet the precise meaning of the term
'controlled release' has often been subject to misunderstanding and has
sometimes been taken to mean 'slow release'. However, pharmacokinetic
studies have shown that very slow release lithium formulations have poor
bioavailability (Birch *et al.*, 1974; Shelley *et al.*, 1986).

   Pharmacokinetic studies of lithium preparations in man are conducted by
oral administration of a test dose of the drug which is then followed by
sequential blood samples at regular, frequent intervals for serum lithium
determinations. These studies are of interest to clinicians and researchers for
the following reasons:

(1)   They can be used to assist the choice of the most appropriate
      formulation for clinical use: this may be on the basis of either
      maximum bioavailability or its release properties (Smith, 1985).
(2)   Transient side effects may be associated with early or relatively high
      peak serum lithium concentrations (Johnson, 1980).

(3) Pharmacokinetic studies can help establish tissue distribution of a drug (Rybakowski *et al.*, 1988) and thus, in the case of lithium, may help elucidate its mode of action.

A number of pharmacokinetic studies of lithium formulations have been carried out in our laboratory (Phillips and Birch, 1987; Phillips, 1989; Phillips *et al.*, 1990). Despite claims of differences in rate of absorption, a preliminary pharmacokinetic comparison of Camcolit 250 ('standard formulation') and Camcolit 400 ('controlled release') showed no significant differences in respect of maximum serum concentrations of lithium or time of the 'peak' (Phillips, 1989). This was a surprising finding in view of observed differences in dissolution characteristics *in vitro* and confirms that dissolution test results should be interpreted with caution unless supported by pharmacokinetic studies.

Comparison of the controlled release formulations Camcolit 400 and Priadel in single dose studies using normal volunteers has shown no difference between the preparations (Shelley and Silverstone, 1986). However, it has been claimed that movement between body compartments at the onset of treatment, and prior to equilibration, may influence serum concentrations of the drug (Hunter, 1988). For this reason it is claimed that single dose studies may not reflect the true clinical situation after prolonged treatment. In one investigation (Phillips and Birch, 1987) we attempted to reflect the situation in lithium-treated patients by using normal subjects stabilized for 21 days at a typical prophylactically effective dose. These results suggested that despite some minor pharmacokinetic differences in intestinal absorption of Priadel and Camcolit 400 in healthy subjects, there was no significant difference between blood levels with the two lithium formulations. This was supported by the results of a similar experiment conducted in a clinical setting using the same production batches of the two drugs (Phillips *et al.*, 1990). Serum lithium profiles of Camcolit 400 and Priadel were compared in a total of 20 patients in two centres at half-hour intervals during the first four hours after the dose, and again at 24 h. There was no significant difference between the formulations in respect of maximum serum lithium concentrations or the concentrations at each time point.

In our experiments there were a number of differences between the normal volunteer and patient groups: mean age of the patients was higher, dosage regimens varied, most were concurrently taking other medication and some had co-existing non-psychiatric medical conditions. The normal volunteers were all free of other medication and took a standard dose of lithium. One major difference was that the normal subjects received their doses in the morning after a light breakfast and the patients took their lithium in the early evening after a meal. The presence of food in the gastrointestinal tract has been shown to affect lithium absorption (Jeppson and Sjogren, 1975), and a diurnal variation in renal lithium clearance has been reported (Lauritsen *et al.*, 1981). The common practice of administering lithium as an early evening dose after a meal may sufficiently delay the lithium peak to reduce the possible discomfort of any transient side-effects and therefore improve patient compliance.

Certain of our pharmacokinetic studies revealed some interesting findings. For example, claims by Hunter (1988) of pharmacokinetic differences between naive and lithium pre-loaded subjects were not substantiated. Hunter's paper compared data for maximum serum lithium concentrations from studies by different groups of workers, in which subjects received different doses of the drug. Our experiments compared both bioavailability and the rise in serum lithium to a maximum concentration from baseline values in the same normal subjects after a single dose, and again after three weeks treatment with the same lithium dose: no significant differences were revealed. This confirmed the findings of an earlier study utilizing the stable isotope [6]Li which showed that the rate of appearance of lithium in the blood after an oral dose was unaffected by the previous state of lithium loading (Birch *et al.*, 1978). These data do not support the suggestion that lithium accumulation in the cellular compartment affects gross pharmacokinetics of the drug in extracellular fluids after chronic exposure (Rybakowsky *et al.*, 1988). An increase in erythrocyte retention of lithium *in vivo* after chronic exposure was confirmed, and was attributed to inhibition of lithium–sodium counter transport rate. We also showed erythrocyte intracellular lithium concentrations to be relatively stable over a 24-h period, in contrast to wide fluctuations in serum levels (Phillips, 1989). In this respect, a highly variable gradient across cell membranes has been suggested to be of importance in the mode of action of lithium (Plenge and Mellerup, 1986, 1988).

It is not known whether there is an increased rate of lithium absorption after pre-treatment with the drug: our pharmacokinetic studies suggest that this is not the case. If such an increase in the rate of intestinal absorption were to occur in lithium treatment, it might be balanced by increased lithium secretion into the gut lumen. It has been shown that lithium is secreted in saliva at a concentration higher than that in plasma, and if this were to occur in other secretory epithelia in the digestive tract, lithium secretion could have a major regulatory role in the control of the overall rate of lithium absorption. In support of this hypothesis, there is a slight flattening of the typical serum lithium pharmacokinetic curve during the elimination phase which has been attributed to the movement of additional lithium to the plasma compartment (Hunter, 1988; Rybakowsky *et al.*, 1988). Some workers have assumed that this is due to efflux of lithium from the intracellular to the extracellular compartment (Rybakowsky *et al.*, 1988). An alternative explanation is that the additional lithium entering the extracellular compartment at this time is due to secondary absorption of lithium which was secreted from the plasma into the gut lumen at the time of the maximum serum lithium concentration (Phillips, 1989).

Overall the pharmacokinetic data indicate that when lithium formulations are similar, other factors tend to have a greater influence on the magnitude and time of peak serum lithium concentrations. When conditions were standardized each of the formulations tested appeared to be absorbed at a similar rate and gave rise to similar serum lithium profiles both in normal subjects and in patients. However, pharmacokinetic studies which employ sampling times at half-hour intervals may not be sufficiently sensitive to differentiate between broadly similar formulations: more frequent blood sampling might be necessary to overcome these difficulties. We are currently evaluating lithium ion-selective electrode analysers in pharmacokinetic experiments. Use of this technology may allow more precise identification of 'peak' serum lithium concentrations at the time of the study.

Conclusions

In conclusion, bioavailability of lithium formulations is most satisfactorily determined by the use of pharmacokinetic studies after administration of a test dose, ideally in a patient population. Blood samples should be taken at frequent intervals: the use of half-hourly samples may not be sufficient to identify the magnitude and time of maximum serum lithium concentrations. Use of new technology in the form of lithium ion-selective electrode analysers to provide instant serum or whole blood lithium concentrations may help to determine the required frequency of blood sampling. Dissolution studies *in vitro* are of less value in comparison of formulations, but may be useful in formulation design.

The long debate continues as to the most appropriate lithium formulation for clinical use. The controlled release formulations Camcolit 400 and Priadel have been shown to be bioequivalent and of equal benefit in the reduction of possible transient side effects. However, dosage regimen is equally relevant and there is some evidence that wide fluctuations in serum lithium concentration may be of considerable importance in its mode of action (Plenge and Mellerup, 1988). This has implications relating to the frequency and timing of the dose and may also influence the incidence of long-term side effects. Further clinical and pharmacokinetic studies will be central to the resolution of this current controversy.

References

Amdisen, A. and Nielsen-Kudsk, F. (1986). Relationships between standardised 12h serum lithium, mean serum lithium of the 24 hour day, dose regimen and therapeutic interval. *Pharmacopsychiatry* **19**, 416–419.

Barnes, R.M. (1978). Recent advances in emission spectroscopy: Inductively coupled plasma discharges for spectrochemical analysis. *C.R.C. Critical Rev. Anal. Chem.* **7**, 203–296.

Birch, N.J. (1982). Lithium in psychiatry. *In* "Metal Ions in Biological Systems", Vol. 14. (ed. H. Sigel), pp. 257–313. Marcel Dekker, New York.

Birch, N.J., Goodwin, J.C., Hullin, R.P. and Tyrer, S.P. (1974). Absorption and excretion of lithium following administration of slow-release and conventional preparations. *Br. J. Clin. Pharmacol.* **1**, 339P.

Birch, N.J., Robinson, D., Inie, R.A. and Hullin, R.P. (1978). ^{6}Li: A stable isotope of lithium determined by atomic absorption spectroscopy and use in human pharmacokinetic studies. *J. Pharm. Pharmacol.* **30**, 683–685.

Blijenberg, B.G. and Leijinse, B. (1968). The determination of lithium in serum by atomic absorption spectroscopy and flame emission spectroscopy. *Clin. Chem. Acta* **19**, 97–99.

Bone, S., Roose, S.P., Dunner, D.L. and Fieve, R.R. (1980). Incidence of side effects in patients on long term lithium therapy. *Amer. J. Psychiat.* **137**, 103–104.

British National Formulary (1988). The British Medical Association and The Pharmaceutical Society, London.

British Pharmacopoea (1980). Her Majesty's Stationery Office, London. p. xix.

Collier, J.L., Storkes, H. and Mulley, B. (1990). Pharmacokinetics of lithium citrate syrup in manic-depressive patients. *Lithium* **1**, 45–48.

Coppen, A., Abou-Saleh, M.T., Milln, P., Bailey, J. and Wood, K. (1983). Reducing lithium dose reduces morbidity and side effects in affective disorders. *J. Affect. Dis.* **6**, 53–66.

Hunter, R. (1988). Steady state pharmacokinetics of lithium carbonate in healthy subjects. *Br. J. Clin. Pharmacol.* **25**, 375–380.

Jeppsson, J. and Sjogren, J. (1975). The influence of food on side effects and absorption of lithium. *Acta Psychiat. Scand.* **51**, 285–288.

Johnson, F.N. (1980). The choice of an appropriate lithium preparation. *In* "Handbook of Lithium Therapy" (ed. F.N. Johnson), pp. 226–236. MTP Press, Lancaster.

Lauritsen, B.J., Mellerup, E.T., Plenge, P., Rasmussen, S., Vestergaard, P. and Schou, M. (1981). Serum lithium concentrations around the clock with different treatment regimens and the diurnal variation of the renal lithium clearance. *Acta Psychiat. Scand.* **64**, 314–319.

Phillips, J.D. (1989). Transport and Inorganic Biochemistry of Lithium and Magnesium. PhD Thesis (CNAA), Wolverhampton Polytechnic, pp. 28–69.

Phillips, J.D. and Birch, N.J. (1987). The effect of lithium formulations on peak serum concentrations. *Med. Sci. Res.* **15**, 943–944.

Phillips, J.D. and Birch, N.J. (1988). Application of a technique for the estimation of lithium tablet dissolution rate. *In* "Lithium: Inorganic Pharmacology and Psychiatric Use" (ed. N.J. Birch), pp. 119–124. IRL Press, Oxford.

Phillips, J.D. and Birch, N.J. (1990). Lithium in Medicine. *In* "Monovalent Cations in Biological Systems" (ed. C.A. Pasternak), pp. 339–355. C.R.C. Press, Boca Raton, FL.

Phillips, J.D., King, J.R., Myers, D.H. and Birch, N.J. (1989). Lithium monitoring close to the patient (letter). *Lancet* **ii**, 1461.

Phillips, J.D., Myers, D.H., King, J.R., Armond, D.A., Derham, C., Puranik, A., Corbett, J.A. and Birch, N.J. (1990). Pharmacokinetics of lithium in patients treated with controlled release lithium formulations. *Int. Clin. Psychopharmacol.* **5**, 65–69.

Plenge, P. and Mellerup, E.T. (1986). Lithium and the kidney: is one daily dose better than two? *Comp. Psychiat.* **27**, 336–342.

Plenge, P. and Mellerup, E.T. (1988). Serum lithium: The importance of peak values, minimum values and mean values. *In* "Lithium: Inorganic Pharmacology and Psychiatric Use" (ed. N.J. Birch), pp. 135–138. IRL Press, Oxford.

Rybakowski, J., Lehmann, W. and Kanarkowski, R. (1988). Erythrocyte lithium-sodium countertransport and total body lithium pharmacokinetics in patients with affective illness. *Human Psychopharmacol.* **3**, 87–93.

Schou, M. (1986). "Lithium Treatment of Manic Depressive Illness: A Practical Guide", 3rd edn, Karger, Basel.

Shelley, R.K., Davidson, R., Silverstone, T., Wickham, A.E. and Reed, J.V. (1986). The relationship between bioavailability and slow release in a series of lithium formulations. *IRCS Med. Sci.* **14**, 1143–1144.

Shelley, R.K. and Silverstone, T. (1986). Single dose pharmacokinetics of five formulations of lithium: a controlled comparison in healthy subjects. *Int. Clin. Psychopharmacol.* **1**, 324–331.

Shelley, R.K. and Silverstone, T. (1987). Lithium preparations. *In* "Depression and Mania: Modern Lithium Therapy" (ed. F.N. Johnson), pp. 94–98. IRL Press, Oxford.

Smith, R.B. (1985). "The Development of a Medicine". Macmillan Press Ltd, London.

Trigger, D.J. and Davies, P.J. (1987a). *In vitro* dissolution of controlled release lithium tablets I. British Pharmacopoea Solution Test Method. *Med. Sci. Res.* **15**, 895–896.

Trigger, D.J. and Davies, P.J. (1987b). *In vitro* dissolution of controlled release lithium tablets II. Alternative test methods. *Med. Sci. Res.* **15**, 897–898.

Tyrer, S.P., Hullin, R.P., Birch, N.J. and Goodwin, J.C. (1976). Absorption of lithium following slow release and conventional preparations. *Psychol. Med.* **6**, 51–58.

Vestergaard, P., Amdisen, A. and Schou, M. (1980). Clinically significant side effects of lithium treatment. *Acta Psychiat. Scand.* **62**, 193–200.

Vestergaard, P., Schou, M. and Thomsen, K. (1982). Monitoring of patients in prophylactic lithium treatment: an assessment based on recent kidney studies. *Brit. J. Psychiat.* **140**, 185–187.

Wadke, D.A. and Jacobson, H. (1980). Preformulation Testing. *In* "Pharmaceutical Dosage Forms" (ed. H. Lieberman and L. Lachman), pp. 1–18. Marcel Dekker, New York.

Wall, B.P., Parkin, J.E. and Sunderland, V.B. (1986). The effect of media and other variables on the BP solution rate test for slow lithium carbonate tablets. *J. Pharm. Pharmacol.* **38**, 633–637.

Wheeling, K. and Christian, G.D. (1984). Spectrofluorometic determination of serum lithium using 1,8 dihydroxyanthraquinone. *Anal. Lett.* **17**, 217–227.

Xie, R.Y. and Christian, G.D. (1986). Serum lithium analysis by coated wire lithium ion selective electrodes in a flow injection analysis dialysis system. *Anal. Chem.* **58**, 1806–1810.

18 The Switch Process and the Effect of Lithium

MICHAEL J. LEVELL[1] and ROY P. HULLIN[2]

[1] *Department of Chemical Pathology, University of Leeds, Leeds LS2 9JT, UK*

[2] *Regional Metabolic Research Unit, High Royds Hospital, Leeds LS29 6AQ, UK*

Affective disorders include all states of abnormal mood but the term is generally applied to the major syndromes of depression and mania where mood disturbance appears to be the central factor. Mania and melancholia have been described for centuries but it is only in this century that the two states were connected. Kraepelin (1921) formulated the concept of manic-depressive psychosis which had a good prognosis for any particular episode but with further episodes likely to recur.

Attacks of mania or depression may be intermittent or periodic with many variations in the course of the illness. Some patients suffer recurrent attacks of depression only; others recurrent attacks of mania. Some have both manic and depressive episodes which may either be isolated or continuous: the patient swinging from mania to depression and back to mania and so on without any intervening periods of normality. Although attacks are usually self-limiting, their duration tends to increase and remissions become shorter as the patient grows older.

Considerable evidence exists for at least two distinct forms of affective disorder: bipolar, where individuals have alternating cycles of high and low mood; and unipolar, where patients have recurrences of depression or mania alone. Unipolar mania is relatively uncommon and is sometimes included under bipolar illness. It may, however, be commoner than once supposed. Nurnberger *et al.* (1979) reported that almost 16% of a bipolar population attending a lithium clinic had unipolar mania by Research Diagnostic Criteria. Abrams and Taylor (1974) found that up to 28% of patients with current mania fell into this category. However, it is unlikely to be a genetically distinct group (Mendlewicz, 1979).

The distinction between bipolar and unipolar disorder was first proposed by Leonhard (1959). Subsequently Dunner *et al.* (1976) proposed a further subdivision of bipolar subjects into two groups I and II. Bipolar II patients differed from bipolar I in that the elated phase was hypomania or mild mania and did not require admission to hospital. The validity of this subdivision is an open question in view of the inconsistent ways in which the terms hypomania and mania are used. The definition and typing of unipolar depression also remains unclear. Perris and d'Elia (1966) originally used the term to describe patients with at least three episodes of depression but other investigators have not defined it so narrowly and a patient hospitalized for a single episode of a relatively severe depression is now usually considered unipolar. The lack of clarity in nomenclature reflects the conflict between clinical practice and research strategies designed to obtain nosologically distinct subgroups. In this chapter we shall confine our attention to the switch process in bipolar patients only, since these patients are more easily recognized and categorized. We shall examine the pathophysiology of bipolar affective disorder in an attempt to resolve how lithium, the accepted treatment of choice in recurrent bipolar affective diseases, acts clinically.

Research effort has tended to concentrate on bipolar patients with short cycles since these patients allow comparisons of successive cycles in longitudinal studies. Such patients are rare and may have very short, or sometimes no, periods of normality between abnormal episodes, but it is from such patients that the concept has originated of a cyclic disease with 'switches' from depression to mania and from mania to depression (Fig. 1). Post *et al.*(1981) have distinguished between patients who switch rapidly and those in whom the change is more gradual. Those with more rapid changes had been ill significantly longer and had suffered a greater number of affective episodes. Ninety-five per cent of bipolar patients do not cycle or switch rapidly: they are usually in normal mood with their episodes of abnormal affect being years apart. In these patients the mood change can be regarded not as a 'switch' but as a relapse.

Genetic and family studies have consistently shown an increased familial prevalence of illness in bipolar Is and usually in bipolar IIs. Bipolars have been demonstrated to have a higher monozygotic:dizygotic ratio than unipolar depressives and to have more frequently two generations of illness. This suggests a dominant mode of transmission but consistent confirmation of this has not been obtained in research studies. Some linkage studies have postulated an X-linked dominant mode of transmission and an association with the XG_a antigen and colour blindness, but other workers have not confirmed this. Moreover Gershon and Bunney (1976) and Gershon *et al.* (1976) concluded from mathematical models that a threshold effect may occur and that bipolar affective disorder is simply a more severe form of the illness than unipolar disorder.

A link between manic-depression and a locus on chromosome 11 was

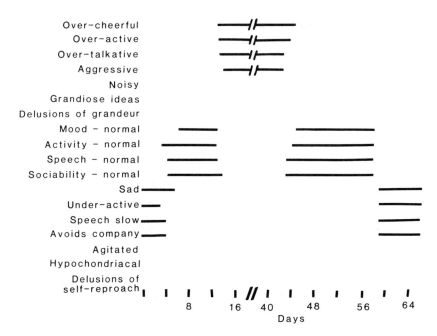

FIG. 1   Twice daily nursing assessment of mood indices in a bipolar patient over a period of months showing an 'up' switch (depression → normal → mania) and a 'down' switch (mania → normal → depression).

reported by Egeland *et al.* (1987). Unfortunately it now appears that this was probably spurious (Robertson, 1989), illustrating the difficulties of work on the genetic mechanisms of psychiatric disease. It would appear that more than one locus is involved, implying various biochemical possibilities, e.g. different enzymes on a single metabolic pathway. Alternatively there could be several different illnesses lumped together because of clinical and nosological ignorance. In any event some variability in bipolar affective disorder is clearly not genetic. For example, episode length and frequency often change as the disease progresses and with age. Thus 'switches' and 'relapses' are not necessarily different, a switch being regarded as a relapse following recovery.

Biological Rhythms

Since in its most dramatic form bipolar disease is a regular cyclical disease, it is natural that there has been extensive study of the possible role of biorhythms in the pathogenesis. A small minority of bipolar patients show a cycle of mania and depression that coincides with a recognized rhythm—daily, weekly, monthly or yearly. Two mechanisms can be envisaged. In one the manic-depressive cycle has a natural frequency that happens to be close to that of a normal biological rhythm and it becomes entrained to it (Silverstone and Romans-Clarkson, 1989). How this entrainment could operate at a biochemical level is a matter of speculation. A second mechanism would involve environmental triggers. In the case of yearly cycles—seasonal affective disorder (SAD)—it is suggested that changing lengths of daylight may provide the triggers. The precise nosological status of SAD patients is in some doubt (Silverstone and Romans-Clarkson, 1989) and it is difficult to generalize from them to all bipolar patients.

Most discussion of biological rhythms and bipolar disorder has centred on abnormal or desynchronized rhythms. The body contains one (or probably two) oscillators which give the well-known circadian periodicities of cortisol, melatonin, electrolyte excretion and many others. These rhythms are normally entrained to the cycle of light and dark and to other social cues.

Studies of bipolar patients have shown various abnormalities of rhythms. Of most interest in the present context are reports of rhythms that are usually linked becoming desynchronized or altered in frequency. A striking example of how this might generate a larger cycle occurred in a patient of Kripke *et al.* (1978). This patient's rhythms of temperature and electrolyte excretion had a cycle length of 21.8 h rather than 24 h. Thus, their relationships to the 24-h environmental cycle recurred every 10 days ($21.8 \times 11 = 24 \times 10$). This kind of mechanism, involving interaction between different rhythms, is a theoretically attractive way of producing longer rhythms. It still begs the question of what acts as a trigger, although it could be something as non-specific as sleep deprivation (Wehr *et al.*, 1987).

Lithium has been shown to alter biological rhythms in many species including man and this has been proposed as the basis of its therapeutic action.Unfortunately, lithium can alter rhythms without necessarily altering mood (Campbell *et al.*, 1989).

Hallonquist *et al.* (1986) conclude that circadian abnormalities may be a factor in bipolar disorder but are not its cause.

Biochemical changes

Many biochemical and physiological differences between mania and depression have been documented. Some of these are expected consequences of the different levels of physical activity, but in two areas—catecholamines and sodium metabolism—the changes may be more fundamental. These biochemical changes are not necessarily causing the behavioural changes but they provide potential 'markers' of the underlying changes that are causative. Of course, even if a particular biochemical change could be shown to 'cause' mania or depression, this would not resolve the problem of switches. However, in the case of sodium, because the daily turnover is only $2-3\%$ of the total in the body, slow changes of the total could theoretically provide a timing mechanism.

Catecholamines

According to the monoamine hypothesis of affective disorders, mania is due to an absolute or relative excess of monoamine neurotransmitters (Schildkraut, 1965) or an increase in receptor sensitivity to those transmitters in the brain (Ashcroft *et al.*, 1972). Conversely depression would result from a deficit of monoamines or a reduced receptor sensitivity to them. Many studies have attempted to investigate this hypothesis, particularly with respect to the catecholamines (noradrenaline and dopamine) and 5-hydroxytryptamine which were early recognized as neurotransmitters in the brain.

Originally noradrenaline was thought to be the neurotransmitter most likely to be involved. Many investigations of the principal brain metabolite of noradrenaline, 3-methoxy-4-hydroxyphenylglycol (MHPG) have been carried out in manic and depressed patients. Despite the variability in the results and imperfections in the methodology in some studies, the majority found an increase in urinary MHPG in mania compared with depression. Bunney *et al.* (1977) reported that, in longitudinal studies of individual patients, increases in urinary MHPG occurred at the time of the switch from depression to mania. Other workers (Annito and Shopsin, 1979) maintained that there was no relationship between mood and excretion of MHPG. Some reports of elevated plasma and cerebrospinal fluid MHPG in the manic phase have also been published, but increased motor activity, which invariably occurs in mania, also causes a rise in the former.

Most antidepressant drugs, including the tricyclics, block the re-uptake of noradrenaline from the synaptic cleft into the pre-synaptic neurone and there have been many examples of mania being produced by antidepressants, especially in bipolar patients. The overall risk of precipitating mania in this way has been estimated at 9% (Bunney, 1978), with some workers reporting that this occurred in nearly 70% of bipolar patients treated with monoamine re-uptake inhibitors.

Randrup *et al.* (1975), Silverstone (1979) and Post (1980) have all suggested that overactivity of dopamine pathways may play an important role in the pathogenesis of mania. The activity of these pathways has usually been studied by determining the cerebrospinal fluid (CSF) level of homovanillic acid (HVA), the major metabolite of dopamine. In most such studies, no elevation of HVA has been found, but several investigations have reported some increase in mania. Bunney *et al.* (1977) described a patient in whom there was a slow increase in CSF HVA before the switch from depression to mania occurred. Furthermore, Murphy (1972) reported that the administration of levadopa, a precursor of dopamine which increases dopamine synthesis in the brain, precipitated mania in eight of nine bipolar patients. Some workers (Van Kammen and Murphy, 1975; Gerner *et al.*, 1976) have also reported manic episodes in bipolar patients produced by a single dose of amphetamine which is believed to act primarily by increasing the release of newly synthesized dopamine from pre-synaptic neurones. The administration of dopamine receptor agonists to bipolar patients suffering from depression has also been found to lead to mania (Gerner *et al.*, 1976).

Kety (1971) and Mendels and Frazer (1975) have postulated an underlying deficiency in 5-hydroxytryptamine activity in both mania and depression. There is little evidence to support this although some of the studies of cerebrospinal fluid 5-hydroxytryptamine metabolites have found normal or reduced levels in manic patients in spite of the demonstration that motor activity may lead to increases.

There have been a number of recent studies of other biochemical changes occurring during the switch from depression to mania. Carney *et al.* (1989) reported that nine out of eleven bipolar patients treated with S-adenosyl-methionine switched into elevated mood state with a tendency for the cerbrospinal fluid HVA to increase during the treatment. Sporadic cases of switching into elevated mood had been earlier linked with S-adenosyl-methionine and folate (Lipinski *et al.*, 1984; Carney, 1986). The authors suggested that S-adenosylmethionine is linked with CSF 5-hydroxyindoleacetic acid and folate metabolism and influences prolactin, thus producing an effect on dopamine metabolism.

Semba *et al.* (1988) reported a single case study in which a rapid-cycle bipolar patient showed an increased excretion of β-phenylethylamine (which is known to increase central catecholamine metabolism) before the switch from mania to depression. The authors suggested that β-phenyl-ethylamine acted as a biochemical trigger in the switch mechanism.

Another single case study (Kennedy *et al.* 1989) showed elevated levels of melatonin during mania and of cortisol during depression. Since melatonin output is mainly under β-noradrenergic influence (Lewy *et al.*, 1981), elevated output during mania would suggest an increase in noradrenergic activity. There was however no reduction in melatonin during the patient's depressed phase in contrast to the report by Brown *et al.* (1985).

Sodium Metabolism

There have been several studies of sodium metabolism in bipolar patients. The simple measurement of plasma sodium concentration gives little information.

Except in extreme pathological situations, whenever the body gains or loses sodium it usually gains or loses a corresponding volume of water, so that extracellular sodium concentration shows little change. It is therefore necessary to measure the mass of sodium in particular compartments of the body, or to estimate changes of mass by balance studies. Neither approach is ideal. Measurements of extracellular fluid and other body compartments depend upon measuring volumes of distribution of marker substances. These measurements are difficult to perform and are theoretically unsound because they assume constant concentrations throughout the compartment. The alternative approach is to measure daily sodium balance. There are difficulties in measuring input and output accurately for weeks or months of patients whose cooperation cannot be guaranteed. Moreover incomplete bladder emptying may occur (Blum and Friedland, 1983). In consequence, few longitudinal studies have been published and ultra-short cycle patients are disproportionately represented.

Sodium balances have been published on three bipolar patients with cycle lengths of less than one week (Klein and Nunn, 1945; Crammer, 1959; Jenner *et al.*, 1967) and we have data on the 72-h cycle patient described by Rees *et al.* (1974). In all cases the patients were on constant sodium intake. The first three showed sodium retention during depression and loss during mania. In contrast our 72-h cycle patient showed sodium retention during mania and loss during depression, although the changes were small. However, the patient of Crammer (1959) also showed this pattern after several months' treatment with chlorpromazine. With such short cycles a lag between a central change and its peripheral manifestation could give altered or variable apparent relationships to mood.

Another factor which appears to have been overlooked hitherto is the acute effect of ACTH on aldosterone. Depressed patients have, on average, higher cortisol levels reflecting higher ACTH levels. ACTH has an acute stimulatory effect on aldosterone, lasting 2–3 days. Thus a patient whose episodes of depression lasted for only this period might be showing this acute stimulation with its associated sodium retention.

Patients with cycle lengths of several weeks have also been studied. Hullin *et al.* (1968) showed balance data on one patient over a period of 33 months during which she also had sodium retention during depression and loss during mania. However, in one of the patients of Crammer (1959) and the patient of Crammer (1986), sodium loss occurred early in depression with retention occurring during much of the remaining time. In a limited study, Allsopp *et al.* (1972) also found sodium and water loss in early depression.

All the above studies have been in patients on constant sodium intake. All showed cyclic changes of sodium balance but the relationships to mood were not identical in all patients. Furthermore, in two studies (Crammer, 1959 and our 72-h cycle patient) high and low sodium diets had little or no effect on mood, nor did they alter the general patterns of sodium retention and loss. Thus a causal connection between sodium and mood is unlikely.

It is difficult to study sodium balance and at the same time allow the patient to choose how much salt to eat. Anderson *et al.* (1964) achieved this and noted a high salt intake during mania in one of two bipolar patients. Crammer (1986) and Penney *et al.* (1987) used the indirect approach of relating body weight to urinary sodium output. In respectively one and three patients they showed increased sodium excretion in mania at a time when rapidly rising body weight indicated fluid retention. This implies increased sodium intake. These results are interesting in the light of early studies of extracellular volume (ECV) and of total exchangeable sodium. Dawson *et al.* (1956) found little difference between mania and depression. However, Coppen *et al.* (1966) found that 'residual' sodium (total − extracellular) was relatively high in mania, and Hullin *et al.* (1967a), in a longitudinal study of three patients, found a higher ECV in mania than depression. Voluntary changes of sodium intake could well account for some of these differences— those patients who chose to eat more salt in mania would expand their ECV.

Aldosterone

Attempts to explain changes in sodium have so far concentrated on aldosterone. When sodium intake was constant, the urinary excretion of aldosterone was higher in manic patients than in depressed patients (Aronoff *et al.*, 1970; Murphy *et al.*, 1969). A longitudinal study in two patients measuring aldosterone production rate confirmed this pattern, showing a progressive fall of aldosterone during the switch from mania to depression at the time of sodium loss (Allsopp *et al.*, 1972). Crammer (1986) found a progressive rise of aldosterone excretion during the switch from depression to mania. This increase was associated with falling sodium excretion and ultimately sodium retention. Thus, longitudinal studies of aldosterone in patients on fixed sodium intake have suggested that it has a causal role in the changes of sodium balance at both up and down switches.

When urinary aldosterone was examined in patients who were not receiving a fixed intake of sodium, a different picture emerged. Levell and Hullin (1988) reported results on three patients whose aldosterone excretion showed little difference between mania and depression. Crammer (1986) also reported little difference in a study in which sodium intake was free. Levell and Hullin (1988) suggested that patients whose sodium intake was not restricted would consume more salt when manic, so tending to block the increase of aldosterone shown by manic patients on a fixed intake.

What causes the variations of urinary aldosterone seen in patients on a fixed sodium intake? Hullin *et al.* (1977) examined plasma renin activity and concluded that aldosterone did not show the same relationship to renin in bipolar patients that it shows in normals. They later found evidence of an inhibitor of aldosterone producton in serum from bipolar patients, especially in early depression (O'Brien *et al.*, 1979; Hullin *et al.*, 1981). They suggested that this inhibitor caused the fall of aldosterone that occurs during the down switch and which is possibly responsible for the loss of sodium that occurs. This inhibitor has never been identified.

The effect on sodium and aldosterone of starting lithium therapy was first examined by Murphy *et al.* (1969) and Aronoff *et al.* (1970). Both groups noted a transient increase of sodium excretion followed by a small increase of urinary aldosterone—presumably a compensatory change. The initial sodium loss appears to be due to a decreased responsiveness to the sodium-retaining action of aldosterone (Thomsen *et al.*, 1976; Stewart *et al.*, 1987). Consistent with this hypothesis, Pederson *et al.* (1977) reported a weak positive correlation between plasma aldosterone and plasma lithium in patients on long-term treatment. In contrast, Hendler (1975), also measuring aldosterone in plasma rather than in urine, found very high values in untreated bipolar patients; lithium treatment lowered aldosterone levels. Miller *et al.* (1979) found no effect of 3–4 weeks treatment on aldosterone or renin.

Our experience of plasma aldosterone suggests that it may be unreliable in bipolar patients whose mood is abnormal. We found occasional very high values that did not correlate with urinary aldosterone (M.J. Levell and R.P. Hullin, unpubl. obsv.). This is in contrast to normal subjects in whom plasma and urinary measurements of aldosterone correlate. We concluded that plasma aldosterone measurements in these patients may be affected by transient stimuli so that they may fail to represent the patient's true aldosterone status.

In patients who are continuing to exhibit episodes of abnormal affect while receiving lithium, the characteristic changes of body sodium continue to occur. There are no obvious differences in pattern between treated and untreated patients. Of the longitudinal studies cited above, the patient of Crammer (1986), and two of the patients of Penney *et al.* (1987) were on lithium. Hullin *et al.* (1968) reported cycles off and on lithium in one patient. It can, of course, be argued that since patients who continue to show abnormal moods are not responding clinically to lithium, one would not expect a biochemical response. There is an obvious difficulty in seeking sodium changes in patients whose behaviour has been completely normalized by lithium (although it might be attempted in ultra-short cycle patients).

The sodium and aldosterone changes in short-cycle bipolar patients may be summarized as follows. An increase of aldosterone during the switch from depression to mania leads to sodium retention. A fall of aldosterone during the down switch leads to sodium loss. Superimposed on these changes, there may be a transient stimulation of aldosterone by ACTH lasting for 2–3 days of depression. In patients on a free salt intake, there will sometimes be an increased consumption of salt in mania which will tend to lower aldosterone. The interaction of these three factors will be variable.

There are interesting interactions between sodium and aldosterone on the one hand and catecholamines on the other. As well as causing sodium retention by the kidney, aldosterone has more widespread effects. High aldosterone levels are associated with high intracellular sodium. Sodium stimulates tyrosine hydroxylase, the rate-limiting enzyme in dopamine and noradrenaline synthesis. This has been studied as part of a hormonal mechanism for sodium regulation (Goldstein et al., 1989), but could account for some of the altered catecholamine levels of bipolar patients. However, it must be stressed that neither aldosterone nor sodium retention, nor the two together, causes any behavioural changes in normal people.

Other interactions between catecholamines and aldosterone occur in aldosterone regulation. Many factors affect aldosterone production. The dominant influences are sodium and potassium. Sodium deprivation and potassium excess stimulate aldosterone. Sodium excess and potassium deprivation inhibit. Potassium acts directly on the adrenal but the sodium effect is mostly indirect. The renin–angiotensin system is one link between sodium and aldosterone, atrial natriuretic peptides are another; there are probably others. ACTH acutely stimulates aldosterone. The degree of stimulation depends upon sodium and potassium status. Moreover, even if ACTH stimulation continues, the aldosterone will return in 2 or 3 days to the level determined by the sodium and potassium.

Catecholamines may influence aldosterone in several ways. Noradrenaline acutely stimulates renin production (and is responsible for the rise of renin and therefore aldosterone, seen on standing up). However, we found no consistent abnormality in the renin response to posture in bipolar patients (Hullin et al., 1977). Dopamine appears to inhibit aldosterone production, although the mechanism is controversial. Increases of dopamine have little effect, but dopamine antagonists cause aldosterone to increase: thus dopamine inhibition of aldosterone is normally maximal. Dopamine agonists have been shown to precipitate mania and antagonists to resolve mania. It would be of interest to measure aldosterone in such studies.

Several neurotransmitters, including noradrenaline and 5-hydroxytryptamine, are involved in ACTH secretion which, as mentioned above, transiently increases aldosterone. Some amines, including 5-hydroxytryptamine and its N-methyl derivative, stimulate aldosterone in vitro but the physiological significance of these effects is unclear.

Antidiuretic Hormone

Arginine vasopressin (AVP), the antidiuretic hormone in humans, is a candidate for involvement in bipolar disease. Not only could it be involved in sodium and water changes, but it also has behavioural effects (Gold *et al.*, 1978). Gold *et al.* (1981) measured AVP in CSF, finding higher concentrations in mania than depression. Gold *et al.* (1978) proposed that high AVP acted as a trigger at some point in the cycle of mania and depression but we failed to find such a point (Penney *et al.*, 1987). Urinary AVP was measured longitudinally in three patients on free intake of water and sodium. From comparison of AVP excretion and urine osmolality, we concluded that there were short-lived spikes of very high AVP excretion but there was no systematic difference in AVP between mania and depression.

Lithium has a pronounced effect on AVP, inhibiting its effect on the kidney. This leads, as in nephrogenic diabetes insipidus, to a compensatory increase of AVP. Since AVP has behavioural effects, this increase could contribute to the therapeutic effect of lithium (Hullin, 1988).

Other Endocrine Changes

The role of neuroendocrine abnormalities in mood disorders has been studied mainly in patients suffering from endogenous depression. This is due to the high incidence of depression in endocrinopathies (e.g. hypothyroidism and Cushing's disease) and because several symptoms of depression, such as changes in sleep pattern, appetite and sex drive, suggest associated hypothalamic dysfunction. Furthermore some neurotransmitters implicated in the etiology of depression, in particular noradrenaline and 5-hydroxytryptamine, also regulate the secretion of hypothalamic cells which control pituitary function. Interpretation of the evidence from research studies in this field presents problems since several neurotransmitters are frequently involved in the regulation of the secretion of each hormone. Other complications include the episodic secretion of some hormones associated with sleep patterns or circadian rhythms and the responsiveness of hormone systems such as ACTH and prolactin to stress.

Notwithstanding the difficulties, significant abnormalities in the secretion of cortisol, growth hormone and thyroid-stimulating hormone (TSH) in endogenous depression have been demonstrated. Of these, cortisol is of greatest interest since bipolar as well as unipolar patients are affected (Hullin *et al.* 1967b). There is well-documented evidence for the hypersecretion of cortisol in certain groups of depressed patients. This seems to be secondary to increased

ACTH secretion. It has been suggested that a noradrenaline deficit may play a role in the mechanism. However, abnormalities of cortisol and other hormones are not confined to depression but occur in a variety of affective disorders (Whalley *et al.*, 1989).

Conclusion

At the switch into mania, several changes occur. Production of dopamine increases, and because dopamine agonists may precipitate mania, this increase is likely to be a cause, rather than a consequence, of the change. Aldosterone rises in patients on restricted salt intake. If salt is not restricted, some patients will eat more salt; this may blunt or prevent the rise of aldosterone. Dopamine inhibited aldosterone so cannot be the cause of the rise. On the other hand, salt may increase dopamine, so that the increased salt appetite or aldosterone-induced sodium retention could cause the rise of dopamine. However, altering dietary salt does not cause mania in normals or in bipolar patients.

Towards the end of mania or at a switch into depression, events are less well documented. Aldosterone production may fall, with sodium loss. Dopaminergic activity must fall, but it is unclear when. ACTH rises, with consequent steroid changes. It is curious that 'down' switches should be poorly documented. Perhaps this is because so much work has been done on unipolar depressives. However, while we might expect some common biochemical changes with depression arising in different diseases, it would be naive to assume a completely common mechanism.

It is difficult to find a single conceptual model that will embrace the extremes of bipolar disease—from the patient who occasionally relapses but is usually normal to the patient who shows a relentless alternation of mania and depression. At the one extreme, environmental triggers seem most likely, at the other extreme an inbuilt rhythm.

There are several cyclic phenomena in humans that might provide models of bipolar disease, though none is universally satisfactory. Circadian rhythms, with a natural periodicity combined with entrainment to environmental cues, provide one model. However, as pointed out earlier, most bipolar patients do not show a regular cycle. A variant of this model is the production of longer rhythms from desynchronized circadian rhythms.

The menstrual cycle provides the most attractive model. Here a series of events, linked in cause and effect relationships, and including structural growth, provide a link between the start of one cycle and the start of the next. Oligomenorrhoea and periods of amenorrhoea provide the parallel to the more common bipolar patient with infrequent episodes. We have failed to find such a coherent series of relationships in bipolar patients, although we suggest that interactions of catecholamines, sodium, and aldosterone may be steps in such a relationship.

A third type of rhythm, determined by environmental or social rhythms, e.g. nutrients or posture, seems not relevant to bipolar disease.

A central feature in bipolar affective disease, but one that is not often considered, is the lack of clear distinction between the 'relapse' of the patient with infrequent episodes and the 'switch' of the regular short-cycle patient. Any mechanism that is proposed must explain both. This problem of switches and relapses might be resolved by postulating that the defect in bipolar patients is not one that *causes* the abnormal mood but rather a deficiency in a mechanism that instantly *corrects* abnormal mood in normal people. Transient hypomania, in particular, is a common occurrence although usually it lasts only seconds, minutes, or for the duration of the party. Most people also experience depression. It is therefore reasonable to postulate the existence of a mechanism that counters abnormal mood. If there is such a mechanism, any defect in it will produce a disease.

Examination of switches offers few clues to the action of lithium. At a clinical level, lithium appears to have at least two actions: an acute antimanic action and a chronic prophylactic action on both mania and depression. The fundamental question of whether lithium corrects the underlying biochemical defects or simply uncouples the behavioural changes from them is unanswered and virtually unanswerable. If the patient is euthymic, when does one look? In some patients, lithium may attenuate the symptoms without changing the pattern of mania and depression, suggesting that it is not affecting the switch process. A 48-h cycle patient had evidence of an underlying cycle even when treated (Hanna *et al.*, 1986). These are but anecdotes, and systematic study is needed. The switches of the short-cycle patient, the relapses of those with a less regular disease, remain as enigmatic as when bipolar disease was first defined (Kraepelin, 1921).

References

Abrams, R. and Taylor, M.A. (1974). Unipolar mania. *Arch. Gen. Psychiat.* **30**, 441–443.

Allsopp, M.N.E., Levell, M.J., Stitch, S.R. and Hullin, R.P. (1972). Aldosterone production rates in manic-depressive psychosis. *Br. J. Psychiat.* **120**, 399–404.

Anderson, W.McC., Dawson, J. and Margerison, J.H. (1964). Serial biochemical, clinical and electroencephalographic studies in affective illness. *Clin. Sci.* **26**, 323–336.

Annito, W. and Shopsin, B. (1979). Neuropharmacology of mania. *In* "Manic Illness" (ed. B. Shopsin), pp. 105–162. Raven Press, New York.

Aronoff, M.S., Evens, R.G. and Durell, J. (1970). Effect of lithium salts on electrolyte metabolism. *J. Psychiat. Res.* **8**, 139–159.

Ashcroft, G.W., Eccleston, D., Murray, L.G., Glen, A.I.M., Crawford, T.B.B., Pullar, I.A. *et al.* (1972). Modified amine hypothesis for the aetiology of affective illness. *Lancet* ii, 573–577.

Blum, A. and Friedland, G.W. (1983). Urinary tract abnormalities due to chronic psychogenic polydipsia. *Amer. J. Psychiat.* **140**, 915–916.

Brown, G.M., Kocsis, J.H., Caroff, S. *et al.* (1985). Differences in nocturnal melatonin secretion between melancholic depressed patients and control subjects. *Amer. J. Psychiat.* **142**, 811–816.

Bunney, W.E. (1978). Psychopharmacology of the switch process in affective illness. *In* "Psychopharmacology, a Generation of Progress" (ed. M.A. Lipton, A. Di Mascio and D.F. Killam), pp. 1249–1259. Raven Press, New York.

Bunney, W.E., Wehr, T.A., Gillin, J.C. *et al.* (1977). The switch process in manic-depressive psychosis. *Ann. Intern. Med.* **87**, 319–335.

Campbell, S.S., Grillin, J.C., Kripke, D.F., Janowsky, D.S. and Pirsch, S.C. (1989). Lithium delays circadian phase of temperature and REM sleep in a bipolar depressive: a case report. *Psychiat. Res.* **27**, 23–29.

Carney, M.W.P. (1986). Neuropharmacology of S-adenosylmethionine. *Clin. Neuropharmacol.* **9**, 235–243.

Carney, M.W.P., Chary, T.K.N., Bottiglieri, T. and Reynolds, E.H. (1989). The switch mechanism and the bipolar/unipolar dichotomy. *Br. J. Psychiat.* **154**, 48–51.

Crammer, J.L. (1959). Water and sodium in two psychotics. *Lancet* i, 1122–1126.

Crammer, J.L. (1986). Disturbance of water and sodium in a manic-depressive illness. *Br. J. Psychiat.* **149**, 337–345.

Coppen, A.J., Shaw, D.M., Malleson, A. and Costain, R. (1966). Mineral metabolism in mania. *Br. Med. J.* i, 71–75.

Dawson, J., Hullin, R.P. and Crockett, B.M. (1956). Metabolic variations in manic-depressive psychosis. *J. Mental Sci.* **102**, 168–177.

Dunner, D.L., Gershon, E.S. and Goodwin, F.K. (1976). Heritable factors in the severity of affective illness. *Biol. Psychiat.* **11**, 31–42.

Egeland, J.A., Gerhard, D.S., Pauls, D.L. *et al.* (1987). Bipolar affective disorders linked to DNA markers on chromosome 11. *Nature* **325**, 783–787.

Gerner, R.H., Post, R.M. and Bunney, W.E. (1976). A dopaminergic mechanism in mania. *Amer. J. Psychiat.* **133**, 1177–1180.

Gershon, E.S. and Bunney, W.E. (1976). The question of X-linkage in bipolar manic-depressive illness. *J. Psychiat. Res.* **13**, 99–117.

Gershon, E.S., Bunney, W.E. and Leckmann, J.F. (1976). The inheritance of affective disorders: a review of data and of hypotheses. *Behav. Genet.* **6**, 227–261.

Gold, P.W., Goodwin, F.K. and Reus, V.I. (1978). Vasopressin in affective illness. *Lancet* i, 1233–1236.

Gold, P.W., Goodwin, F.K., Post, R.M. and Robertson, G.L. (1981). Vasopressin function in depression and mania. *Pharmacol. Bull.* **17**, 7–9.

Goldstein, D.S., Stull, R., Eisenhofer, G. and Gill, J.R. (1989). Urinary excretion of dihydroxyphenylalanine and dopamine during alterations of dietary salt intake in humans. *Clin. Sci.* **76**, 517–522.

Hallonquist, J.D., Goldberg, M.A. and Brandes, J.S. (1986). Affective disorders and circadian rhythms. *Can. J. Psychiat.* **31**, 259–271.

Hanna, S.M., Jenner, F.A. and Souster, L.P. (1986). Electro-oculogram changes at the switch in a manic-depressive plant. *Br. J. Psychiat.* **149**, 229–232.

Hendler, N. (1975). Lithium-responsive hyperaldosteronism in manic patients. *J. Nerv. Ment. Dis.* **161**, 49–54.

Hullin, R.P. (1988). Closing address: a review of the congress. *In* "Lithium: Inorganic

Pharmacology and Psychiatric Use" (ed. N.J. Birch), pp. 335–338. IRL Press, Oxford.

Hullin, R.P., Bailey, A.D., McDonald, R., Dransfield, G.A. and Milne, H.B. (1967a). Body water variations in manic-depressive psychosis. *Br. J. Psychiat.* **113**, 584–592.

Hullin, R.P., Bailey, A.D., McDonald, R., Dransfield, G.A. and Milne, H.B. (1967b). Variations in 11-hydroxycorticosteroids in depression and manic-depressive psychosis. *Br. J. Psychiat.* **113**, 593–600.

Hullin, R.P., Swinscoe, J.C., McDonald, R. and Dransfield, G.A. (1968). Metabolic balance studies on the effect of lithium salts in manic-depressive psychosis. *Br. J. Psychiat.* **114**, 1561–1573.

Hullin, R.P., Jerram, T.C., Lee, M.R., Levell, M.J. and Tyrer, S.P. (1977). Renin and aldosterone relationships in manic-depressive psychosis. *Br. J. Psychiat.* **131**, 575–581.

Hullin, R.P., Levell, M.J., O'Brien, M.J. and Toumba, K.J. (1981). Inhibition of *in vitro* production of aldosterone by manic-depressive sera. *Br. J. Psychiat.* **138**, 373–380.

Jenner, F.A., Gjessing, L.R., Cox, J.R., Davies-Jones, A., Hullin, R.P. and Hanna, S.M. (1967). A manic-depressive psychotic with a persistent 48-hour cycle. *Br. J. Psychiat.* **113**, 895–910.

Kennedy, S.H., Tighe, S., McVey, G. and Brown, G.M. (1989). Melatonin and cortisol "switches" during mania, depression, and euthymia in a drug-free bipolar patient. *J. Nerv. Ment. Dis.* **177**, 300–303.

Kety, S.S. (1971). Brain amines and affective disorder. *In* "Brain Biochemistry and Mental Disease" (ed. B.T. Ho and W.M. McIsaac), pp. 237–244. Plenum Press, New York.

Klein, R. and Nunn, R.F. (1945). Clinical and biochemical analysis of a case of manic-depressive psychosis showing regular weekly cycles. *J. Ment. Sci.* **91**, 79–88.

Kraepelin, E. (1921). "Manic-depressive Insanity and Paranoia" (translated by R.M. Barclay). E & S Livingstone, Edinburgh.

Kripke, D.F., Mullaney, D.J., Atkinson, M. and Wolf, S. (1978). Circadian rhythm disorders in manic-depressives. *Biol. Psychiat.* **13**, 335–351.

Leonhard, K. (1959). "Aufeiling der Endogenen Psychosen", 2nd edn. Akademie-Verlag, Berlin.

Levell, M.J. and Hullin, R.P. (1988). The switch process in periodic affective disorder: aldosterone and AVP. *In* "Lithium: Inorganic Pharmacology and Psychiatric Use" (ed. N.J. Birch), pp. 169–172. IRL Press, Oxford.

Lewy, A.J., Wehr, T.A., Goodwin, F.K. *et al.* (1981). Manic-depressive patients may be supersensitive to light. *Lancet* **i**, 383–384.

Lipinski, J.F., Cohen, B.M., Frankenburg, F. *et al.* (1984). Open trial of S-adenosylmethionine for treatment of depression. *Amer. J. Psychiat.* **141**, 448–450.

Mendels, J. and Frazer, A. (1975). Reduced central serotonergic activity in mania. *Br. J. Psychiat.* **126**, 241–248.

Mendlewicz, J. (1979). Genetic forms of manic illness and the question of atypical mania. *In* "Manic Illness" (ed. B. Shopsin), pp. 49–55. Raven Press, New York.

Miller, P.D., Dubovsky, S.L., McDonald, K.M., Katz, F.H., Robertson, G.L. and Schrier, R.W. (1979). Central, renal and adrenal effects of lithium in man. *Amer. J. Med.* **66**, 797–803.

Murphy, D.L. (1972). L-dopa, behavioural activation and psychopathology. *Res. Publ. Assoc. Nerv. Ment. Dis.* **50**, 472–493.

Murphy, D.L., Goodwin, F.K. and Bunney, W.E. (1969). Aldosterone and sodium response to lithium administration in man. *Lancet* **ii**, 458–460.

Nurnberger, J., Roose, S.P., Dunner, D.L. and Fieve, R.R. (1979). Unipolar mania: a distinct clinical entity? *Amer. J. Psychiat.* **136**, 1420–1423.

O'Brien, M.J., Levell, M.J. and Hullin, R.P. (1979). Inhibition of aldosterone production in adrenal cell suspensions by serum from patients with manic-depressive psychosis. *J. Endocrinol.* **80**, 41–50.

Pederson, E.B., Darling, S., Kierkegaard-Hansen, A. *et al.* (1977). Plasma aldosterone during lithium treatment. *Neuropsychobiology* **3**, 153–159.

Penney, M.D., Levell, M.J. and Hullin, R.P. (1987). Arginine vasopressin in manic-depressive psychosis. *Psychol. Med.* **17**, 861–867.

Perris, C. and d'Elia, G. (1966). Therapy and prognosis. *Acta Psychiatr. Scand. (Suppl.)* **194**, 153–171.

Post, R.M. (1980). Biochemical theories of mania. *In* "Mania" (ed. R.H. Belmaker and H.M. VanPraag), pp. 217–265. MTP, Lancaster.

Post, R.M., Ballenger, J.C., Rey, A.C. and Bunney, W.E. (1981). Slow and rapid onset of manic episodes: implications for underlying biology. *Psychiat. Res.* **4**, 229–237.

Randrup, A., Munkv, I., Fog, R., Gerlach, J., Molander, L., Kjelberg, B. and Scheel-Kruger, J. (1975). Mania, depression and brain dopamine. *In* "Current Developments in Psychopharmacology", Vol. 2 (ed. W.B. Essman and L. Valzelli), pp. 205–248. Spectrum, New York.

Rees, J.R., Allsopp, M.N.E. and Hullin, R.P. (1974). Plasma concentrations of tryptophan and other amino-acids in manic-depressive patients. *Psychol. Med.* **4**, 334–337.

Reynolds, E.H. (1981). *In* "Advances in Epileptology" (ed. M. Dam, L. Gram and J.K. Perry). Raven Press, New York.

Robertson, M. (1989). False start on manic depression. *Nature* **342**, 222.

Schildkraut, J.J. (1965). The catecholamine hypothesis of affective disorders: a review of supporting evidence. *Amer. J. Psychiat.* **122**, 509–522.

Semba, J., Nankai, M., Maruyama, Y., Kaneno, S., Watanabe, A. and Takahashi, R. (1988). Increase in urinary β-phenylethylamine preceding the switch from mania to depression: a "rapid cycler". *J. Nerv. Ment. Dis.* **176**, 116–119.

Silverstone, T. (1979). Psychopharmacology of manic-depressive illness. *In* "Current Themes in Psychiatry 2" (ed. R.N. Gaind and B.L. Hudson), pp. 271–282. Macmillan, London.

Silverstone, T. and Romans-Clarkson, S. (1989). Bipolar affective disorder: causes and prevention of relapse. *Br. J. Psychiat.* **154**, 321–335.

Stewart, P.M., Grieve, J., Nairn, P.L., Padfield, P.L. and Edwards, C.R.W. (1987). Lithium inhibits the action of fludrocortisone on the kidney. *Clin. Endocrinol.* **27**, 63–68.

Thomsen, K., Jensen, J. and Olesen, O.V. (1976). Effect of prolonged lithium ingestion on the response to mineralocorticoids in rats. *J. Pharm. Exp. Ther.* **196**, 463–468.

Van Kammen, D.P. and Murphy, D.L. (1975). Attenuation of the euphoriant and activating effects of *d*- and *l*-amphetamine by lithium carbonate treatment. *Psychopharmacologia* **44**, 215–224.

Wehr, J.A., Sack, D.A. and Rosenthal, N.E. (1987). Sleep reduction as a final common pathway in the genesis of mania. *Amer. J. Psychiat.* **144**, 201–204.

Whalley, L.J., Christie, J.E., Blackwood, D.H.R. *et al.* (1989). Disturbed endocrine function in the psychoses. I: Disordered homeostasis or disease process? *Br. J. Psychiat.* **155**, 455–461.

Index